Lecture Notes in Control and Informati 439

For further volumes:
http://www.springer.com/series/642

Eli Gershon and Uri Shaked

Advanced Topics in Control and Estimation of State-Multiplicative Noisy Systems

Springer

Dr. Eli Gershon
Department of Electrical
and Electronics Engineering
Holon Institute of Technology
Holon
Israel

Prof. Uri Shaked
School of Electrical Engineering
Tel Aviv University
Ramat Aviv
Israel

ISSN 0170-8643 ISSN 1610-7411 (electronic)
ISBN 978-1-4471-5069-5 ISBN 978-1-4471-5070-1 (eBook)
DOI 10.1007/978-1-4471-5070-1
Springer London Heidelberg New York Dordrecht

Library of Congress Control Number: 2013932638

Printed on acid-free paper

Springer is part of Springer Science+Business Media (www.springer.com)

To the memory of Professor Nadav Berman, our friend and colleague whose impact on the field of stochastic nonlinear control is most significant.

Eli Gershon and Uri Shaked

Preface

In this monograph the theory that has been developed in our previous monograph entitled "H_∞ Control and Estimation of State-multiplicative Linear Systems", Lecture Notes in Control and Information Sciences, LNCIS, Springer, vol. 318, is extended to various systems with state dependent noise such as: linear systems with time delay and nonlinear systems with multiplicative noise. Similar to our previous book, this monograph is addressed to engineers that are engaged in control systems research and development, to graduate students specializing in stochastic linear and nonlinear control theory and to applied mathematicians that are interested in control problems.

The reader of this book will benefit from our previous monograph which is solely devoted to delay-free state-multiplicative systems. He should be familiar with the theory of linear and nonlinear systems, state-space methods, and optimal control theory. Some knowledge of stochastic processes, by large, would be an advantage, although the monograph provides a brief background of the subject. Some of the stochastic tools needed to master the subject material are given in the Introduction chapter, where principle concepts are introduced and explained.

The book consists of five parts. The first part includes an introduction and extensive literature survey for both: delayed linear and delay-free nonlinear continuous-time and discrete-time systems with state dependent noise. This part also treats some special topics in stochastic processes which are also complemented in Appendix A.

The second part of the book deals with time delay stochastic systems and it includes the following four chapters: In Chapter 2 the formulation of the stochastic version of the continuous-time Bounded Real Lemma (BRL) is presented for delayed nominal and uncertain systems and the solutions to the state and the measurement-feedback control problems are given, where the solution to the latter utilizes a general-type controller. Chapter 2 also presents a solution to the zero-order (static) output-feedback control problem. In Chapter 3 a treatment of the continuous-time reduced-order control problem is given and in Chapter 4 various preview-patterns are considered within the stochastic tracking control problem for delayed systems. Chapter 5 deals with the stability, BRL, and control and filtering problems of the discrete-time counterpart of the delayed stochastic systems of Chapter 2. We note that Chapters 2,3,and 5 deal also with the robust uncertain cases for polytopic uncertain systems. In Chapter 2 also the robust norm-bounded case is covered. In Part 3 (Chapters 6-10), we deal with the problem of stability, control and estimation of stochastic nonlinear systems in the H_∞ sense, for both continuous-time (Chapters 6-8) and discrete-time systems (Chapters 9,10).

In Part 4 (Chapters 11, 12), we deal with special topics that include stochastic switched systems with dwell time (stability and control) and L_∞ (peak to peak) filtering of nonlinear stochastic systems. The last part (Part 5) introduces six practical examples of noisy state-multiplicative control and filtering problems for linear and nonlinear systems. They are taken from various fields of control engineering, and they include problems of reduced order control and tracking for retarded systems.

The Appendix part of the book contains some basic features of stochastic control processes. In Appendix A, a self-contained description of the σ-algebra, which is needed for the study of state-dependent noisy systems, is brought. A brief description of the LMI method which has become an essential technique in system theory is given in Appendix B. In Appendix C, we bring the Matlab codes of some of the files used to solve Example 11.6 (in Chapter 11) and Example 13.3 (in Chapter 13) which concern stochastic L_2-gain and state-feedback control, respectively, of switched systems with dwell time.

Few words about the numbering scheme used in the book are in order. Each chapter is divided into sections. Thus, Section 2.3 refers to the third section within the second chapter. In each chapter, theorems, lemmas, corollaries, examples and figures are numbered consecutively within the chapter.

Acknowledgments

We owe a debt of gratitude to our colleague Professor E. Fridman from Tel Aviv University for sharing with us her experience in the field of delayed processes and for her valuable comments that contributed to the rigor concerning the input-output approach.

We are also grateful to Dr. I. Yaesh of IMI whose comments on some of the aviation-oriented examples were crucial and to L. Allerhand for his help in the solution of the example in Chapter 13, Section 13.3. Finally, we thank C&M Maus Chair at Tel Aviv University for supporting the research reported in this book.

Tel Aviv, Israel
December 2012

Eli Gershon
Uri Shaked

Contents

1 **Introduction** 1
- 1.1 Stochastic State-Multiplicative Time Delay Systems 1
- 1.2 The Input–Output Approach for Delayed Systems 2
  - 1.2.1 Continuous-Time Case 3
  - 1.2.2 Discrete-Time Case 4
- 1.3 Non Linear Control of Stochastic State-Multiplicative Systems 5
  - 1.3.1 The Continuous-Time Case 5
  - 1.3.2 Stability 8
  - 1.3.3 Dissipative Stochastic Systems 9
  - 1.3.4 The Discrete-Time-Time Case 9
  - 1.3.5 Stability 11
  - 1.3.6 Dissipative Discrete-Time Nonlinear Stochastic Systems 11
- 1.4 Stochastic Processes – Short Survey 12
- 1.5 Mean Square Calculus 13
- 1.6 White Noise Sequences and Wiener Process 14
  - 1.6.1 Wiener Process 14
  - 1.6.2 White Noise Sequences 14
- 1.7 Stochastic Differential Equations 16
- 1.8 Ito Lemma 17
- 1.9 Nomenclature 18
- 1.10 Abbreviations 19

2 **Time Delay Systems – H_∞ Control and General-Type Filtering** 21
- 2.1 Introduction 21
- 2.2 Problem Formulation and Preliminaries 22

2.2.1 The Nominal Case ... 22
2.2.2 The Robust Case – Norm-Bounded Uncertain Systems ... 23
2.2.3 The Robust Case – Polytopic Uncertain Systems ... 24
2.3 Stability Criterion ... 25
2.3.1 The Nominal Case – Stability ... 25
2.3.2 Robust Stability – The Norm-Bounded Case ... 28
2.3.3 Robust Stability – The Polytopic Case ... 29
2.4 Bounded Real Lemma ... 30
2.4.1 BRL for Delayed State-Multiplicative Systems – The Norm-Bounded Case ... 31
2.4.2 BRL – The Polytopic Case ... 32
2.5 Stochastic State-Feedback Control ... 32
2.5.1 State-Feedback Control – The Nominal Case ... 32
2.5.2 Robust State-Feedback Control – The Norm-Bounded Case ... 34
2.5.3 Robust Polytopic State-Feedback Control ... 35
2.5.4 Example – State-Feedback Control ... 37
2.6 Stochastic Filtering for Delayed Systems ... 37
2.6.1 Stochastic Filtering – The Nominal Case ... 37
2.6.2 Robust Filtering – The Norm-Bounded Case ... 41
2.6.3 Robust Polytopic Stochastic Filtering ... 42
2.6.4 Example – Filtering ... 43
2.7 Stochastic Output-Feedback Control for Delayed Systems ... 44
2.7.1 Stochastic Output-Feedback Control – The Nominal Case ... 44
2.7.2 Example – Output-Feedback Control ... 47
2.7.3 Robust Stochastic Output-Feedback Control – The Norm-Bounded Case ... 49
2.7.4 Robust Polytopic Stochastic Output-Feedback Control ... 50
2.8 Static Output-Feedback Control ... 53
2.9 Robust Polytopic Static Output-Feedback Control ... 57
2.10 Conclusions ... 59

3 Reduced-Order H_∞ Output-Feedback Control ... 61
3.1 Introduction ... 61
3.2 Problem Formulation ... 62
3.3 The Delayed Stochastic Reduced-Order H_∞ Control ... 64
3.4 Conclusions ... 74

4 Tracking Control with Preview ... 75
4.1 Introduction ... 75
4.2 Problem Formulation ... 76
4.3 Stability of the Delayed Tracking System ... 78

4.4 The State-Feedback Tracking ... 80
4.5 Example ... 85
4.6 Conclusions ... 85
4.7 Appendix ... 86

5 H_∞ Control and Estimation of Retarded Linear Discrete-Time Systems ... 95
5.1 Introduction ... 95
5.2 Problem Formulation ... 96
5.3 Mean-Square Exponential Stability ... 98
5.3.1 Example – Stability ... 101
5.4 The Bounded Real Lemma ... 101
5.4.1 Example – BRL ... 103
5.5 State-Feedback Control ... 103
5.5.1 Example – Robust State-Feedback ... 105
5.6 Delayed Filtering ... 106
5.6.1 Example – Filtering ... 112
5.7 Conclusions ... 113

6 H_∞-Like Control for Nonlinear Stochastic Systems ... 115
6.1 Introduction ... 115
6.2 Stochastic H_∞ SF Control ... 117
6.3 The Infinite-Time Horizon Case: A Stabilizing Controller ... 121
6.3.1 Example ... 124
6.4 Norm-Bounded Uncertainty in the Stationary Case ... 125
6.4.1 Example ... 127
6.5 Conclusions ... 128

7 Non Linear Systems – H_∞-Type Estimation ... 129
7.1 Introduction ... 129
7.2 Stochastic H_∞ Estimation ... 130
7.2.1 Stability ... 132
7.3 Norm-Bounded Uncertainty ... 134
7.3.1 Example ... 136
7.4 Conclusions ... 138

8 Non Linear Systems – Measurement Output-Feedback Control ... 139
8.1 Introduction and Problem Formulation ... 139
8.2 Stochastic H_∞ OF Control ... 140
8.2.1 Example ... 144
8.2.2 The Case of Nonzero G_2 ... 145
8.3 Norm-Bounded Uncertainty ... 147
8.4 Infinite-Time Horizon Case: A Stabilizing H_∞ Controller ... 151
8.5 Conclusions ... 154

9 l_2-Gain and Robust SF Control of Discrete-Time NL Stochastic Systems 155
9.1 Introduction 155
9.2 Sufficient Conditions for l_2- Gain$\leq \gamma$: A Special Case 159
9.3 Norm-Bounded Uncertainty 161
9.4 Conclusions 163

10 H_∞ Output-Feedback Control of Discrete-Time Systems 165
10.1 Sufficient Conditions for l_2-Gain$\leq \gamma$: A Special Case 166
10.1.1 Example 168
10.2 The OF Case 170
10.2.1 Example 175
10.3 Conclusions 175

11 H_∞ Control of Stochastic Switched Systems with Dwell Time 177
11.1 Introduction 177
11.2 Problem Formulation 178
11.3 Stochastic Stability 179
11.4 Stochastic L_2-Gain 182
11.5 H_∞ State-Feedback Control 187
11.6 Example – Stochastic L_2-Gain Bound 190
11.7 Conclusions 190

12 Robust L_∞-Induced Control and Filtering 193
12.1 Introduction 193
12.2 Problem Formulation and Preliminaries 194
12.3 Stability and P2P Norm Bound of Multiplicative Noisy Systems 195
12.4 P2P State-Feedback Control 197
12.5 P2P Filtering 199
12.6 Conclusions 202

13 Applications 203
13.1 Reduced-Order Control 203
13.2 Terrain Following Control 205
13.3 State-Feedback Control of Switched Systems 206
13.4 Non Linear Systems: Measurement Output-Feedback Control 208
13.5 Discrete-Time Non Linear Systems: l_2-Gain 211
13.6 L_∞ Control and Estimation 212

A Appendix: Stochastic Control Processes – Basic Concepts ... 219
A.1 Basics of σ-Algebra ... 219
A.1.1 The Continuous-Time Case ... 219
A.1.2 The Discrete-Time Case ... 220
A.2 Application of Ito Lemma ... 221
A.3 Stochastic Stability ... 222
A.4 Simulation of Stochastic Differential Equations ... 224

B The LMI Optimization Method ... 227

C Stochastic Switching with Dwell Time – Matlab Scripts ... 229
C.1 Example 11.6 ... 229
C.1.1 Code for the Result of Theorem 11.2 in Table 11.1 ... 229
C.1.2 Code for the Result of Corollary 11.4.1 in Table 11.1 ... 232
C.2 Example 13.3 ... 236
C.2.1 Code for the Result of Theorem 11.3 in Table 13.3 ... 236

References ... 243

Index ... 251

1

Introduction

1.1 Stochastic State-Multiplicative Time Delay Systems

The field of stochastic state-multiplicative control and filtering has greatly matured since, its emergence in the 60's of the last century [72], [91], [92], [93], [97], [100], [102], [121], [122], [124], [125], (see also [16], [53], [80] for extensive review). The linear quadratic optimization problems (stochastic H_2) that were treated in the first two decades cleared the way, in the mid 80s, to the H_∞ worst case control strategy, resulting in a great expansion of research effort, aimed at the solution of related problems such as state-feedback control, estimation, dynamic output-feedback control, preview tracking control and zero-order control among other problems, for various types of nominal and uncertain systems [13], [24]-[28], [44]-[54], [94], [95], [99], [115] (see also [53] for extensive review). In the last decade a great deal of the research in this field has been centered around time-delay systems for both: linear systems and nonlinear ones, where solutions have been obtained for various problems in the H_∞ context. These problems mainly concern continuous-time delayed systems and include various stability issues, state-feedback control, estimation and dynamic output-feedback control [18]- [20], [31], [32], [55]-[64], [81], [85], [88], [89], [116], [117], [119], [120], [126], [127], [128], [130], [134], [138], similarly to the ones encountered in the delay-free stochastic systems and indeed, in the deterministic delayed systems counterpart [12], [33]-[36], [71], [96], [111]. The discrete-time setting has been also tackled but to a lesser extent (see [39], [38], [118] and the references therein).

The stability, and control of deterministic delayed systems of various types (i.e., constant time delay, slow and fast varying delay, etc.) has been a central field within the system theory sciences. In the last two decades, systems with uncertain time-delay have been a subject of recurring interest, especially due to the emergence of the H_∞ control theory in the early 80s. Most of the research, within the control and estimation community, is based on application of different types of Lyapunov Krasovskii Functionals (LKFs) (see for example, [33], [79]). Also continuous-time systems with *fast varying delays* (i.e.,

E. Gershon & U. Shaked: *Advanced Topics in Control & Estimation*, LNCIS 439, pp. 1–19.
DOI: 10.1007/978-1-4471-5070-1_1

without any constraints on the delay derivative) were treated via descriptor type LKF [33], where the derivative of the LKF along the trajectories of the system depends on the state and the state derivative. The results that have been obtained for the stability of deterministic retarded systems have been naturally applied also to stochastic systems by large and in the last decade to state-multiplicative noisy systems.

In the this monograph we extend our previous results, concerning linear stochastic systems that are summed up in [53] to linear-delayed stochastic systems. We apply an *input–output* approach [77] for mostly **delay-dependent** solutions of various control and the filtering stochastic problems. By this approach, the system is replaced by one with norm-bounded operators without delays, which is treated by the standard theory of non-retarded systems with state-multiplicative noise [53]. In our system we allow for a slowly varying delay, (i.e. delays with the derivative less than 1) where the uncertain stochastic parameters multiply both the delayed and the non delayed states in the state space model of the system. In the filtering and dynamic output-feedback problems of Chapter 2, we incorporate also a stochastic multiplicative noise in the measurement equation. We address first the stability issue of the stochastic state-multiplicative delayed systems and we then formulate and find the conditions for the corresponding Bounded Real Lemma (BRL). Based on the BRL we then solve the various control and the filtering problems in Chapters 2–5. We also demonstrate the applicability and tractability of our results via many examples, some of which are given in the latter chapters and some, the more advanced ones, appear in Chapter 13 (Applications). We start this monograph by introducing first, in the following section, the input–output approach for the the design of control and estimation of linear time-delayed stochastic systems.

1.2 The Input–Output Approach for Delayed Systems

In Chapters 2–5 we apply the input-output approach to linear time invariant delayed systems. This approach transforms a given delayed system to a norm-bounded uncertain system that can be treated, in the stochastic context, by the various solutions methods that can be found in [53]. The major advantage of the input-output approach lies in its simplicity of use such that the resulting inequalities that emerge are relatively tractable and simple, for both: delay-dependent and delay-independent solutions. However, this technique entails some degree of conservatism that can be compensated by a clever choice of the Lyapunov function that is involved in the solution method. In the following two subsections we introduce the input-output approach for continuous-time, and discrete-time stochastic systems.

1.2.1 Continuous-Time Case

We consider the following system:

$$dx(t) = [A_0x(t) + A_1x(t - \tau(t))]dt + Hx(t - \tau(t))d\zeta(t) \\ +Gx(t)d\beta(t), \quad x(\theta) = 0, \theta \leq 0, \tag{1.1}$$

where $x(t) \in R^n$ is the state vector and A_0, A_1 and G, H are time invariant matrices and where $\beta(t)$, $\zeta(t)$ are zero-mean real scalar Wiener processes satisfying:

$$\mathcal{E}\{\beta(t)\beta(s)\} = min(t, s),\ \mathcal{E}\{\zeta(t)\zeta(s)\} = min(t, s),$$

$$\mathcal{E}\{\beta(t)\zeta(s)\} = \bar{\alpha} \cdot min(t, s),\ |\bar{\alpha}| \leq 1.$$

In (1.1), $\tau(t)$ is an unknown time-delay which satisfies:

$$0 \leq \tau(t) \leq h, \quad \dot{\tau}(t) \leq d < 1. \tag{1.2}$$

In the input–output approach, we use the following operators:

$$(\Delta_1 g)(t) \triangleq g(t - \tau(t)), \quad (\Delta_2 g)(t) \triangleq \int_{t-\tau(t)}^{t} g(s)ds. \tag{1.3}$$

In what follows we use the fact that the induced L_2-norm of Δ_1 is bounded by $\frac{1}{\sqrt{1-d}}$, and, similarly to [77], the fact that the induced L_2-norm of Δ_2 is bounded by h. Using the above operator notations, the system (1.1) becomes a special case of the following system:

$$dx(t) = [A_0 + m]x(t)dt + (A_1 - m)w_1(t)dt - mw_2(t)dt + Gx(t)d\beta(t) \\ +Hw_1(t)d\zeta(t) - \Gamma_\beta dt - \Gamma_\zeta dt, \\ \bar{y}(t) = [A_0 + m]x(t) + (A_1 - m)w_1(t) - mw_2(t) - \Gamma_\beta - \Gamma_\zeta, \tag{1.4}$$

where

$$\Gamma_\beta = m\int_{t-\tau}^{t} Gx(s)d\beta(s), \text{ and } \Gamma_\zeta = m\int_{t-\tau}^{t} Hw_1(s)d\zeta(s), \tag{1.5}$$

and where

$$w_1(t) = (\Delta_1 x)(t), \quad \text{and} \quad w_2(t) = (\Delta_2 \bar{y})(t). \tag{1.6}$$

Remark 1.1. The dynamics of (1.1) is a special case of that of (1.4) as follows: Noting (1.6) and applying the operators of (1.3), Equation (1.4a) can be written as:

$$dx(t) = [A_0 + m]x(t)dt + (A_1 - m)w_1(t)dt - m\{\int_{t-\tau}^{t} \bar{y}(t')dt'\}dt$$

$$+Gx(t)d\beta(t) + Hw_1(t)d\zeta, \quad w_1(t) = x(t - \tau(t)).$$

Now, recalling $\bar{y}$ of (1.4b) one can write:

$$dx(t) = \bar{y}(t)dt + Gx(t)d\beta(t) + Hw_1(t)d\zeta(t)$$

and therefore

$$\bar{y}(t')dt' = dx(t') - Gx(t')d\beta(t') - Hw_1(t')d\zeta(t').$$

Hence,

$$-mw_2(t) \triangleq -m\int_{t-\tau}^{t} \bar{y}(t')dt' = -m\int_{t-\tau}^{t} \{dx(t') - Gx(t')d\beta(t') - Hw_1(t')d\zeta(t')\}$$

$$= -mx(t) + mx(t-\tau) + \Gamma_\beta + \Gamma_\zeta = -mx(t) + mw_1(t) + \Gamma_\beta + \Gamma_\zeta,$$

where Γ_β and Γ_ζ are defined in (1.5a,b) respectively. Replacing the right hand side of the latter equation for $-mw_2(t)$ in (1.4a), the dynamics of (1.1a) is recovered.

We note that the matrix m is a $n \times n$ unknown constant matrix to be determined. This matrix is introduced into the dynamics of (1.4) in order to achieve additional degree of freedom in the design of the various controllers in all the linear continuous-time delayed systems considered in this book. Using the fact that $||\Delta_1||_\infty \leq \frac{1}{\sqrt{1-d}}$ and $||\Delta_2||_\infty \leq h$, (1.4) may be cast into what is entitled: the norm-bounded uncertain model, by introducing into (1.1) the above new variables of (1.6) where $||\Delta_1||_\infty \leq \frac{1}{\sqrt{1-d}}$ and $||\Delta_2||_\infty \leq h$ are diagonal operators having identical scalar operators on the main diagonal.

1.2.2 Discrete-Time Case

We consider the following linear retarded system:

$$\begin{aligned} x_{k+1} &= (A_0 + D\nu_k)x_k + (A_1 + F\mu_k)x_{k-\tau(k)}, \\ x_l &= 0,\ l \leq 0, \end{aligned} \tag{1.7}$$

where $x_k \in \mathcal{R}^n$ is the system state vector and where the time delay is denoted by the integer τ_k and it is assumed that $0 \leq \tau_k \leq h$, $\forall k$. The variables $\{\mu_k\}$ and $\{\nu_k\}$ are zero-mean real scalar white-noise sequences that satisfy:

$$E\{\nu_k\nu_j\} = \delta_{kj},\ E\{\mu_k\mu_j\} = \delta_{kj},$$
$$E\{\mu_k\nu_j\} = 0,\ \ \forall k,j\ \ \geq 0.$$

The matrices in (1.7) are constant matrices of appropriate dimensions.

In order to tackle the stability (and hence the BRL) of the retarded discrete-time system, we introduce the following scalar operators which are needed, in the sequel, for transforming the delayed system to an equivalent norm-bounded nominal system:

$$\Delta_1(g_k) = g_{k-h}, \quad \Delta_2(g_k) = \sum_{j=k-h}^{k-1} g_j. \tag{1.8}$$

Denoting

$$\bar{y}_k = x_{k+1} - x_k$$

and using the fact that

$$\Delta_2(\bar{y}_k) = x_k - x_{k-h},$$

the following state space description of the system is obtained:

$$x_{k+1} = (A_0 + D\nu_k + M)x_k + (A_1 - M + F\mu_k)\Delta_1(x_k) - M\Delta_2(\bar{y}_k)$$

$$x_l = 0, \ l \leq 0,$$

where the matrix M is a free decision variable to be determined. Similarly to the continuous-time case, this matrix is introduced into the dynamics of (1.7) in order to achieve additional degree of freedom in the design of the state-feedback controller and the estimator of Chapter 5 in this book. We consider then the following auxiliary system:

$$x_{k+1} = (A_0 + D\nu_k + M)x_k + (A_1 - M + F\mu_k)w_{1,k} - Mw_{2,k} \tag{1.9}$$

with the feedback

$$w_{1,k} = \Delta_1(x_k), \quad w_{2,k} = \Delta_2(\bar{y}_k). \tag{1.10}$$

In Chapter 5, it is shown how the norm-boundness of the operators of (1.8) together with the auxiliary system of (1.9) enable the solution of the stability issue of discrete-time stochastic retarded systems.

1.3 Non Linear Control of Stochastic State-Multiplicative Systems

1.3.1 The Continuous-Time Case

In recent years there has been a growing interest, as reflects from the various published research works, in the extension of H_∞ control and estimation theory to accommodate stochastic systems [13], [24]-[28], [44]-[54] (see also [53] for extensive review).

The main thrust for these efforts stems from the attempt to model system uncertainties as a stochastic process, in particular, as a white noise, or formally as a Wiener process. This has led to the development of an H_∞ theory for stochastic linear systems with multiplicative noise. There has been also some work done in the direction of extending the linear stochastic H_∞

control theory to accommodate stochastic nonlinear systems (see, e.g., [5], [6], [7], [17] and [137]). As [6], [7], [137] have developed theories for the state feedback case, [17] and [5] considered the output feedback case. The latter adopted a differential game point of view, and stressed the connection to the risk-sensitive control (see, e.g. [29] and [76]), while the former took the differential game approach and considered (for technical reasons) finitely additive measure space as a foundation for the measurement process structure. The objective of the present chapter is to develop an H_∞-like control theory for nonlinear stochastic systems with output feedback, exhibiting a combination of deterministic and stochastic uncertainties, together with a presence of what is called norm bounded uncertainty. In particular, we consider the following nonlinear stochastic system.

$$\begin{aligned} dx_t = {} & f(x_t,t)dt+g(x_t,t)u_t dt+g_1(x_t,t)v_t dt+\bar{g}(x_t,t)u_t dW_t \\ & +g_2(x_t,t)v_t dW_t^2+G(x_t,t)dW_t^1 \end{aligned} \tag{1.11}$$

$$dy_t = h_2(x_t,t)dt + g_3(x_t,t)v_t dt + G_2(x_t,t)dW_t^3 \tag{1.12}$$

where $\{x_t\}_{t\geq 0}$ is a solution to (1.11) with: an initial condition x_0, an exogenous disturbance $\{v_t\}_{t\geq 0}$, a control signal $\{u_t\}_{t\geq 0}$, and Wiener processes $\{W_t\}_{t\geq 0}$, $\{W_t^1\}_{t\geq 0}$, $\{W_t^2\}_{t\geq 0}$, $\{W_t^3\}_{t\geq 0}$. Also, y_t is an R^p-valued observation vector which is corrupted by noise (Wiener processes $\{W_t^3\}_{t\geq 0}$), and contains an uncertain component (a stochastic process $\{v_t\}_{t\geq 0}$).

This type of system may be viewed, on one hand, as an extension of the linear case, which has been extensively treated in the literature (see, e.g., [53], and the references therein, or [69] where a motivation for considering such systems is provided), and on the other hand it may be considered as a result of an expansion of a general nonlinear stochastic system in terms of its states and its uncertain variables up to the second order terms, or Volterra type stochastic systems.

The following will be assumed to hold throughout this work.

1. Let $(\Omega, F, \{F_t\}_{t\geq 0}, P)$ be a filtered probability space where $\{F_t\}_{t\geq 0}$ is the family of sub σ- algebras generated by $\{W_t\}_{t\geq 0}$, $\{W_t^1\}_{t\geq 0}$, $\{W_t^2\}_{t\geq 0}$, $\{W_t^3\}_{t\geq 0}$, which are all Wiener processes taken to be R^1- valued, R^l- valued, R^1-valued, R^{m_3}-valued, respectively.

2. All the functions below are assumed to be Borel measurable on $R^n \times [0,\infty)$. $f : R^n \times [0,\infty) \to R^n$, $g : R^n \times [0,\infty) \to R^{n\times m}$, $g_1 : R^n \times [0,\infty) \to R^{n\times m_1}$, $g_2 : R^n \times [0,\infty) \to R^{n\times m_2}$, $G : R^n \times [0,\infty) \to R^{n\times l}$. In aaddition $G_2 : R^n \times [0,\infty) \to R^{p\times m_3}$, $h_2 : R^n \times [0,\infty) \to R^p$, $g_3 : R^n \times [0,\infty) \to R^{p\times m_1}$. It is also assumed that $f(0,t) = 0$, $G(0,t) = 0$, $h_2(0,t) = 0$, $G_2(0,t) = 0$ for all $t \geq 0$.

3. $\{v_t\}_{t\geq 0}$ is a non-anticipative R^{m_1}-valued stochastic process defined on $(\Omega, F, \{F_t\}_{t\geq 0}, P)$, which satisfies $\mathcal{E}\{\int_0^t \|v_s\|^2 ds\} < \infty$ for all $t \in [0,\infty)$, where E denotes expectation: $\mathcal{E}\{x\} = \int_\Omega x(\omega)dP(\omega)$.

4. $\{u_t\}_{t\geq 0}$ is a non-anticipative R^m-valued stochastic process defined on $(\Omega, F, \{F_t\}_{t\geq 0}, P)$.

5. x_0 is assumed to be F_0-measurable, and to satisfy $\mathcal{E}\{\|x_0\|^2\} < \infty$

In this chapter we consider the case for which the solution x_t satisfies $\mathcal{E}\{||x_t||^2\} < \infty, \quad \forall t \geq 0$. For the pertaining conditions which guarantee this see e.g. [65].

Definition 1.3.1 *The pair $\{u_t, v_t\}_{t\in[(0,\infty)}$, or in short $\{u, v\}$, is said to be admissible if the stochastic differential equation (1.11) possesses a unique strong solution relative to the filtered probability space $(\Omega, F, \{F_t\}_{t\geq 0}, P)$ so that $\mathcal{E}\{\|x_t\|^2\} < \infty$ for all $t \in [(0, \infty)$.*

Remark 1.2. The family of all admissible pairs $\{u, v\}$ will be denoted by $\mathcal{A}$. The notation $\mathcal{A}_u$ will be used for all admissible pairs $\{u, v\}$ with fixed u. We note that $\mathcal{A}_u$ may be empty for some non-anticipative u.

Let $\alpha(\cdot, \cdot)$ be positive Borel function on $R^n \times [0, \infty)$ (where $R^n \times [0, \infty)$ is endowed with the Borel σ-algebra). In what follows it will be assumed that $\mathcal{E}\{\alpha(x, t)\} < \infty$ for all $t \in [0, \infty)$ and for all F-measurable, R^n-valued random variables which satisfy $\mathcal{E}\{||x||^2\} < \infty$. The control objective is now stated as follows. Consider the controlled output:

$$z_t = \begin{bmatrix} h(x_t, t) \\ u_t \end{bmatrix}, \quad t \in [0, \infty) \tag{1.13}$$

where $h : R^n \times [0, \infty) \to R^r$ is a Borel measurable function, and let $\mathcal{Y}_t = \{y_s : s \leq t\}$. Find an output-feedback controller $u_t = u(\mathcal{Y}_t, t)$ such that, for a given $\gamma > 0$, the following H_∞ criterion is satisfied.

$$\mathcal{E}\{\int_{t_1}^{t_2} \|z_t\|^2 dt\} \leq \gamma^2 \mathcal{E}\{\alpha(x_{t_1}, t_1) + \int_{t_1}^{t_2} \|v_t\|^2 dt\} \tag{1.14}$$

for all $0 \leq t_1 < t_2$, for all F_0-measurable x_0 with $\mathcal{E}\{||x_0||^2\} < \infty$, and for all disturbances v_t in $\mathcal{A}_u$ (provided $\mathcal{A}_u$ is nonempty). Whenever the system (1.11) satisfies the above inequality, it is said to possess an L_2-gain that is less than or equal to γ. Note that for the infinite time-horizon $t_2 = \infty$, and it is required that v satisfies: $\int_0^\infty ||v_t||^2 dt < \infty$.

As we adopt the stochastic dissipativity point of view in dealing with this control problem, we first recall the concept of stochastic dissipative systems and then discuss some properties of these systems. In addition, we introduce what we call the Bounded Real Lemma (BRL) for nonlinear stochastic systems which was first developed in [6]. This is done in Section 2. In Section 3, we develop the H_∞ output-feedback control for nonlinear stochastic systems of the type described by (1.11). We introduce there a pair of Hamilton-Jacobi Inequalities (HJI), the solution of which yields an output-feedback controller that renders the underlying closed-loop system L_2-gain$\leq \gamma$. The approach taken here is analogous to the one introduced by Isidori ([73]) in the framework of the deterministic counterpart. Section 4 deals with the special case of systems with norm-bounded uncertainties where we introduce certain matrix inequalities whose solution yields a robust output-feedback. In Section 5, we consider the infinite-time horizon where we introduce sufficient conditions for the synthesis of a stabilizing (in both mean square and in probability senses) output-feedback controller. This section is concluded with a simple example of a single-degree-of-freedom inverted pendulum.

1.3.2 Stability

We recall few facts from the theory of stochastic stability (see e.g. [67]). We remark that in what follows global stability is considered. Obviously, local stability results may also be achieved, in a similar way.

Definition 1.3.2 *Consider the stochastic system*

$$dx_t = f(x_t,t)dt + G(x_t,t)dW_t \tag{1.15}$$

with $f(0,t) = G(0,t) = 0$ *for all* $t \geq 0$, *and assume that* f *and* G *are such that (1.15) possesses a unique strong solution for all* $t \geq 0$ *relative to the filtered probability space* $(\Omega, F, \{F_t\}_{t\geq 0}, P)$, *where* $\{F_t\}$ *is generated by the Wiener process* $\{W_t\}_{t\geq 0}$. *The solution* x_t *is said to be stable in probability if for any* $\varepsilon > 0$ $\lim_{x\to 0} P\{\sup_{t\geq 0} \|x_t\| > \varepsilon\} = 0$.

Definition 1.3.3 *The solution* x_t *of (1.15) is said to be globally asymptotically stable in probability if it is stable in probability, and if* $P\{\lim_{t\to\infty} x_t = 0\} = 1$ *for any initial state* $x_0 \in R^n$.

A sufficient condition for a global stability in probability is given by the following theorem.

Theorem 1.3.1 *([67]) Assume there exists a positive storage function* $V(x,t) \in C^{2,1}$, *with* $V(0,t) = 0$. *Let* $L(x,t)$ *be the infinitesimal generator of the process* x_t, *that is*

$$LV(x,t) = V_x(x,t)f(x,t) + \frac{1}{2}Tr\{G^T(x,t)V_{xx}(x,t)G(x,t)\},$$

so that $(LV)(x,t) < 0$ *for all* $x \in R^n$ *and for all* $t \geq 0$. *Assume also that* $\inf_{t>0} V(x,t) \to \infty$ *as* $\|x\| \to \infty$. *Then, the system of (1.15) is globally asymptotically stable in probability.*

Definition 1.3.4 *[stability in the mean-square sense]*
The system (1.15) is said to be globally exponentially stable in the mean-square sense if $\mathcal{E}\{||x_t||^2\} \leq k\mathcal{E}\{\|x_s\|^2\}\exp\{-\alpha(t-s)\}$ *for all* $0 \leq s \leq t$, *and for some positive numbers* k *and* α.

Theorem 1.3.2 *([67]) Assume there exists a positive function* $V(x,t) \in C^{2,1}$, *with* $V(0,t) = 0$. *Then, the system of (1.15) is globally exponentially stable if there are positive numbers* k_1, k_2, k_3 *such that the following hold.*

$$k_1||x||^2 \leq V(x,t) \leq k_2||x||^2, \quad (LV)(x,t) \leq -k_3||x||^2 \text{ for all } t \geq 0. \tag{1.16}$$

1.3.3 Dissipative Stochastic Systems

In this section we summarize various results pertaining what is called Stochastic Bounded Real Lemma (SBRL) which serve as a basis for the theory to be developed in the sequel. A full account of what follows may be found in ([113], [7], and [11]).

Let $S : R^m \times R^{r+m_1} \rightarrow R$, be a Borel measurable function which will be entitled supply rate.

Definition 1.3.5 *Consider the system (1.11) together with the controlled output $z(t)$ as defined in (1.13), and let S be a supply rate as defined above. Let u be such that $\mathcal{A}_u$ is nonempty. Then, the system (1.11) is said to be dissipative with respect to the supply rate S if there is a function $V : R^n \times [0,\infty) \rightarrow R$, with $V(x,t) \geq 0$ for all $x \in R^n$ and $t \in [0,\infty)$, so that $V(0,t) = 0 \ \forall t \in [0,\infty)$ satisfies $\mathcal{E}\{V(x,t)\} < \infty$ for all t and for all F-measurable r.v.s satisfying $\mathcal{E}\{||x||^2\} < \infty$, such that:*

$$\mathcal{E}\{V(x_t,t)\} \leq \mathcal{E}\{V(x_s,s)\} + \mathcal{E}\{\int_s^t S(v_\sigma, z_\sigma) d\sigma\} \tag{1.17}$$

for all $t \geq s \geq 0$ and for all admissible disturbances $\{v_t\}_{t\geq 0}$ in $\mathcal{A}_u$, where x_t is the solution to the differential equation (1.11). V is then called a storage function for the system (1.11).

Similar to the deterministic theory of dissipative systems, the theorem below establishes conditions under which the system (1.11) possesses a storage function. First we introduce a candidate for a storage function.

Definition 1.3.6 *Consider the system (1.11) and let x be an R^n valued random variable defined on the probability space (Ω, F, P) with $E\{\|x\|^2\} < \infty$. Assume also that x is F_t measurable. Let u be such that $\mathcal{A}_u$ is nonempty. Given $t \in [0,\infty)$ and let $x_t = x$. Define*

$$V_a(x,t) = \sup_{T \geq t, \ v \in \mathcal{A}_u} [-E\{[\int_t^T S(v_s, z_s) ds]/x\}]. \tag{1.18}$$

1.3.4 The Discrete-Time-Time Case

H_∞ control for discrete-time deterministic nonlinear systems has been considered by numerous researchers, see e.g. [87], which utilize, in part, the dissipativity concept. In order to develop an analogous theory for the stochastic counterpart, we setup, in the present paper, some theory of stochastic dissipativity. A related topic is the risk sensitive control problems (see, e.g. [29] and references therein) which, in general, deal with optimization of stochastic systems where the cost function involves an exponential.

In Chapter 9 an effort is made to extend the work presented in [6], in order to include discrete-time nonlinear stochastic systems. The contribution of this paper is in that it provides means (the Bounded Real Lemma-BRL) for synthesizing a state-feedback H_∞ controller for a large class of nonlinear stochastic systems. In addition, the BRL facilitates, in a natural way, the utilization of the Linear Matrix Inequality (LMI) techniques to achieve l_2-gain$\leq \gamma$ for a large class of uncertain nonlinear systems with norm bounded uncertainties. Another contribution of this paper is the introduction of what we call stochastic dissipation, which serves as a basis for the H_∞ control theory developed in the sequel.

We consider the following stochastic system

$$x_{k+1} = f_k(x_k, v_k, u_k, \omega_k) \tag{1.19}$$

where $\{x_k\}_{k\geq 0}$ is a solution to (1.19), with: an initial condition x_0, an exogenous disturbances $\{v_k\}_{k\geq 0}$, a control signal $\{u_k\}_{k\geq 0}$, and a white noise sequence $\omega = \{\omega_k\}_{k\geq 0}$ defined on a probability space (Ω, F, P). In the sequel $\omega = \{\omega_k\}_{k\geq 0}$ describes both exogenous random inputs and parameter uncertainty of the system. The following will be assumed to hold throughout this work.

1. Let $(\Omega, F, \{F_k\}_{k\geq 0}, P)$ be a filtered probability space where $\{F_k\}_{k\geq 0}$ is the family of sub σ-algebras of F generated by $\{\omega_k\}_{k\geq 0}$, where ω_k are assumed to be R^l-valued. In fact, each F_k is assumed to be the minimal σ-algebra generated by $\{\omega_i\}_{0\leq i\leq k-1}$ while F_0 is assumed to be some given sub σ-algebra of F, independent of F_i for all $i > 0$.

2. $\{u_k\}_{k\geq 0}$ and $\{v_k\}_{k\geq 0}$ are non-anticipative (that is u_k, v_k are independent of $\{F_i, i > k\}$), R^m and R^{m_1} valued, respectively, stochastic processes defined on $(\Omega, F, \{F_k\}_{k\geq 0}, P)$. The vectors x_k are R^n valued, and for each k, $f_k : R^n \times R^{m_1} \times R^m \times R^l \to R^n$ is assumed to be continuous on $R^n \times R^{m_1} \times R^m \times R^l$.

3. $\{v_k\}_{k\geq 0}$ satisfies $E\{\sum_{k=0}^{N} ||v_k||^2\} < \infty \; \forall N \geq 0$, where E stands for the expectation operation.

4. x_0 is assumed to be F_0-measurable and to satisfy $E\{||x_0||^2\} < \infty$.

5. The following notation will be used in the sequel. Let X, Y be R^{n_1} and R^{n_2}-valued random variables defined on (Ω, F, P) and let $V : R^{n_1} \times R^{n_2} \to R$. Let P_y be the probability distribution of Y. Then the random variable $E_y\{V(X,Y)\}$ is defined by $E_y\{V(X,Y)\} \doteq \int_{R^{n_2}} V(X,y)dP_y(y)$.

In this work we investigate the problem of stochastic H_∞ state-feedback control which is formulated as follows. Given a controlled output:

$$z_k(x,u) = \begin{bmatrix} h_k(x) \\ u \end{bmatrix} \tag{1.20}$$

where $h_k : R^n \to R^r$, synthesize a controller $u_k = K_k(x_k)$ such that, for a given $\gamma > 0$, the following H_∞ criterion is satisfied.

$$E\{\textstyle\sum_{i=j}^{k-1} ||z_i(x_i,u_i)||^2\} \leq E\{\beta_j(x_j)\} + \textstyle\sum_{i=j}^{k-1} \gamma^2 E\{||v_i||^2\} \tag{1.21}$$

for all $0 \leq j \leq k$, for all $\{v_k\}_{k \geq 0}$ such that $\sum_{i=0}^{k} E\{||v_i||^2\} < \infty$, where $\beta_j : R^n \rightarrow R^+$ are positive functions satisfying $E\{\beta_j(x)\} < \infty$ whenever x is an F-measurable random variable satisfying $E\{||x||^2\} < \infty$. A system that satisfies this property is said to be l_2-gain$\leq \gamma$. We also denote the family of all exogenous disturbances v which satisfy $\sum_{i=0}^{N} E\{||v_i||^2\} < \infty$, and such that (1.19) possesses a solution $\{x_k\}_{k \geq 0}$ with $E\{||x_k||^2 < \infty\}$ for some $\{u_k\}_{k \geq 0}$, by $\mathcal{A}_u$.

1.3.5 Stability

Consider the following nonlinear stochastic system

$$x_{k+1} = f_k(x_k, w_k) \tag{1.22}$$

where $x_k \in R^n$ and $\{w_k\}_k$ is an R^l-valued stochastic process defined on the probability space $(\Omega, F, \{F_k\}_{k \geq 0}, P)$ as defined above. It will be assumed, in what follows, that (1.22) has a unique solution for all $k \geq 0$. The following is standard.

Definition 1.3.7 *The (1.22) is said to be globally exponential mean square stable if there is a positive constant $a < 1$ so that the solution of (1.22) satisfies*

$$\mathcal{E}\{||x_k||^2\} \leq a^k \mathcal{E}\{||x_0||^2\} \quad \forall k \geq 0, \quad x_0 \in L_2(\Omega, F_0, P, R^n)$$

The following sufficient conditions are obtained for the system (1.22) to be exponentially mean square stable; these are a straightforward adjustment of the continuous-time counterparts.

Theorem 1.3.3 *Consider the system (1.22). Let $\{V_k(x)\}_{k \geq 0}$ be a family of positive functions defined on R^n such that the following conditions are satisfied. There are positive constants k_1, k_2, k_3 so that:*

1. $k_1||x||^2 \leq V_k(x) \leq k_2||x||^2 \quad \forall k \geq 0 \quad \forall x \in R^n$
2. $\mathcal{E}_{w_k}\{V_{k+1}(f_k(x, w_k))\} - V_k(x) \leq -k_3||x||^2 \quad \forall k \geq 0 \quad \forall x \in R^n$.

Then, the system (1.22) is exponentially mean square stable.

1.3.6 Dissipative Discrete-Time Nonlinear Stochastic Systems

In analogy to the continuous-time case, we define the notion of stochastic dissipative system.

Definition 1.3.8 *Consider the discrete-time stochastic system (1.19) defined on the discrete time interval $[0, \infty)$, with the associated controlled output (1.20) and let $S : R^{r+m} \times R^{m_1} \rightarrow R$ be a Borel measurable function on $R^{r+m} \times R^{m_1}$. Then, the system (1.19) is said to be dissipative with respect* *to S if, for some admissible control sequence* $\{u_k\}_k$, *there exists a family of functions* $V_k : (R^n \times N^+) \to R$ *with* $V_k(x) \geq 0 \quad \forall x \in R^n, k \in N^+$ *(*N^+ *denotes the positive integers), so that* $V_k(0) = 0 \quad \forall k \in N^+$, $E\{V_0(x_0)\} < \infty$, *and*

$$E\{V_k(x_k)\} \leq E\{V_j(x_j)\} + E\{\sum_{i=j}^{k-1} S(z_i, v_i)\} \tag{1.23}$$

for all $k : k \geq j \geq 0$, *with the convention:* $\sum_{i=j}^{l-1} S(z_i, v_i) = 0$ *whenever* $l = j$, *and for all non-anticipative* $v = \{v_i\}_{i \geq 0}$ *with* $E\{\sum_{i=0}^{\infty} ||v_i||^2\} < \infty$. *We call S supply rate (see e.g. Willems [122] for the deterministic case), and the family* $V = \{V_k\}_k$ *is said to be a storage function of the system.*

We have now the following trivial result which connects the l_2-gain property to the notion of stochastic dissipation.

Lemma 1.3.1 *Consider the system (1.19) and let the supply rate be* $S(z, v) = \gamma^2||v||^2 - ||z||^2$. *Assume the system possesses a storage function V so that the system is dissipative with respect to this S. Then, the system has* l_2*-gain* $\leq \gamma$.

1.4 Stochastic Processes – Short Survey

Stochastic processes are a family of random variables parameterized by time $t \in \mathcal{T}$. Namely, at each instant t, $x(t)$ is a random variable. When t is continuous (namely $\mathcal{T} = \mathcal{R}$), we say that $x(t)$ is a continuous-time stochastic process, and if t is discrete (namely $\mathcal{T} = \{1, 2, ...\}$), we say that $x(t)$ is a discrete-time variable. For any finite set of $\{t_1, t_2, ...t_n\} \in \mathcal{T}$, we can define the joint distribution $F(x(t_1), x(t_2), ..., x(t_n))$ and the corresponding joint density $p(x(t_1), x(t_2), ..., x(t_n))$.

The first and the second order distribution functions, $p(x(t))$ and $p(x(t), x(\tau))$, respectively, play an important role in our discussion. Also the mean $m_x(t) \triangleq E\{x(t)\}$ and the autocorrelation $\gamma_x(t, \tau) \triangleq E\{x(t)x(\tau)\}$ are useful characteristics of the stochastic process $x(t)$. When $x(t)$ is vector valued, the autocorrelation is generalized to be $\Gamma_x(t, \tau) = E\{x(t)x(\tau)^T\}$. The covariance matrix of a vector valued stochastic process $x(t)$ is a measure of its perturbations with respect to its mean value and is defined by $P_x(t) \triangleq E\{(x(t) - m_x(t))(x(t) - m_x(t))^T\}$.

A process $x(t)$ is said to be stationary if

$$p(x(t_1), x(t_2), ..., x(t_n)) = p(x(t_1 + \tau), x(t_2 + \tau), ..., x(t_n + \tau))$$

for all n and τ. If the latter is true only for $n = 1$, then the process $x(t)$ is said to be stationary of order 1 and then $p(x(t))$ does not depend on t. Consequently, the mean $m_x(t)$ is constant and $p(x(t), x(\tau))$ depends only on $t - \tau$. Also in such a case, the autocorrelation function of two time instants depends only on the time difference, namely $\gamma_x(t, t - \tau) = \gamma_x(\tau)$.

An important class of stochastic processes is one of Markov processes. A stochastic process $x(t)$ is called a Markov process if for any finite set of time instants $t_1 < t_2 < ... < t_{n-1} < t_n$ and for any real λ it satisfies

$$Pr\{x(t_n) < \lambda | x(t_1), x(t_2), ..., x(t_{n-1}), x(t_n)\} = Pr\{x(t_n) | x(t_{n-1}\}.$$

Stochastic processes convergence properties of a process $x(t)$ to a limit x can be analyzed using different definitions. The common definitions are almost sure or with probability 1 convergence (namely $x(t) \to x$ almost surely, meaning that this is satisfied except for an event with a zero probability), convergence in probability (namely for all $\epsilon > 0$, the probability of $|x(t) - x| \geq \epsilon$ goes to zero), and mean square convergence, where given that $E\{x(t)^2\}$ and $E\{x^2\}$ are both finite, $E\{(x(t) - x)^2\} \to 0$. In general, almost sure convergence neither implies nor it is implied by mean square convergence, but both imply convergence in probability. In the present book we adopt the notion of mean square convergence and the corresponding measure of stability, namely mean square stability.

1.5 Mean Square Calculus

Working with continuous-time stochastic processes in terms of differentiation, integration, etc. is similar to the analysis of deterministic functions, but it requires some extra care in evaluation of limits. One of the most useful approaches to calculus of stochastic processes is the so called mean square calculus where mean square convergence is used when evaluating limits.

The full scope of mean square calculus is covered in [75] and [105] but we bring here only a few results that are useful to our discussion.

The notions of mean square continuity and differentiability are key issues in our discussion. A process $x(t)$ is said to be mean square continuous if $lim_{h \to 0} x(t + h) = x(t)$. It is easy to see that if $\gamma_x(t, \tau)$ is continuous at (t, t) then also $x(t)$ is mean square continuous. Since the converse is also true, then mean square continuity of $x(t)$ is equivalent to continuity of $\gamma(t, \tau)$ in (t, t). Defining mean square derivative by the mean square limit as $h \to 0$ of $(x(t + h) - x(t))/h$, then it is similarly obtained that $x(t)$ is mean square differentiable (i.e. its derivative exists in the mean square sense) if and only if $\gamma_x(t, \tau)$ is differentiable at (t, t). A stochastic process is said to be mean square integrable, whenever $\sum_{i=0}^{n-1} x(\tau_i)(t_{i+1} - ti)$ is mean square convergent where $a = t_0 < t_1 < ... < t_n = b$, where $\tau_i \in [t_i, t_{i+1}]$ and where $|t_{i+1} - ti| \to 0$. In such a case, the resulting limit is denoted by $\int_a^b x(t)dt$. It is important to know that $x(t)$ is mean square integrable on $[a, b]$ if and only if $\gamma_x(t, \tau)$ is integrable on $[a, b] \times [a, b]$. The fundamental theorem of mean square calculus then states that if $\dot{x}(t)$ is mean square integrable on $[a, b]$, then for any $t \in [a, b]$, we have

$$x(t) - x(a) = \int_a^t \dot{x}(\tau) d\tau.$$

The reader is referred to [75] for a more comprehensive coverage of mean square calculus.

1.6 White Noise Sequences and Wiener Process

In this section we consider both: discrete-time white noise type stochastic processes (Chapters 5,9,10) and continuous-time Wiener-type stochastic processes (Chapters 2–4, 6–8, 11). We start with the description of the Wiener type stochastic processes.

1.6.1 Wiener Process

A process $\beta(t)$ is said to be a Wiener Process (also referred to as Wiener-Levy process or Brownian motion) if it has the initial value of $\beta(0) = 0$ with probability 1, has stationary independent increments, and is normally distributed with zero mean for all $t \geq 0$. The Wiener process has then the following properties : $\beta(t) - \beta(\tau)$ is normally distributed with zero mean and variance $\sigma^2(t-\tau)$ for $t > \tau$ where σ^2 is an empirical positive constant. Consider now for $t > \tau$ the autocorrelation

$$\begin{aligned}\gamma_\beta(t,\tau) &= E\{\beta_t\beta_\tau)\} = E\{(\beta(t) - \beta(\tau) + \beta(\tau))\beta(\tau)\} \\ &= E\{(\beta(t) - \beta(\tau))\beta(\tau)\} + E\{\beta^2(\tau)\}.\end{aligned}$$

Since the first term is zero, due to the independent increments property of the Wiener process, it is readily obtained that $\gamma_\beta(t,\tau) = \sigma^2\tau$. Since we have assumed that $t > \tau$ we have in fact that $\gamma_\beta(t,\tau) = \sigma^2 min(t,\tau)$. Since the latter is obviously continuous at (t,t), it follows that $\beta(t)$ is mean square continuous. However, a direct calculation (see [75]) of the second order derivative of $\gamma_\beta(t,\tau)$, with respect to t and τ at (t,t), shows that

$$\frac{min(t+h, t+h') - min(t,t)}{hh'} = 1/max(h,h')$$

which is clearly unbounded as h and h' tend to zero. Therefore, $\gamma_\beta(t,\tau)$ is not differentiable at any (t,t) and consequently $\beta(t)$ is not mean square differentiable anywhere. It is, therefore, concluded that the Wiener process is continuous but not differentiable in the mean square sense. In fact, it can be shown that the latter conclusion holds also in the sense of almost sure convergence.

1.6.2 White Noise Sequences

A discrete-time process is said to be white if it is a Markov process and if all $x(k)$ are mutually independent. Such a process is said to be a white Gaussian noise if, additionally, its samples are normally distributed. The mutual independence property leads, in the vector valued case, to $E\{x(n)x^T(m)\} = Q_n\delta_{n,m}$ where $\delta_{n,m}$ is the Kronecker delta function (1 for equal arguments and zero otherwise) and where $Q_n \geq 0$. The discrete-time white noise is a useful approximation of measurement noise in many practical cases. Its continuous-time analog also appears to be useful. Consider a stationary process $x(t)$ whose samples are mutually independent, taken at large enough intervals. Namely,

$$\gamma(\tau) = E\{x(t+\tau)x(t)\} = \sigma^2\frac{\rho}{2}e^{-\rho|\tau}$$

where $\rho >> 1$. As ρ tends to infinity $\gamma(\tau)$ rapidly decays as a function of τ, and therefore the samples of $x(t)$ become virtually independent and the process becomes white. Noting that for ρ that tends to infinity, $\frac{\rho}{2}e^{-\rho|\tau} \rightarrow \delta(\tau)$ where δ is the Dirac delta function [75], a vector valued white process $x(t)$ is formally considered to have the autocorrelation of $\gamma(\tau) = Q(t)\delta(\tau)$ where $Q(t) \geq 0$. Namely, $E\{x(t)x(\tau)\} = Q(t)\delta(t-\tau)$. Defining the spectral density of $x(t)$ by the Fourier transform of its autocorrelation, namely by

$$f(\omega) = \int_{-\infty}^{\infty} e^{-i\tau\omega}\sigma^2\frac{\rho}{2}e^{-\rho|\tau|}d\tau = \frac{\sigma^2}{1+\omega^2/\rho^2}$$

we see that this spectral density is constant and has the value of σ^2 up to about the frequency ρ where it starts dropping to zero. Namely, the spectrum of $x(t)$ is nearly flat independently of the frequency, which is the source of the name "white" noise, in analogy to white light including all frequencies or wavelengths. We note that for finite $\rho >> 1$, $x(t)$ is said to be a wide-band noise (where 1 may represent the measured process bandwidth and ρ the measurement noise bandwidth). In such a case, modelling $x(t)$ as a white noise is a reasonable approximation. We note, however, that constant $f(\omega)$ or white spectrum corresponds to infinite energy by Parseval's theorem . Alternatively, looking at the autocorrelation at $\tau = 0$, we see that

$$\gamma(0) = E\{x^2(t)\} = \frac{1}{2\pi}\int_{-\infty}^{\infty} f(\omega)d\omega \rightarrow \infty.$$

Therefore, white noise is not physically realizable but is an approximation to wide band noise. To allow mathematical manipulations of white noise, we relate it to Wiener processes which are well defined. To this end we recall that the autocorrelation of a Wiener process $\beta(t)$ is given by

$$\gamma(t,\tau) = E\{x(t)x(\tau)\} = \sigma^2 min(t,\tau).$$

Since expectation and derivatives can be interchanged, namely

$$E\{\frac{d\beta(t)}{dt}\frac{d\beta(\tau)}{d\tau}\} = \frac{d^2}{dtd\tau}E\{\beta_t\beta_\tau\},$$

it follows that the autocorrelation of $\dot{\beta(t)}$ is given by $\sigma^2\frac{d}{d\tau}[\frac{d}{dt}min(t,\tau)]$. However, $min(t,\tau)$ is τ for $\tau < t$ and t otherwise; therefore, its partial derivative with respect to t is a step function of τ rising from 0 to 1 at $\tau = t$. Consequently, the partial derivative of this step function is just $\sigma^2\delta(t-\tau)$. The autocorrelation of $\dot{\beta}(t)$ is thus $\sigma^2\delta(t-\tau)$, just as the autocorrelation of white noise, and we may, therefore, formally conclude that white noise is the derivative, with respect to time, of a Wiener process.

1.7 Stochastic Differential Equations

Many stochastic processes are formally described (see [75] and [105]) by the Langevin's equation:

$$\frac{dx}{dt} = f(x(t),t) + g(x(t),t)\dot{\beta}(t)$$

where $\dot{\beta}(t)$ is a white noise process. For example, in this monograph, the so called state-multiplicative process is obtained when $f(x(t),t) = Ax$ and $g(x(t),t) = Dx$ leading to

$$\frac{dx}{dt} = Ax + Dx\dot{\beta}(t).$$

When we write the latter in terms of differentials rather than in terms of derivatives, we obtain the following equation

$$dx = Axdt + Dxd\beta$$

where the physically unrealizable $\dot{\beta}(t)$ no longer appears but instead the differential $d\beta$ of $\beta(t)$ drives the equation. Note that

$$d\beta(t) \triangleq \beta(t) - \beta(t-dt)$$

is normally distributed with zero mean and $\sigma^2(t-(t-dt)) = \sigma^2 dt$. When $\sigma^2 = 1$ we say that the Wiener process $\beta(t)$ and the corresponding white noise process $\dot{\beta}(t)$ are standard. Back to Langevin's equation, we may realize that it can also be written in terms of differentials as

$$dx(t) = f(x(t),t)dt + g(x(t),t)d\beta(t).$$

This equation is, in fact, interpreted by

$$x(t) - x(t_0) = \int_{t_0}^{t} f(x(\tau,\tau)d\tau + \int_{t_0}^{t} g(x(\tau,\tau)d\beta(\tau),$$

where the first term is a Lebesgue-Stieltges integral and the second term is an Ito integral with respect to the Wiener process $\beta(t)$. Namely, this integral is defined via approximation by the sum:

$$\sum_{i=0}^{n-1} g_{t_i}[\beta(t_{i+1}) - \beta(t_i)], \quad \text{where} \quad a = t_0 < t_1 < ... < t_n = b,$$

where g_{t_i} is a random variable at the fixed time t_i which is $\mathcal{F}_{t_i}$ measurable (see Section 1.4.1). It is assumed that g_{t_i} is independent on future increments $\beta(t_k)-\beta(t_l)$ for $t_i \le t_l \le t_k \le b$ of $\beta(t)$. The stochastic integral is then defined by choosing a series $g^n$of piecewise step functions which converge to g, in the sense that the mean square of the integral of $g^n - g$ tends to zero as n tends to infinity. Whenever g_t is mean square integrable and is independent of future increments of $\beta(t)$, the stochastic Ito sense integral exists. Furthermore, it satisfies two useful identities:

$$E\{\int g_t d\beta(t)\} = 0, \quad \text{and} \quad E\{\int g_t d\beta(t) \int f_t d\beta(t)\} = \sigma^2 \int E\{g_t f_t\} dt.$$

In fact, we have defined in the above the first order stochastic integral. The second order stochastic integral in the Ito sense, $\int_0^t g_t d\beta^2(t)$, is similarly defined by taking the limit of n to infinity in $\sum_{i=0}^{n-1} g_{t_i}[\beta(t_{i+1}) - \beta(t_i)]$. It can be shown [75] that the latter converges, in mean square, to just $\int_{t_0}^t \sigma^2 g_t dt$.

We consider next

$$dx(t) = x(t)d\beta(t) + \frac{1}{2}x(t)d\beta^2(t)$$

where $x(0) = 1$ almost surely. Integrating the latter yields

$$x(t) - 1 = \int_0^t x(t)d\beta(t) + \frac{1}{2}\int_0^t x(t)d\beta^2(t) = \int_0^t x(t)d\beta(t) + \frac{\sigma^2}{2}\int_0^t x(t)dt.$$

The latter is simply the integral form in the Ito sense of:

$$dx(t) = x(t)d\beta(t) + \frac{\sigma^2}{2}x(t)dt,$$

meaning that in stochastic differential equations, $d\beta^2(t)$ can be replaced in the mean square sense by $\sigma^2 dt$.

1.8 Ito Lemma

In this monograph, Ito lemma is a key lemma which is widely used in the various chapters to evaluate differentials of nonlinear scalar valued functions $\varphi(x(t))$ of solutions $x(t)$ of Ito type stochastic differential equations.

Consider a scalar process $x(t)$ which satisfies

$$\frac{dx}{dt} = f(x(t), t) + g(x(t), t)\dot{\beta}(t).$$

Then, using Taylor expansion, we have

$$d\varphi = \varphi_t dt + \varphi_x dx + \frac{1}{2}\varphi_{xx} dx^2 + \frac{1}{3}\varphi_{xxx} dx^3 + ...$$

Discarding terms of the order $o(dt)$, recalling that $d\beta^2(t)$ is of the order of dt, and substituting for dx in the above Taylor expansion, it is found that [75]:

$$d\varphi = \varphi_t dt + \varphi_x dx + \frac{1}{2}\varphi_{xx} g^2 d\beta^2(t).$$

Substituting $\sigma^2 dt$ for $d\beta^2(t)$ we obtain

$$d\varphi = \varphi_t dt + \varphi_x dx + \frac{\sigma^2}{2}\varphi_{xx} g^2 dt.$$

For vector valued $x(t)$, where $Qdt = E\{d\beta d\beta^T\}$, the latter result reads:

$$d\varphi = \varphi_t dt + \varphi_x dx + \frac{1}{2} Tr\{gQg^T \varphi_{xx}\} dt$$

where φ_{xx} is the Hessian of φ with respect to x.

1.9 Nomenclature

$(\cdot)^T$	matrix transposition.
$\mathcal{R}^n$	the n dimensional Euclidean space.
$\mathcal{R}^{n\times m}$	the set of all $n \times m$ real matrices.
$P>0, (P\geq 0)$	the matrix $P \in \mathcal{R}^{n\times n}$ is symmetric and positive definite (respectively, semi-definite).
$\|\|.\|\|_2^2$	the standard l_2-norm: $\|\|d\|\|_2^2 = (\Sigma_{k=0}^{N-1} d_k^T d_k)$.
$l_2[0 \quad N-1]$	the space of square summable functions over $[0 \quad N-1]$.
$\|\|f_k\|\|_R^2$	the product $f_k^T R f_k$.
$\|\|f\|\|$	the Euclidean norm of f.
$\underset{v}{\mathcal{E}}\{\cdot\}$	the expectation operator with respect to v.
$[Q_k]_+$	the causal part of a sequence $\{Q_i,\ i = 1, 2, ..., N\}$.
$[Q_k]_-$	the anti causal part of a sequence $\{Q_i,\ i = 1, 2, ..., N\}$.
$Tr\{\cdot\}$	the trace of a matrix.
δ_{ij}	the Kronecker delta function.
$\delta(t)$	the Dirac delta function.
$\mathcal{N}$	the set of natural numbers.
Ω	the sample space.
$\mathcal{F}$	$\sigma-$algebra of subsets of Ω called events.
$\mathcal{P}$	the probability measure on $\mathcal{F}$.
$Pr(\cdot)$	probability of $(\cdot)$.
$l^2(\Omega, \mathcal{R}^n)$	the space of square-summable $\mathcal{R}^n-$ valued functions.

	on the probability space $(\Omega, \mathcal{F}, \mathcal{P})$.
$(\mathcal{F}_k)_{k\in\mathcal{N}}$	an increasing family of $\sigma-$algebras $\mathcal{F}_k \subset \mathcal{F}$.
$\tilde{l}^2([0,N];\mathcal{R}^n)$	the space of nonanticipative stochastic processes. $\{f_k\}$=$\{f_k\}_{k\in[0,N]}$ in $\mathcal{R}^n$ with respect to $(\mathcal{F}_k)_{k\in[0,N)}$ satisfying $\|\|f_k\|\|^2_{\tilde{l}_2} = E\{\sum_0^N \|\|f_k\|\|^2\} = \sum_0^N E\{\|\|f_k\|\|^2\} < \infty$, $f_k \in \tilde{l}_2([0,\infty);\mathcal{R}^n)$.
$\tilde{l}^2([0,\infty);\mathcal{R}^n)$	the above space for $N \to \infty$
$\tilde{L}^2([0,T);\mathcal{R}^k)$	the space of non anticipative stochastic processes. $f(\cdot) = (f(t))_{t\in[0,T]}$ in $\mathcal{R}^k$ with respect to $(\mathcal{F}_t)_{t\in[0,T)}$ satisfying $\|\|f(\cdot)\|\|^2_{\tilde{L}_2} = \mathcal{E}\{\int_0^T \|\|f(t)\|\|^2 dt\} = \int_0^T E\{\|\|f(t)\|\|^2\}dt < \infty$.
$\begin{bmatrix} P & R \\ * & Q \end{bmatrix}$	$= \begin{bmatrix} P & * \\ R^T & Q \end{bmatrix}$,for symmetric P and Q, is the matrix $\begin{bmatrix} P & R \\ R^T & Q \end{bmatrix}$.
$diag\{A,\ B\}$	the block diagonal matrix $\begin{bmatrix} A & 0 \\ 0 & B \end{bmatrix}$.
$col\{a,b\}$	the matrix (vector) $\begin{bmatrix} a \\ b \end{bmatrix}$.

1.10 Abbreviations

BRL	Bonded Real Lemma
BLS	Bilinear System
BLSS	Bilinear Stochastic System
BRL	Bounded Real Lemma
DBLS	Deterministic Bilinear System
DRE	Difference Riccati Equation
GBM	Geometrical Brownian Motion
LF	Lyapunov Function
LKF	Lyapunov Krasovskii Functional
LMI	Linear Matrix Inequality
LPD	Lyapunov Parameter Dependent
LTI	Linear Time Invariant
LTV	Linear Time Variant
OF	Output Feedback
P2P	Peak to Peak
SDN	State Dependent Noise
SF	State Feedback
SNR	Signal to Noise Ratio
SOF	Static Output-Feedback

2

Time Delay Systems – H_∞ Control and General-Type Filtering

2.1 Introduction

In this chapter we consider state-multiplicative LTI stochastic systems that may encounter parameter uncertainties. We adopt here (and also in Chapters 3–5) the input–output approach that is identical to the one presented in the Introduction (Chapter 1, Section 1.2.1) and in [59]. That is, the system's delay action is represented by linear operators, with no delay, which in turn allow us to replace the underlying system with a nonretarded one that possesses norm-bounded uncertainty. The latter system may, therefore, be treated by the theory of non-retarded systems with state-multiplicative noise and norm-bounded uncertainties [53]. In all the systems that are considered in the present chapter, we allow for time-varying delay where the uncertain stochastic parameters multiply both the delayed and the nondelayed states in the state space model.

The present chapter is organized as follows: We first investigate the stability of nominal delayed systems using the input–output approach and we then treat the robust stability problem for the resulting nonretarded systems that encounter either norm-bounded or polytopic type parameter uncertainties. We continue by deriving bounds on the L_2-gain of the uncertain delayed systems. Following the latter derivation, we obtain solutions to the problems of: state-feedback control, general-type filtering, and full-order measurement control for nominal systems and for uncertain ones. The latter solutions are followed by the solution to the SOF control problem. The latter problem is also solved later, in Chapter 3, as a special case of the reduced-order control problem. Various examples are given that demonstrate the applicability of the theory involved.

E. Gershon & U. Shaked: *Advanced Topics in Control & Estimation,* LNCIS 439, pp. 21–60.
DOI: 10.1007/978-1-4471-5070-1_2

2.2 Problem Formulation and Preliminaries

2.2.1 *The Nominal Case*

We consider the following system:

$$\begin{aligned}
&dx = [A_0x(t) + A_1x(t-\tau(t)) + B_1w(t)]dt + Hx(t-\tau(t))d\nu(t) + Gx(t)d\beta(t)\\
&+B_2u(t)dt, \qquad x(\theta)=0, \quad \theta \le 0\\
&y(t) = [C_2x(t) + \bar{C}_2x(t-\tau(t)) + D_{21}n(t)] + Fx(t)d\zeta(t),\\
&z(t) = C_1x(t) + D_{12}u(t),
\end{aligned} \tag{2.1}$$

where $x(t) \in R^n$ is the state vector, $w(t) \in \tilde{L}^2_{\mathcal{F}_t}([0,\infty);\mathcal{R}^q)$ is an exogenous disturbance, $y(t) \in R^m$ is the measurement vector, $n(t) \in \tilde{L}^2_{\mathcal{F}_t}([0,\infty);\mathcal{R}^p)$ is an additive measurement noise, $z(t) \in R^r$ is the objective vector, and $u(t) \in R^\ell$ is the control input signal, A_0, A_1, B_1, B_2, C_1, C_2, $\bar{C}_2$, D_{12}, D_{21} and F, G, H are time-invariant matrices of the appropriate dimension. $\tau(t)$ is an unknown time-delay which satisfies:

$$0 \le \tau(t) \le h, \quad \dot{\tau}(t) \le d < 1. \tag{2.2}$$

The zero-mean real scalar Wiener process $\beta(t)$, $\nu(t)$, $\zeta(t)$ satisfy:

$$\mathcal{E}\{\zeta(t)\zeta(s)\} = min(t,s),\ \mathcal{E}\{\beta(t)\beta(s)\} = min(t,s),\ \mathcal{E}\{\nu(t)\nu(s)\} = min(t,s),$$
$$\mathcal{E}\{\beta(t)\nu(s)\} = \bar{\alpha}\cdot min(t,s),\ |\bar{\alpha}| \le 1,\ \mathcal{E}\{\nu(t)\zeta(t)\} = 0,$$
$$\mathcal{E}\{\beta(t)\zeta(s)\} = 0$$

We treat the following problems:

i) H_∞ state-feedback control:
We consider the system of (2.1a,c) and the following performance index:

$$J_E \triangleq \mathcal{E}\{\int_0^\infty ||z(t)||^2 dt - \gamma^2 \int_0^\infty ||w(t)||^2 dt\}. \tag{2.3}$$

Our objective is to find a state-feedback control law $u(t) = Kx(t)$ that achieves $J_E < 0$, for the worst-case of the process disturbance $w(t) \in \tilde{L}^2_{\mathcal{F}_t}([0,\infty);\mathcal{R}^q)$ and for the prescribed scalar $\gamma > 0$.

ii) H_∞ filtering:
We consider the system of (2.1a–c) where $B_2 = 0$, $D_{12} = 0$, $\bar{C}_2 = 0$ and consider the following general-type estimator:

$$\begin{aligned}
d\hat{x}(t) &= A_c\hat{x}(t)dt + B_cy(t),\\
\hat{z} &= C_c\hat{x}(t).
\end{aligned} \tag{2.4}$$

We denote

$$e(t) = x(t) - \hat{x}(t) \quad \text{and} \quad \bar{z}(t) = z(t) - \hat{z}(t) \tag{2.5}$$

and we consider the following cost function:

$$J_F \stackrel{\Delta}{=} \mathcal{E}\{\int_0^{\infty} ||\bar{z}(t)||^2 dt - \gamma^2 [\int_0^{\infty} ||w(t)||^2 dt + \int_0^{\infty} ||n(t)||^2 dt]\}. \tag{2.6}$$

Given $\gamma > 0$, we seek an estimate $C_c\hat{x}(t)$ of $C_1 x(t)$ over the infinite time horizon $[0, \infty)$ such that J_F given by (2.6) is negative for all nonzero $w(t), n(t)$ where $w(t) \in \tilde{L}^2_{\mathcal{F}_t}([0, \infty); \mathcal{R}^q)$, $n(t) \in \tilde{L}^2_{\mathcal{F}_t}([0, T]; \mathcal{R}^p)$.

iii) H_∞ output-feedback control:
We consider the system of (2.1a–c). Our objective is to find a controller of the following general-type structure:

$$\begin{aligned} d\hat{x}(t) &= A_c\hat{x}(t)dt + B_c y(t)dt, \quad \hat{x}(0) = 0, \\ u(t) &= C_c\hat{x}(t), \end{aligned} \tag{2.7}$$

that achieves $J_E < 0$, where J_E is given in (2.3), for the worst-case disturbance $w(t) \in \tilde{L}^2_{\mathcal{F}_t}([0, \infty); \mathcal{R}^q)$ and measurement noise $n(t) \in \tilde{L}^2_{\mathcal{F}_t}([0, T]; \mathcal{R}^p)$, and for the prescribed scalar $\gamma > 0$.

iv) H_∞ static output-feedback control:
We consider the system of (2.1a–c). We seek a controller of the following form:

$$u = Ky, \tag{2.8}$$

and we consider the following performance index:

$$J_{static} \stackrel{\Delta}{=} \mathcal{E}\{\int_0^{\infty} ||z(t)||^2 dt - \gamma^2 \int_0^{\infty} ||w(t)||^2 dt\}. \tag{2.9}$$

Assuming that the exogenous disturbance signal is energy bounded, a static control gain is sought which, for a prescribed scalar γ and for all nonzero $w(t) \in \tilde{L}^2_{\mathcal{F}_t}([0, \infty); \mathcal{R}^q)$, guarantees $J_{static} < 0$.

2.2.2 The Robust Case – Norm-Bounded Uncertain Systems

We consider the system of (2.1a,c) where the matrices A_0, A_1, and B_2 are replaced by $\tilde{A}_0$, $\tilde{A}_1$, and $\tilde{B}_2$ respectively. The latter matrices encounter the following norm-bounded uncertainties:

$$\tilde{A}_0 = A_0 + E_0 F_0 \bar{H}_0, \; \tilde{A}_1 = A_1 + E_1 F_1 \bar{H}_1, \; \tilde{B}_2 = B_2 + E_0 F_0 \bar{H}_2, \tag{2.10}$$

where

$$F_i^T F_i \leq I, \quad E_i \in \mathcal{R}^{n \times r_i}, \; F_i \in \mathcal{R}^{r_i \times \bar{r}_i}, \; \bar{H}_i \in \mathcal{R}^{\bar{r}_i \times n}, \; i = 0, 1, \quad \bar{H}_2 \in \mathcal{R}^{\bar{r}_0 \times l} \tag{2.11}$$

and where the matrices $E_0,\ E_1,\ \bar{H}_i,\ i=0,1,2$ are constant matrices. Similarly to the nominal case, we treat the following problems:

i) Robust H_∞ state-feedback control:
We consider the system of (2.1a,c) where the matrices $A_0,\ A_1$, and B_2 are replaced by $\tilde{A}_0,\ \tilde{A}_1$, and $\tilde{B}_2$ respectively, and the performance index of (2.3).

Our objective is to find a state-feedback control law $u(t) = Kx(t)$ that achieves $J_E < 0$, for the worst-case of the process disturbance $w(t) \in \tilde{L}^2_{\mathcal{F}_t}([0,\infty);\mathcal{R}^q)$ and for the prescribed scalar $\gamma > 0$.

ii) Robust H_∞ filtering:
We consider the system of (2.1a–c) where $B_2 = 0,\ D_{12} = 0,\ \bar{C}_2 = 0$ where the matrices $A_0,\ A_1$ are replaced by $\tilde{A}_0,\ \tilde{A}_1$ respectively, and consider the estimator of (2.4) and the performance index of (2.6). Given $\gamma > 0$, we seek an estimate $C_c\hat{x}(t)$ of $C_1x(t)$ over the infinite time horizon $[0,\infty)$ such that J_F given by (2.6) is negative for all nonzero $w(t), n(t)$ where $w(t) \in \tilde{L}^2_{\mathcal{F}_t}([0,\infty);\mathcal{R}^q)$, $n(t) \in \tilde{L}^2_{\mathcal{F}_t}([0,T];\mathcal{R}^p)$.

iii) Robust H_∞ output-feedback control:
We consider the system of (2.1a–c) where the matrices $A_0,\ A_1$, and B_2 are replaced by $\tilde{A}_0,\ \tilde{A}_1$, and $\tilde{B}_2$ respectively, and the general-type controller (2.7). Our objective is to find a controller of the type (2.7) that achieves $J_E < 0$, where J_E is given in (2.3), for the worst-case disturbance $w(t) \in \tilde{L}^2_{\mathcal{F}_t}([0,\infty);\mathcal{R}^q)$ and measurement noise $n(t) \in \tilde{L}^2_{\mathcal{F}_t}([0,T];\mathcal{R}^p)$, and for the prescribed scalar $\gamma > 0$.

2.2.3 The Robust Case – Polytopic Uncertain Systems

In the robust polytopic case, we assume that the system parameters lie within the following polytope

$$\bar{\Omega} \stackrel{\Delta}{=} \left[A_0\ A_1\ B_1\ B_2\ C_1\ C_2\ \bar{C}_2\ D_{12}\ D_{21}\ H\ G\ F\right], \tag{2.12}$$

which is described by its vertices:

$$\bar{\Omega} = Co\{\bar{\Omega}_1, \bar{\Omega}_2, ..., \bar{\Omega}_N\}, \tag{2.13}$$

where $\bar{\Omega}_i \stackrel{\Delta}{=}$

$$\left[A_0^{(i)}\ A_1^{(i)}\ B_1^{(i)}\ B_2^{(i)}\ C_1^{(i)}\ C_2^{(i)}\ \bar{C}_2^{(i)}\ D_{12}^{(i)}\ D_{21}^{(i)}\ H^{(i)}\ G^{(i)}\ F^{(i)}\right], \tag{2.14}$$

and where N is the number of vertices. In other words:

$$\bar{\Omega} = \sum_{i=1}^{N} \bar{\Omega}_i f_i \quad , \sum_{i=1}^{N} f_i = 1 \quad , f_i \geq 0. \tag{2.15}$$

We treat the following three problems:

i) Robust Polytopic H_∞ state-feedback control:
We consider the system of (2.1a,c) where the system matrices lie within the polytope $\bar{\Omega}$ of (2.12), and J_E of (2.3). Our objective is to find a state-feedback control law $u(t) = Kx(t)$ that achieves $J_E < 0$, for the worst-case of the process disturbance $w(t) \in \tilde{L}^2_{\mathcal{F}_t}([0,\infty);\mathcal{R}^q)$ and for a prescribed scalar $\gamma > 0$.

ii) Robust Polytopic H_∞ filtering:
We consider the system of (2.1a–c) where $B_2 = 0,\ D_{12} = 0,\ \bar{C}_2 = 0$ where the system matrices lie within the polytope $\bar{\Omega}$ of (2.12), and consider the estimator of (2.4) and the performance index of (2.6). Given $\gamma > 0$, we seek an estimate $C_c\hat{x}(t)$ of $C_1x(t)$ over the infinite time horizon $[0,\infty)$ such that J_F given by (2.6) is negative for all nonzero $w(t), n(t)$ where $w(t) \in \tilde{L}^2_{\mathcal{F}_t}([0,\infty);\mathcal{R}^q)$, $n(t) \in \tilde{L}^2_{\mathcal{F}_t}([0,T];\mathcal{R}^p)$.

iii) Robust polytopic H_∞ output-feedback control:
We consider the system of (2.1a–c) where the system matrices lie within the polytope $\bar{\Omega}$ of (2.12), and J_E of (2.3). Our objective is to find a controller of the general-type structure of (2.7) that achieves $J_E < 0$, for the worst-case disturbance $w(t) \in \tilde{L}^2_{\mathcal{F}_t}([0,\infty);\mathcal{R}^q)$ and measurement noise $n(t) \in \tilde{L}^2_{\mathcal{F}_t}([0,T];\mathcal{R}^p)$, and for a prescribed scalar $\gamma > 0$.

iv) Robust polytopic H_∞ static output-feedback control:
We consider the system of (2.1a–c) where the system matrices lie within the polytope $\bar{\Omega}$ of (2.12), and J_{static} of (2.9). Our objective is to find a static controller of the type of (2.8) that achieves $J_{static} < 0$, for the worst-case disturbance $w(t) \in \tilde{L}^2_{\mathcal{F}_t}([0,\infty);\mathcal{R}^q)$ and for a prescribed scalar $\gamma > 0$.

2.3 Stability Criterion

We first consider the issue of stability of the stochastic delayed nominal autonomous system and then the stability of the norm-bounded and polytopic uncertain systems. Once the criteria for stability is found, we formulate and obtain the nominal and robust H_∞ BRL for the latter system. We use the resulting BRL to solve the state-feedback, filtering, dynamic output-feedback, and static output-feedback control problems that were defined above.

2.3.1 The Nominal Case – Stability

We consider the following system:

$$\begin{aligned} dx(t) &= [A_0x(t) + A_1x(t-\tau(t)) + B_1w(t)]dt + Hx(t-\tau(t))d\nu(t) \\ &\quad + Gx(t)d\beta(t), \quad x(\theta) = 0, \theta \leq 0, \end{aligned} \tag{2.16}$$

where $x(t) \in R^n$ is the state vector, $w(t) \in \tilde{L}^2_{\mathcal{F}_t}([0,\infty);\mathcal{R}^q)$ is an exogenous disturbance, and $A_0,\ A_1,\ B_1$ and $G,\ H$ are time invariant matrices and where $\beta(t),\ \nu(t)$ are zero-mean real scalar Wiener processes satisfying:

$$\mathcal{E}\{\beta(t)\beta(s)\} = min(t,s),\ \mathcal{E}\{\nu(t)\nu(s)\} = min(t,s),$$
$$\mathcal{E}\{\beta(t)\nu(s)\} = \bar{\alpha}\cdot min(t,s),\ |\bar{\alpha}| \le 1.$$

In (2.16), $\tau(t)$ is an unknown time-delay which satisfies:

$$0 \le \tau(t) \le h, \quad \dot{\tau}(t) \le d < 1. \tag{2.17}$$

In order to solve the above problem, we introduce the following operators:

$$(\Delta_1 g)(t) \triangleq g(t-\tau(t)), \quad (\Delta_2 g)(t) \triangleq \int_{t-\tau(t)}^{t} g(s)ds. \tag{2.18}$$

In what follows we use the fact that the induced L_2-norm of Δ_1 is bounded by $\frac{1}{\sqrt{1-d}}$, and, similarly to [77], the fact that the induced L_2-norm of Δ_2 is bounded by h. Using the above operator notations, the system (2.16a) becomes a special case of the following system (see Chapter 1, Remark 1.1):

$$\begin{aligned} dx(t) &= [A_0+m]x(t)dt + (A_1-m)w_1(t)dt - mw_2(t)dt + B_1w(t)dt \\ &\quad +Gx(t)d\beta(t) + Hw_1(t)d\nu - \Gamma_\beta dt - \Gamma_\nu dt, \\ \bar{y}(t) &= [A_0+m]x(t) + (A_1-m)w_1(t) - mw_2(t) + B_1w(t) - \Gamma_\beta - \Gamma_\nu, \end{aligned} \tag{2.19}$$

where

$$\Gamma_\beta = m\int_{t-\tau}^{t} Gx(s)d\beta(s), \text{ and } \Gamma_\nu = m\int_{t-\tau}^{t} Hw_1(s)d\nu(s),$$

and where $w_1(t) = (\Delta_1 x)(t), \quad$ and $\quad w_2(t) = (\Delta_2\bar{y})(t)$.

The matrix m is a $n\times n$ unknown constant matrix to be determined. We note that this matrix is introduced into the dynamics of (2.16) in order to achieve additional degree of freedom in the design of the controller. Using the fact that $||\Delta_1||_\infty \le \frac{1}{\sqrt{1-d}}$ and $||\Delta_2||_\infty \le h$, (2.19) may be cast into what is entitled: the norm-bounded uncertain model, by introducing into (2.16) the following new variables:

$$w_1(t) = \Delta_1 x(t), \quad and \quad w_2(t) = \Delta_2\bar{y}(t), \tag{2.20}$$

where $||\Delta_1||_\infty \le \frac{1}{\sqrt{1-d}}$ and $||\Delta_2||_\infty \le h$ are *diagonal* operators having identical scalar operators on the main diagonal.

The conditions for the stability of the system (2.16) with the 'feedback' that is described in (2.20) are usually derived using the small gain theorem. It is well known, however, (see for example [15]) that these conditions are identical to those obtained applying the Lyapunov approach to the same system with feedback that is described by the diagonal 'feedback' operators $\bar{\Delta}_1$ and $\bar{\Delta}_2$ that, for all x and $\bar{y}$ in $\mathcal{R}^n$, satisfy the following:

$$x^T(t)\bar{\Delta}_1^T\bar{\Delta}_1 x(t) \le (1-d)^{-1}\|x(t)\|^2$$

and

$$\bar{y}^T(t)\bar{\Delta}_2^T\bar{\Delta}_2\bar{y}(t) \le h^2\|\bar{y}(t)\|^2.$$

In order to derive the above conditions for stability, we replace Δ_i by $\bar{\Delta}_i$, $i = 1,2$ and we consider the system of (2.19), with $B_1 = 0$, and the following positive definite function:

$$V(x) = x^T Q x,$$

where $Q > 0$ is a constant matrix. Noting that $\mathcal{E}\{\Gamma_\beta\} = 0$ and $\mathcal{E}\{\Gamma_\nu\} = 0$, we apply the Ito lemma (see, e.g. [133]) and taking expectation we obtain:

$$\begin{aligned}\mathcal{E}\{(LV)(t)\} = \mathcal{E}\{\langle Qx(t), [(A_0+m)x(t) + (A_1-m)\bar{\Delta}_1 x(t) - m\bar{\Delta}_2\bar{y}(t)]\rangle\}\\ +\mathcal{E}\{Tr\{Q[Gx(t)\ \ Hw_1(t)]\bar{P}[Gx(t)\ \ Hw_1(t)]^T\}\},\end{aligned}$$

where $\bar{P} \triangleq \begin{bmatrix} 1 & \bar{\alpha} \\ \bar{\alpha} & 1 \end{bmatrix}$ is the covariance matrix of the augmented Wiener process vector $col\{\beta(t)\ \ \nu(t)\}$, that is $\mathcal{E}\{col\{\beta(t)\ \ \nu(t)\}\{\beta(t)\ \ \nu(t)\}\} = \bar{P}t$.

We also have the following:

$$Tr\{Q[Gx(t)\ \ Hw_1(t)]\bar{P}[Gx(t)\ \ Hw_1(t)]^T\} = Tr\{\begin{bmatrix} x^T(t)G^T \\ w_1^T(t)H^T \end{bmatrix} Q[Gx(t)\ \ Hw_1(t)]\bar{P}\}$$

$$= Tr\{\begin{bmatrix} x^T(t)G^TQGx(t) & x^T(t)G^TQHw_1(t) \\ w_1^T(t)H^TQGx(t) & w_1^T(t)H^TQHw_1(t) \end{bmatrix}\begin{bmatrix} 1 & \bar{\alpha} \\ \bar{\alpha} & 1 \end{bmatrix}\}$$

$$= x^T(t)G^TQGx(t) + 2\bar{\alpha}x^T(t)G^TQHw_1(t) + w_1^T(t)H^TQHw_1(t).$$

In the attempt to establish $\mathcal{E}\{(LV)(t)\} \le -k||x(t)||^2$, for some $k > 0$ (uniformly in t), we obtain the following inequality:

$$\begin{aligned}2x^TQ[(A_0+m)x(t) + (A_1-m)w_1(t) - m\bar{\Delta}_2\bar{y}(t)] + kx^T(t)x(t)\\ +x^T(t)G^TQGx(t) + 2\bar{\alpha}x^T(t)G^TQHw_1(t) + w_1^T(t)H^TQHw_1(t) < 0.\end{aligned} \tag{2.21}$$

Adding the following term, which is nonnegative due to the diagonality of Δ_1, to (2.21):

$$x^T(t)(\frac{1}{1-d}R_1 - \bar{\Delta}_1^T R_1\bar{\Delta}_1)x(t) = x^T(t)\frac{1}{1-d}R_1 x(t) - w_1^T(t)R_1 w_1(t)$$

and using the fact that

$$-2x^T(t)Qm\bar{\Delta}_2\bar{y}(t) \le h^2\bar{y}^T(t)R_2\bar{y}(t) + x^T(t)QmR_2^{-1}m^TQx(t), \tag{2.22}$$

where R_1 and R_2 are constant positive definite matrices, we obtain that (2.21) holds if the following inequality is satisfied:

$$\begin{array}{c}2x^T(t)Q[(A_0+m)x(t)+(A_1-m)w_1(t)]+x^T(t)QmR_2^{-1}m^TQx(t)\\ +h^2\bar{y}^T(t)R_2\bar{y}(t)+x^T(t)G^TQGx(t)+2\bar{\alpha}x^T(t)G^TQHw_1(t)+x^T(t)\frac{1}{1-d}R_1x(t)\\ -w_1^T(t)R_1w_1(t)+w_1^T(t)H^TQHw_1(t)+kx^T(t)x(t)\le 0.\end{array} \tag{2.23}$$

Denoting $\zeta(t)=col\{x(t),w_1(t)\}$, we obtain that (2.23) holds if the following inequality holds:

$$\zeta^T(t)\begin{bmatrix}\bar{\Psi}_{11} & Q(A_1-m)+\bar{\alpha}G^TQH\\ * & -R_1+H^TQH\end{bmatrix}\zeta(t)+h^2\bar{y}^T(t)R_2\bar{y}(t)<0,$$

where

$$\bar{\Psi}_{11}=Q(A_0+m)+(A_0+m)^TQ+\frac{1}{1-d}R_1+G^TQG+QmR_2^{-1}m^TQ.$$

Applying Schur complement, denoting $Q_m=Qm$ and taking $R_2=\epsilon_f Q$ where $\epsilon_f>0$ is a tuning parameter, the following inequality is obtained:

$$\Psi=\begin{bmatrix}\Psi_{11} & QA_1-Q_m+\bar{\alpha}G^TQH & Q_m & h\epsilon_f(A_0^TQ+Q_m^T)\\ * & -R_1+H^TQH & 0 & h\epsilon_f(A_1^TQ-Q_m^T)\\ * & * & -\epsilon_fQ & -h\epsilon_fQ_m^T\\ * & * & * & -\epsilon_fQ\end{bmatrix}<0, \tag{2.24}$$

where

$$\Psi_{11}=QA_0+Q_m+A_0^TQ+Q_m^T+\frac{1}{1-d}R_1+G^TQG.$$

We thus arrive at the following theorem:

Theorem 2.1. *The exponential stability in the mean square sense of the system (2.16) is guaranteed if there exist for a given tuning parameter $\epsilon_f>0$, matrices $Q>0,\ R_1>0, Q_m$ that satisfy the LMI condition of (2.24).*

Remark 2.1. A similar attempt to derive conditions for the mean square stability of the system of the type in (2.16) has been done in [55] for the nominal case (with no deterministic norm-bounded uncertainties). However, since the BRL obtained there applies a conservative bound on the operator D_1 of (2.18), we bring the corrected result of (2.24).

2.3.2 *Robust Stability – The Norm-Bounded Case*

We consider the system of (2.16) where $A_0,\ A_1$ are replaced by $\tilde{A}_0,\ \tilde{A}_1$, respectively and where the latter matrices are given in (2.10). Noting that

$$\bar{H}_i^T\epsilon_j\bar{H}_i+QE_i\epsilon_j^{-1}E_i^TQ\ge QE_iF_i\bar{H}_i+\bar{H}_i^TF_i^TE_i^TQ,\quad \epsilon_j>0,\ i=0,1,\ j=i+1$$

we arrive at the following theorem:

Theorem 2.2. *The exponential stability in the mean square sense of the system (2.16), (2.10), and (2.11) is guaranteed if there exist for a given tuning parameter* $\epsilon_f > 0$*, matrices* $Q > 0,\ R_1 > 0, Q_m$ *and scalars* $\epsilon_1,\ \epsilon_2$ *that satisfy the following LMI condition:*

$$\begin{bmatrix} \hat{\Psi}_{11} & \hat{\Psi}_{12} & Q_m & \hat{\Psi}_{14} & QE_0 & QE_1 \\ * & \hat{\Psi}_{22} & 0 & \hat{\Psi}_{24} & 0 & 0 \\ * & * & -\epsilon_f Q & -h\epsilon_f Q_m^T & 0 & 0 \\ * & * & * & -\epsilon_f Q & h\epsilon_f QE_0 & h\epsilon_f QE_1 \\ * & * & * & * & -\epsilon_1 I_n & 0 \\ * & * & * & * & * & -\epsilon_2 I_n \end{bmatrix} < 0 \tag{2.25}$$

where

$$\begin{aligned} \hat{\Psi}_{11} &= QA_0 + Q_m + A_0^T Q + Q_m^T + \tfrac{1}{1-d} R_1 + G^T Q G + \epsilon_1 \bar{H}_0^T \bar{H}_0, \\ \hat{\Psi}_{12} &= QA_1 - Q_m + \bar{\alpha} G^T Q H, \\ \hat{\Psi}_{14} &= h\epsilon_f A_0^T Q + h\epsilon_f Q_m^T, \\ \hat{\Psi}_{22} &= -R_1 + H^T Q H + \epsilon_2 \bar{H}_1^T \bar{H}_1 \\ \hat{\Psi}_{24} &= h\epsilon_f A_1^T Q - h\epsilon_f Q_m^T. \end{aligned}$$

Remark 2.2. The above result provides a delay dependent stability condition. A corresponding delay independent (but rate dependent) result is readily obtained by choosing $m = 0$ and $\epsilon_f \to 0$. This choice is obtained when only the operator Δ_1 is applied to the system (2.16).

Corollary 2.3.1 *A delay independent exponential stability in the mean square sense of the system (2.16) is guaranteed if there exist matrices* $Q > 0,\ R_1 > 0$ *and scalars* $\epsilon_1,\ \epsilon_2$ *that satisfy the following LMI.*

$$\begin{bmatrix} \tilde{\Gamma}_{11} & QA_1 + \bar{\alpha} G^T Q H & QE_0 & QE_1 & G^T Q \\ * & -R_1 + H^T Q H + \epsilon_2 \bar{H}_1^T \bar{H}_1 & 0 & 0 & 0 \\ * & * & -\epsilon_1 I_n & 0 & 0 \\ * & * & * & -\epsilon_2 I_n & 0 \\ * & * & * & * & -Q \end{bmatrix} < 0. \tag{2.26}$$

where $\tilde{\Gamma}_{11} = QA_0 + A_0^T Q + \frac{1}{1-d} R_1 + \epsilon_1 \bar{H}_0^T \bar{H}_0$.

2.3.3 Robust Stability – The Polytopic Case

We consider the system of (2.16) where the system matrices lie within the polytope $\bar{\Omega}$ of (2.12). We obtain the following theorem:

Theorem 2.3. *The exponential stability in the mean square of the system (2.16), where the system matrices lie within the polytope* $\bar{\Omega}$ *of (2.12) is guaranteed if there exist matrices* $Q > 0,\ R_1 > 0,\ Q_m$ *and a positive scalar* ϵ_f *that satisfy the following set of LMIs:*

$$\hat{\Psi}_i = \begin{bmatrix} \hat{\Psi}_{11,i} & QA_1^i - Q_m & Q_m & h\epsilon_f({A_0^i}^T Q + Q_m^T) \\ * & -R_1 + {H^i}^T Q H^i & 0 & h\epsilon_f({A_1^i}^T Q - Q_m^T) \\ * & * & -\epsilon_f Q & -h\epsilon_f Q_m^T \\ * & * & * & -\epsilon_f Q \end{bmatrix} < 0, \qquad (2.27)$$

$\forall i,\ i = 1, 2, ..., N$, where

$$\hat{\Psi}_{11,i} = QA_0^i + Q_m + {A_0^i}^T Q + Q_m^T + \frac{1}{1-d} R_1 + {G^i}^T Q G^i.$$

2.4 Bounded Real Lemma

We consider the system of (2.1a,c) with $B_2 = 0$ and $D_{12} = 0$ and the following index of performance

$$J_B \triangleq \mathcal{E}\{\int_0^\infty ||z(t)||^2 dt - \gamma^2 \int_0^\infty ||w(t)||^2 dt\}. \qquad (2.28)$$

Assuming that the stability condition of (2.24) is satisfied, we seek a condition that guarantees the following:

$$\mathcal{E} \int_0^\infty [LV + z^T(t)z(t) - \gamma^2 w^T(t)w(t)]dt < 0,$$

where in the expression for $\mathcal{E}\{LV\}$ the operators Δ_1 and Δ_2 are used. Using arguments similar to those in subsection 2.3.1, the latter requirement is described by the following theorem:

Theorem 2.4. *Consider the system (2.1a,c) with $B_2 = 0$ and $D_{12} = 0$. The system is exponentially stable in the mean square sense and, for a prescribed scalar $\gamma > 0$ and a given tuning parameter $\epsilon_f > 0$, the requirement of $J_B < 0$ is achieves for all nonzero $w \in \tilde{L}^2_{\mathcal{F}_t}([0,\infty); \mathcal{R}^q)$, if there exist matrices $Q > 0$, $R_1 > 0$, Q_m that satisfy the following LMI condition:*

$$\Gamma = \begin{bmatrix} \bar{\Psi}_{11} & \bar{\Psi}_{12} & Q_m & QB_1 & \bar{\Psi}_{15} \\ * & \bar{\Psi}_{22} & 0 & 0 & \bar{\Psi}_{25} \\ * & * & -\epsilon_f Q & 0 & -h\epsilon_f Q_m^T \\ * & * & * & -\gamma^2 I_q & h\epsilon_f B_1^T Q \\ * & * & * & * & -\epsilon_f Q \end{bmatrix} < 0, \qquad (2.29)$$

where

$$\begin{aligned}
\bar{\Psi}_{11} &= QA_0 + Q_m + A_0^T Q + Q_m^T + \tfrac{1}{1-d} R_1 + G^T Q G + C_1^T C_1\\
\bar{\Psi}_{12} &= QA_1 - Q_m + \bar{\alpha} G^T Q H,\\
\bar{\Psi}_{15} &= h\epsilon_f A_0^T Q + h\epsilon_f Q_m^T,\\
\bar{\Psi}_{22} &= -R_1 + H^T Q H,\\
\bar{\Psi}_{25} &= h\epsilon_f A_1^T Q - h\epsilon_f Q_m^T.
\end{aligned}$$

2.4.1 BRL for Delayed State-Multiplicative Systems – The Norm-Bounded Case

We consider the system (2.1a,c) with $B_2 = 0$ and $D_{12} = 0$, where A_0, A_1 are replaced by $\tilde{A}_0$, $\tilde{A}_1$, respectively and where the latter matrices are given in (2.10). Noting that

$$\bar{H}_i^T \epsilon_j \bar{H}_i + Q E_i \epsilon_j^{-1} E_i^T Q \geq Q E_i F_i \bar{H}_i + \bar{H}_i^T F_i^T E_i^T Q, \quad \epsilon_j > 0,\ i = 0,1,\ j = i+1$$

and applying similar arguments to those of Subsection 2.4, we arrive at the following theorem:

Theorem 2.5. *Consider the system (2.1a,c) with $B_2 = 0$ and $D_{12} = 0$. The system is exponentially stable in the mean square sense and, for a prescribed scalar $\gamma > 0$ and a given tuning parameter $\epsilon_f > 0$, the requirement of $J_B < 0$ is achieves for all nonzero $w \in \tilde{L}^2_{\mathcal{F}_t}([0,\infty); \mathcal{R}^q)$, if there exist matrices $Q > 0$, $R_1 > 0$, Q_m and scalars ϵ_1, ϵ_2 that satisfy the following LMI:*

$$\Gamma = \begin{bmatrix}
\tilde{\Psi}_{11} & \bar{\Psi}_{12} & Q_m & QB_1 & \bar{\Psi}_{15} & QE_0 & QE_1\\
* & \tilde{\Psi}_{22} & 0 & 0 & \bar{\Psi}_{25} & 0 & 0\\
* & * & -\epsilon_f Q & 0 & -h\epsilon_f Q_m^T & 0 & 0\\
* & * & * & -\gamma^2 I & h\epsilon_f B_1^T Q & 0 & 0\\
* & * & * & * & -\epsilon_f Q & h\epsilon_f Q E_0 & h\epsilon_f Q E_1\\
* & * & * & * & * & -\epsilon_1 I_n & 0\\
* & * & * & * & * & * & -\epsilon_2 I_n
\end{bmatrix} < 0, \qquad (2.30)$$

where

$$\tilde{\Psi}_{11} = \bar{\Psi}_{11} + C_1^T C_1 + \epsilon_1 \bar{H}_0^T \bar{H}_0, \quad \tilde{\Psi}_{22} = \bar{\Psi}_{22} + \epsilon_2 \bar{H}_1^T \bar{H}_1$$

and $\bar{\Psi}_{11}$, $\bar{\Psi}_{12}$, $\bar{\Psi}_{15}$ and $\bar{\Psi}_{25}$, are given in Theorem 2.4.

Remark 2.3. The delay independent BRL is readily obtained from the latter LMI by deleting the 3rd and 5th column and row blocks in Γ and by choosing $Q_m = 0$ in $\tilde{\Psi}_{11}$ and $\bar{\Psi}_{12}$.

2.4.2 BRL – The Polytopic Case

We consider the system (2.1a,c) with $B_2 = 0$ and $D_{12} = 0$, where the system matrices lie within the polytope $\bar{\Omega}$ of (2.12). Assuming that the stability condition of Theorem 2.3 is satisfied, we arrive at the following theorem:

Theorem 2.6. *Consider the system (2.1a,c) with $B_2 = 0$ and $D_{12} = 0$,, where the system matrices lie within the polytope $\bar{\Omega}$ of (2.12). The system is exponentially stable in the mean square sense and, for a prescribed $\gamma > 0$ and a given tuning parameter $\epsilon_f > 0$, the requirement of $J_B < 0$ is achieved for all nonzero $w \in \tilde{L}^2_{\mathcal{F}_t}([0,\infty);\mathcal{R}^q)$, if there exist matrices $Q > 0$, $R_1 > 0$ and Q_m that satisfy the following set of LMIs:*

$$\tilde{\Psi}_i = \begin{bmatrix} \tilde{\Psi}_{11,i} & \tilde{\Psi}_{12,i} & Q_m & QB_1^i & \tilde{\Psi}_{15,i} \\ * & \tilde{\Psi}_{22,i} & 0 & 0 & \tilde{\Psi}_{25,i} \\ * & * & -\epsilon_f Q & 0 & -h\epsilon_f Q_m^T \\ * & * & * & -\gamma^2 I_q & h\epsilon_f {B^i_1}^T Q \\ * & * & * & * & -\epsilon_f Q \end{bmatrix} < 0, \quad (2.31)$$

$\forall i,\ i = 1, 2, ..., N$, where

$$\begin{aligned} \tilde{\Psi}_{11,i} &= \hat{\Psi}_{11,i} + {C^i_1}^T C^i_1, \\ \tilde{\Psi}_{12,i} &= QA_1^i - Q_m, \\ \tilde{\Psi}_{22,i} &= -R_1 + {H^i}^T QH^i, \\ \tilde{\Psi}_{15,i} &= h\epsilon_f({A^i_0}^T Q + Q_m^T), \\ \tilde{\Psi}_{25,i} &= h\epsilon_f({A^i_1}^T Q - Q_m^T). \end{aligned}$$

2.5 Stochastic State-Feedback Control

2.5.1 State-Feedback Control – The Nominal Case

In this section we address the problem of finding the following state-feedback control law

$$u(t) = Kx(t), \quad (2.32)$$

that stabilizes the system and achieves a prescribed level of attenuation. We consider the system of (2.1a,c) and we apply the control law of (2.32), where A_0 is replaced by $A_0 + B_2 K$, C_1 is replaced by $C_1 + D_{12}K$ and where we assume, for simplicity, that $\bar{\alpha} = 0$. We obtain the following inequality:

$$\begin{bmatrix} \Upsilon_{11} & QA_1 - Q_m & Q_m & QB_1 & \Upsilon_{15} & \Upsilon_{16} & G^T Q \\ * & -R_1 + H^T Q H & 0 & 0 & \tilde{\Psi}_{25} & 0 & 0 \\ * & * & -\epsilon_f Q & 0 & -h\epsilon_f Q_m^T & 0 & 0 \\ * & * & * & -\gamma^2 I_q & h\epsilon_f B_1^T Q & 0 & 0 \\ * & * & * & * & -\epsilon_f Q & 0 & 0 \\ * & * & * & * & * & -I_r & 0 \\ * & * & * & * & * & * & -Q \end{bmatrix} < 0,$$

where

$$\Upsilon_{11} = QB_2K + K^T B_2^T Q + QA_0 + Q_m + A_0{}^T Q + Q_m^T + \tfrac{1}{1-d} R_1,$$
$$\Upsilon_{15} = h\epsilon_f([A_0 + B_2K]^T Q + Q_m^T),$$
$$\Upsilon_{16} = (C_1 + D_{12}K)^T,$$
$$\tilde{\Psi}_{25} = h\epsilon_f(A_1{}^T Q - Q_m^T).$$

Multiplying the above inequality by $diag\{Q^{-1}, Q^{-1}, Q^{-1}, I, Q^{-1}, I, Q^{-1}\}$, from the left and the right, denoting $P \triangleq Q^{-1}$, $m_p = PQ_mP$, $\bar{R}_p = PR_1P$ and $\hat{Y} = KP$, we obtain the following inequality:

$$\begin{bmatrix} \hat{\Upsilon}_{11} & A_1P - m_p & m_p & B_1 & \hat{\Upsilon}_{15} & \hat{\Upsilon}_{16} & PG^T & 0 \\ * & -\bar{R}_p & 0 & 0 & \hat{\Upsilon}_{25} & 0 & 0 & PH^T \\ * & * & -\epsilon_f P & 0 & -h\epsilon_f m_p^T & 0 & 0 & 0 \\ * & * & * & -\gamma^2 I_q & h\epsilon_f B_1^T & 0 & 0 & 0 \\ * & * & * & * & -\epsilon_f P & 0 & 0 & 0 \\ * & * & * & * & * & -I_r & 0 & 0 \\ * & * & * & * & * & * & -P & 0 \\ * & * & * & * & * & * & * & -P \end{bmatrix} < 0, \qquad (2.33)$$

where

$$\hat{\Upsilon}_{11} = B_2\hat{Y} + \hat{Y}^T B_2^T + m_p + m_p^T + PA_0^T + A_0P + \tfrac{1}{1-d}\bar{R}_p,$$
$$\hat{\Upsilon}_{15} = \epsilon_f h[PA_0^T + m_p^T + \hat{Y}^T B_2^T],$$
$$\hat{\Upsilon}_{16} = PC_1^T + \hat{Y}^T D_{12}^T,$$
$$\hat{\Upsilon}_{25} = \epsilon_f h[PA_1^T - m_p^T].$$

We thus arrive at the following theorem:

Theorem 2.7. *Consider the system (2.1a,c) and J_E of (2.3). For a prescribed $\gamma > 0$ and a given tuning scalar $\epsilon_f > 0$, there exists a state-feedback gain that achieves negative $J_E < 0$ for all nonzero $w \in \tilde{L}^2_{\mathcal{F}_t}([0,\infty); \mathcal{R}^q)$, if there exist* *matrices* $P > 0$, $\bar{R}_p$, m_p *and* $\hat{Y}$ *that satisfy (2.33). In the latter case the state-feedback gain is given by:*

$$K = \hat{Y}P^{-1}. \tag{2.34}$$

2.5.2 Robust State-Feedback Control – The Norm-Bounded Case

We consider the system (2.1a,c) where A_0, A_1, B_2 are replaced by $\tilde{A}_0$, $\tilde{A}_1$, $\tilde{B}_2$ respectively and where the latter matrices are given in (2.10). Noting that

$$\bar{H}_i^T \bar{\epsilon}_j^{-1} \bar{H}_i + QE_i\bar{\epsilon}_j E_i^T Q \geq QE_i F_i \bar{H}_i + H_i^T F_i^T E_i^T Q,\ \bar{\epsilon}_j > 0,\ i = 0, 1, j = i+1, \tag{2.35}$$

and applying similar arguments to those of Subsection 2.4.1, we arrive at the following theorem:

Theorem 2.8. *Consider the system (2.1a,c) and* J_E *of (2.3). For a prescribed* $\gamma > 0$ *and a given tuning scalar* $\epsilon_f > 0$, *there exists a robust state-feedback gain that achieves negative* $J_E < 0$ *for all nonzero* $w \in \tilde{L}^2_{\mathcal{F}_t}([0,\infty);\mathcal{R}^q)$, *if there exist matrices* $P > 0$, $\bar{R}_p$, m_p *and* $\hat{Y}$ *and scalars* $\bar{\epsilon}_1$, $\bar{\epsilon}_2$ *that satisfy (2.37). In the latter case the state-feedback gain is given by:*

$$K = \hat{Y}P^{-1}. \tag{2.36}$$

$$\begin{bmatrix}
\hat{\Upsilon}_{11} & \hat{\Upsilon}_{12} & m_p & B_1 & \hat{\Upsilon}_{15} & \hat{\Upsilon}_{16} & PG^T & 0 & \bar{\epsilon}_1 E_0 & \hat{\Upsilon}_{1,10} & \bar{\epsilon}_2 E_1 & 0 \\
* & -\bar{R}_p & 0 & 0 & \hat{\Upsilon}_{25} & 0 & 0 & PH^T & 0 & 0 & 0 & \hat{\Upsilon}_{2,12} \\
* & * & -\epsilon_f P & 0 & -h\epsilon_f m_p^T & 0 & 0 & 0 & 0 & 0 & 0 & 0 \\
* & * & * & -\gamma^2 I_q & h\epsilon_f B_1^T & 0 & 0 & 0 & 0 & 0 & 0 & 0 \\
* & * & * & * & -\epsilon_f P & 0 & 0 & 0 & 0 & 0 & 0 & 0 \\
* & * & * & * & * & -I_r & 0 & 0 & \tilde{\epsilon}_1 E_0 & 0 & \tilde{\epsilon}_2 E_1 & 0 \\
* & * & * & * & * & * & -P & 0 & 0 & 0 & 0 & 0 \\
* & * & * & * & * & * & * & -P & 0 & 0 & 0 & 0 \\
* & * & * & * & * & * & * & * & -\bar{\epsilon}_1 I_n & 0 & 0 & 0 \\
* & * & * & * & * & * & * & * & * & -\bar{\epsilon}_1 I_n & 0 & 0 \\
* & * & * & * & * & * & * & * & * & * & -\bar{\epsilon}_2 I_n & 0 \\
* & * & * & * & * & * & * & * & * & * & * & -\bar{\epsilon}_2 I_n
\end{bmatrix} < 0, \tag{2.37}$$

where

$$\hat{\Upsilon}_{11} = B_2\hat{Y} + \hat{Y}^T B_2^T + m_p + m_p^T + PA_0^T + A_0P + \tfrac{1}{1-d}\bar{R}_p,$$
$$\hat{\Upsilon}_{12} = A_1P - m_p,$$
$$\hat{\Upsilon}_{15} = \epsilon_f h[PA_0^T + m_p^T + \hat{Y}^T B_2^T],$$
$$\hat{\Upsilon}_{16} = PC_1^T + \hat{Y}^T D_{12}^T,$$
$$\hat{\Upsilon}_{1,10} = [\bar{H}_0P + \bar{H}_2\hat{Y}]^T,$$
$$\hat{\Upsilon}_{25} = \epsilon_f h[PA_1^T - m_p^T],$$
$$\hat{\Upsilon}_{2,12} = P\bar{H}_1^T,$$
$$\tilde{\epsilon}_1 = h\epsilon_f\bar{\epsilon}_1,$$
$$\tilde{\epsilon}_2 = h\epsilon_f\bar{\epsilon}_2.$$

Proof:
Noting that

$$\tilde{A}_0 = A_0 + E_0F_0\bar{H}_0,\ \tilde{A}_1 = A_1 + E_1F_1\bar{H}_1,\ \tilde{B}_2 = B_2 + E_0F_0\bar{H}_2,$$

we obtain the following inequality where Γ_1 is the left side of the inequality of (2.33)

$$\Gamma_1 + \Gamma_2F_0\bar{\Gamma}_2 + [\Gamma_2F_0\bar{\Gamma}_2]^T + \Gamma_3F_1\bar{\Gamma}_3 + [\Gamma_3F_1\bar{\Gamma}_3]^T < 0, \qquad (2.38)$$

where

$$\Gamma_2^T = [E_0^T\ 0\ 0\ 0\ 0\ [\epsilon_f hE_0^T]\ 0\ 0\ 0\ 0\ 0\ 0],$$
$$\bar{\Gamma}_2 = [[\bar{H}_0P + \bar{H}_2\hat{Y}]\ 0\ 0\ 0\ 0\ 0\ 0\ 0\ 0\ 0\ 0\ 0],$$
$$\Gamma_3^T = [E_1^T\ 0\ 0\ 0\ 0\ [\epsilon_f hE_1^T]\ 0\ 0\ 0\ 0\ 0\ 0],$$
$$\bar{\Gamma}_3 = [0\ \bar{H}_1P\ 0\ 0\ 0\ 0\ 0\ 0\ 0\ 0\ 0\ 0].$$

Applying the inequality of (2.35) and using Schur complement, the LMI of (2.37) is obtained.

2.5.3 Robust Polytopic State-Feedback Control

We consider the system of (2.1a,c) where A_0 is replaced by $A_0 + B_2K$, C_1 is replaced by $C_1 + D_{12}K$ and where the system matrices lie within the polytope $\bar{\Omega}$ of (2.12). Assuming, for simplicity, that $\bar{\alpha} = 0$, we obtain the following inequality:

$$\begin{bmatrix} \Upsilon_{11,i} & \tilde{\Psi}_{12,i} & Q_m & QB_1^i & \Upsilon_{15,i} & \Upsilon_{16,i} & G^{i,T}Q \\ * & -R_1+H^{i,T}QH^i & 0 & 0 & \tilde{\Psi}_{25,i} & 0 & 0 \\ * & * & -\epsilon_f Q & 0 & -h\epsilon_f Q_m^T & 0 & 0 \\ * & * & * & -\gamma^2 I_q & h\epsilon_f B_1^{i,T} Q & 0 & 0 \\ * & * & * & * & -\epsilon_f Q & 0 & 0 \\ * & * & * & * & * & -I_r & 0 \\ * & * & * & * & * & * & -Q \end{bmatrix} < 0,$$

$\forall i,\ i=1,2,....,N$, where

$$\Upsilon_{11,i} = QB_2^i K + K^T B_2^{i,T} Q + QA_0^i + Q_m + A_0^{i,T} Q + Q_m^T + \tfrac{1}{1-d} R_1,$$
$$\Upsilon_{12,i} = QA_1^i - Q_m,$$
$$\Upsilon_{15,i} = h\epsilon_f([A_0^{i,T} + K^T B_2^{i,T}]Q + Q_m^T),$$
$$\Upsilon_{16,i} = (C_1^i + D_{12}^i K)^T,$$
$$\Upsilon_{25,i} = h\epsilon_f({A_1^i}^T Q - Q_m^T).$$

Multiplying the above inequality by $diag\{Q^{-1}, Q^{-1}, Q^{-1}, I, Q^{-1}, I, Q^{-1}\}$, from the left and the right, denoting $P \stackrel{\Delta}{=} Q^{-1}$, $m_p = mP$, $\bar{R}_p = PR_1P$ and $\hat{Y} = KP$, we obtain the following inequalities:

$$\hat{\Upsilon}_i = \begin{bmatrix} \hat{\Upsilon}_{11,i} & \hat{\Upsilon}_{12,i} & m_p & B_1^i & \hat{\Upsilon}_{15,i} & \hat{\Upsilon}_{16,i} & PG^{i,T} & 0 \\ * & -\bar{R}_p & 0 & 0 & \hat{\Upsilon}_{25,i} & 0 & 0 & PH^{i,T} \\ * & * & -\epsilon_f P & 0 & -h\epsilon_f m_p^T & 0 & 0 & 0 \\ * & * & * & -\gamma^2 I_q & h\epsilon_f B_1^{i,T} & 0 & 0 & 0 \\ * & * & * & * & -\epsilon_f P & 0 & 0 & 0 \\ * & * & * & * & * & -I_r & 0 & 0 \\ * & * & * & * & * & * & -P & 0 \\ * & * & * & * & * & * & * & -P \end{bmatrix} < 0, \quad (2.39)$$

$\forall i,\ i=1,2,....,N$, where

$$\hat{\Upsilon}_{11,i} = B_2^i \hat{Y} + \hat{Y}^T B_2^{i,T} + m_p + m_p^T + PA_0^{i,T} + A_0^i P + \tfrac{1}{1-d}\bar{R}_1,$$
$$\hat{\Upsilon}_{12,i} = A_1^i P - m_p,$$
$$\hat{\Upsilon}_{15,i} = \epsilon_f h[PA_0^{i,T} + \bar{m}_p^T + \hat{Y}^T B_2^{I,T}],$$
$$\hat{\Upsilon}_{16,i} = PC_1^{i,T} + \hat{Y}^T D_{12}^{I,T},$$
$$\hat{\Upsilon}_{25,i} = \epsilon_f h[PA_1^{i,T} - m_p^T].$$

We thus arrive at the following theorem:

Theorem 2.9. *Consider the system (2.1a,c) where the system matrices lie within the polytope $\bar{\Omega}$ of (2.12). For a prescribed scalar $\gamma > 0$ and a given tuning scalar $\epsilon_f > 0$, there exists a robust state-feedback gain that achieves negative $J_E < 0$ for all nonzero $w \in \tilde{L}^2_{\mathcal{F}_t}([0,\infty);\mathcal{R}^q)$, if there exist matrices $P > 0$, $\bar{R}_p$, m_p and $\hat{Y}$ that satisfy (2.39). In the latter case the state-feedback gain is given by (2.34).*

2.5.4 Example – State-Feedback Control

We consider the system of (2.1a,c) , with $h = 0.6$ and $d = 0$ and with the following objective function:

$$J = \mathcal{E}\int_0^\infty \{||C_1x||^2 + 0.01||u||^2 - \gamma^2||w||^2 d\tau.\}$$

It is assumed that there is an access to the two states of the system and that

$$A = \begin{bmatrix} 0 & 1 \\ -1 & -0.4 \end{bmatrix}, \ A_1 = \begin{bmatrix} 0 & 0.1 \\ -0.10 & -0.04 \end{bmatrix}, \ B_1 = \begin{bmatrix} 1 \\ -1 \end{bmatrix},$$

$$B_2 = \begin{bmatrix} 0 \\ 1 \end{bmatrix} \quad G = \begin{bmatrix} 0 & 0.3 \\ -0.2 & -0.04 \end{bmatrix}, \ H = \begin{bmatrix} 0 & 0.18 \\ -0.09 & -0.15 \end{bmatrix}, \quad C_1 = \begin{bmatrix} -0.5 & 4 \\ 0 & 0 \end{bmatrix}.$$

It is further assumed that the stochastic multiplicative noise processes in both the delayed and the nondelayed states are uncorrelated (i.e $\bar{\alpha} = 0$).

Applying Corollary 2 of [129], a minimum attenuation level of $\gamma = 4.855$ is obtained. Using Theorem 2.7 above we obtain a near minimum attenuation level of $\gamma = 4.656$. The corresponding state-feedback gain is $K = -\begin{bmatrix} 4.2218 & 6.0887 \end{bmatrix}$.

2.6 Stochastic Filtering for Delayed Systems

2.6.1 Stochastic Filtering – The Nominal Case

In this section we address the filtering problem of the delayed state-multiplicative noisy system. We consider the system of (2.1a–c) with $B_2 = 0, \ D_{12} = 0, \ \bar{C}_2 = 0$ and the general type filter of (2.4). Denoting $\xi^T(t) \stackrel{\Delta}{=} [x(t)^T \ \ \hat{x}(t)^T]$, $\bar{w}^T(t) \stackrel{\Delta}{=} [w(t)^T \ \ n(t)^T]$ we obtain the following augmented system:

$$\begin{aligned} &d\xi(t) = [\tilde{A}_0\xi(t) + \tilde{B}\bar{w}(t)]dt + \tilde{A}_1\xi(t-\tau(t))dt + \tilde{H}\xi(t-\tau(t))d\nu(t) \\ &+\tilde{G}\xi(t)d\beta(t) + \tilde{F}\xi(t)d\zeta(t), \quad \xi(\theta) = 0, \ over[-h \ 0], \\ &\tilde{z}(t) = \tilde{C}\xi(t), \end{aligned} \tag{2.40}$$

where

$$\tilde{A}_0 = \begin{bmatrix} A_0 & 0 \\ B_c C_2 & A_c \end{bmatrix}, \ \tilde{B} = \begin{bmatrix} B_1 & 0 \\ 0 & B_c D_{21} \end{bmatrix}, \tilde{A}_1 = \begin{bmatrix} A_1 & 0 \\ 0 & 0 \end{bmatrix}, \ \tilde{H} = \begin{bmatrix} H & 0 \\ 0 & 0 \end{bmatrix},$$

$$\tilde{G} = \begin{bmatrix} G & 0 \\ 0 & 0 \end{bmatrix}, \ \tilde{F} = \begin{bmatrix} 0 & 0 \\ B_c F & 0 \end{bmatrix}, \ \tilde{C} = [C_1 \ - C_c]. \tag{2.41}$$

Using the BRL result, taking for simplicity $\bar{\alpha} = 0$ and applying Schur complement we obtain that (2.6) is satisfied if there exist Q, R_1, R_2, and m that satisfy the following inequality

$$\begin{bmatrix} \Upsilon_{11} & \Upsilon_{12} & Qm & Q\tilde{B} & \Upsilon_{15} & \tilde{C}^T & \tilde{G}^T Q & \Upsilon_{18} & 0 \\ * & -R_1 & 0 & 0 & \Upsilon_{25} & 0 & 0 & 0 & \Upsilon_{29} \\ * & * & -R_2 & 0 & -hm^T R_2 & 0 & 0 & 0 & 0 \\ * & * & * & -\gamma^2 I_{q+p} & h\tilde{B}^T R_2 & 0 & 0 & 0 & 0 \\ * & * & * & * & -R_2 & 0 & 0 & 0 & 0 \\ * & * & * & * & * & -I_r & 0 & 0 & 0 \\ * & * & * & * & * & * & -Q & 0 & 0 \\ * & * & * & * & * & * & * & -Q & 0 \\ * & * & * & * & * & * & * & * & -Q \end{bmatrix} < 0, \tag{2.42}$$

where

$$\begin{aligned} \Upsilon_{11} &= Q(\tilde{A}_0 + m) + (\tilde{A}_0 + m)^T Q + \tfrac{1}{1-d} R_1, \\ \Upsilon_{12} &= Q(\tilde{A}_1 - m), \\ \Upsilon_{15} &= h\tilde{A}_0^T R_2 + hm^T R_2, \\ \Upsilon_{18} &= \tilde{F}^T Q, \\ \Upsilon_{25} &= h\tilde{A}_1^T R_2 - hm^T R_2, \\ \Upsilon_{29} &= \tilde{H}^T Q. \end{aligned}$$

Defining $\tilde{Q} = Q^{-1}$, denoting the following partitions:

$$\tilde{Q} = \begin{bmatrix} X & M^T \\ M & T \end{bmatrix}, \quad Q = \begin{bmatrix} Y & N^T \\ N & W \end{bmatrix}$$

and defining

$$J = \begin{bmatrix} X^{-1} & Y \\ 0 & N \end{bmatrix},$$

we multiply (2.42) by $\hat{J} = diag\{\tilde{Q}J, \tilde{Q}J, \tilde{Q}J, I, \tilde{Q}J, I, \tilde{Q}J, \tilde{Q}J, \tilde{Q}J,\}$ from the right and by $\hat{J}^T$, from the left.

We obtain:

$$\begin{bmatrix} \tilde{\Upsilon}_{11} & J^T(\tilde{A}_1 - m)\tilde{Q}J & J^T m\tilde{Q}J & J^T\tilde{B} & \tilde{\Upsilon}_{15} & J^T\tilde{Q}\tilde{C}^T & J^T\tilde{Q}\tilde{G}^T J & J^T\tilde{Q}\tilde{F}^T J & 0 \\ * & -\bar{R}_1 & 0 & 0 & \tilde{\Upsilon}_{25} & 0 & 0 & 0 & J^T\tilde{Q}\tilde{H}^T J \\ * & * & -\bar{R}_2 & 0 & \tilde{\Upsilon}_{35} & 0 & 0 & 0 & 0 \\ * & * & * & -\gamma^2 I_{q+p} & h\epsilon_f\tilde{B}^T J & 0 & 0 & 0 & 0 \\ * & * & * & * & -\bar{R}_2 & 0 & 0 & 0 & 0 \\ * & * & * & * & * & -I_r & 0 & 0 & 0 \\ * & * & * & * & * & * & -J^T\tilde{Q}J & 0 & 0 \\ * & * & * & * & * & * & * & -J^T\tilde{Q}J & 0 \\ * & * & * & * & * & * & * & * & -J^T\tilde{Q}J \end{bmatrix} < 0, \tag{2.43}$$

where:

$$\begin{aligned} \tilde{\Upsilon}_{11} &= J^T\tilde{Q}\Upsilon_{11}\tilde{Q}J, \\ \tilde{\Upsilon}_{15} &= J^T\tilde{Q}\Upsilon_{15}\tilde{Q}J = J^T\tilde{Q}[h(\tilde{A}_0^T R_2 + m^T R_2)]\tilde{Q}J, \\ \tilde{\Upsilon}_{25} &= J^T\tilde{Q}\Upsilon_{25}\tilde{Q}J = J^T\tilde{Q}[h(\tilde{A}_1^T R_2 - m^T R_2)]\tilde{Q}J, \\ \tilde{\Upsilon}_{35} &= -hJ^T\tilde{Q}m^T R_2\tilde{Q}J, \end{aligned}$$

and $\bar{R}_i = J^T\tilde{Q}R_i\tilde{Q}J,\ i = 1, 2$. Carrying out the various multiplications contained in (2.43), taking

$$R_2 = \epsilon_f Q$$

and defining

$$\begin{aligned} \bar{X} = X^{-1}, \quad \bar{X}_Y = \begin{bmatrix} \bar{X} & \bar{X} \\ \bar{X} & Y \end{bmatrix}, \quad K_0 = N^T A_c M\bar{X}, \\ U = N^T B_c, \quad \tilde{M} = J^T m\tilde{Q}J,\ Z = C_c M\bar{X}, \end{aligned} \tag{2.44}$$

we obtain:

$$\begin{bmatrix} \tilde{\Upsilon}_{11} & \tilde{\Upsilon}_{12} & \tilde{M} & \tilde{\Upsilon}_{14} & \tilde{\Upsilon}_{15} & \tilde{\Upsilon}_{16} & \tilde{\Upsilon}_{17} & \tilde{\Upsilon}_{18} & 0 \\ * & -\bar{R}_1 & 0 & 0 & \tilde{\Upsilon}_{25} & 0 & 0 & 0 & \tilde{\Upsilon}_{29} \\ * & * & -\epsilon_f \bar{X}_Y & 0 & -\epsilon_f h \tilde{M}^T & 0 & 0 & 0 & 0 \\ * & * & * & -\gamma^2 I_{q+p} & \tilde{\Upsilon}_{45} & 0 & 0 & 0 & 0 \\ * & * & * & * & -\epsilon_f \bar{X}_Y & 0 & 0 & 0 & 0 \\ * & * & * & * & * & -I_r & 0 & 0 & 0 \\ * & * & * & * & * & * & -\bar{X}_Y & 0 & 0 \\ * & * & * & * & * & * & * & -\bar{X}_Y & 0 \\ * & * & * & * & * & * & * & * & -\bar{X}_Y \end{bmatrix} < 0, \quad (2.45)$$

where

$$\begin{aligned}
\tilde{\Upsilon}_{11} &= \begin{bmatrix} \bar{X}A_0 + A_0^T \bar{X} & \bar{X}A_0 + A_0^T Y + C_2^T U^T + K_0^T \\ YA_0 + UC_2 + K_0 + A_0^T \bar{X} & YA_0 + UC_2 + A_0^T Y + C_2^T U^T \end{bmatrix} \\
&+ \tilde{M} + \tilde{M}^T + \tfrac{1}{1-d}\bar{R}_1, \\
\tilde{\Upsilon}_{12} &= \begin{bmatrix} \bar{X}A_1 & \bar{X}A_1 \\ YA_1 & YA_1 \end{bmatrix} - \tilde{M}, \\
\tilde{\Upsilon}_{14} &= \begin{bmatrix} \bar{X}B_1 & 0 \\ YB_1 & UD_{21} \end{bmatrix}, \\
\tilde{\Upsilon}_{15} &= \epsilon_f h \begin{bmatrix} A_0^T \bar{X} & A_0^T Y + C_2^T U^T + K_0^T \\ A_0^T \bar{X} & C_2^T U^T \end{bmatrix} + \epsilon_f h \tilde{M}^T, \\
\tilde{\Upsilon}_{16} &= \begin{bmatrix} C^T - Z^T \\ -C^T \end{bmatrix}, \\
\tilde{\Upsilon}_{17} &= \begin{bmatrix} G^T \bar{X} & G^T Y \\ G^T \bar{X} & G^T Y \end{bmatrix}, \\
\tilde{\Upsilon}_{18} &= \begin{bmatrix} 0 & F^T U^T \\ 0 & F^T U^T \end{bmatrix}, \\
\tilde{\Upsilon}_{25} &= \epsilon_f h \begin{bmatrix} A_1^T \bar{X} & A_1^T Y \\ A_1^T \bar{X} & A_1^T Y \end{bmatrix} + \epsilon_f h \tilde{M}^T, \\
\tilde{\Upsilon}_{29} &= \begin{bmatrix} H^T \bar{X} & H^T Y \\ H^T \bar{X} & H^T Y \end{bmatrix}, \\
\tilde{\Upsilon}_{45} &= \epsilon_f h \tilde{\Upsilon}_{14}.
\end{aligned} \quad (2.46)$$

We thus arrive at the following theorem:

Theorem 2.10. *Consider the system of (2.1a-c) with* $B_2 = 0,\ D_{12} = 0$ *and* $\bar{C}_2 = 0$. *For a prescribed* $\gamma > 0$ *and tuning scalar* ϵ_f, *there exists a filter of the structure (2.4) that achieves (2.6) for all nonzero* $w \in \tilde{L}^2([0,\infty);\mathcal{R}^q),\ n \in \tilde{L}^2([0,\infty);\mathcal{R}^p)$, *if there exist matrices* $\bar{X} > 0,\ Y > 0,\ K_0,\ U,\ \bar{R}_1,\ \tilde{M},\ Z$, *that satisfy (2.45). In the latter case the filter parameters can be extracted using the following equations:*

$$A_c = N^{-T}K_0\bar{X}M^{-1},\ B_c = N^{-T}U,\ C_c = Z\bar{X}M^{-1}. \tag{2.47}$$

Namely, the transfer function matrix of the filter is given by $T(s) = Z(s(\bar{X} - Y) - K_0)^{-1}U$.

2.6.2 Robust Filtering – The Norm-Bounded Case

We consider the system of (2.1a–c) with $B_2 = 0,\ D_{12} = 0,\ \bar{C}_2 = 0$ and the general type filter of (2.4), where $A_0,\ A_1$ are replaced by $\tilde{A}_0,\ \tilde{A}_1$, respectively, and where the latter matrices are given in (2.10). Noting that

$$\bar{H}_i^T\epsilon_j\bar{H}_i + QE_i\epsilon_j^{-1}E_i^TQ \geq QE_iF_i\bar{H}_i + \bar{H}_i^TF_i^TE_i^TQ, \quad \epsilon_j > 0,\ i = 0,1, j = i+1,$$

we arrive at the following theorem:

Theorem 2.11. *Consider the system of (2.1a-c) with* $B_2 = 0,\ D_{12} = 0$ *and* $\bar{C}_2 = 0$ *and the general type filter of (2.4), where* $A_0,\ A_1$ *are replaced by* $\tilde{A}_0,\ \tilde{A}_1$ *respectively and where the latter matrices are given in (2.10). For a prescribed* $\gamma > 0$ *and tuning scalar* ϵ_f, *there exists a filter of the structure (2.4) that achieves (2.6) for all nonzero* $w \in \tilde{L}^2([0,\infty);\mathcal{R}^q),\ n \in \tilde{L}^2([0,\infty);\mathcal{R}^p)$, *if there exist scalars* $\epsilon_1,\ \epsilon_2$ *and matrices* $\bar{X} > 0,\ Y > 0,\ K_0,\ U,\ \bar{R}_1,\ \tilde{M},\ Z$, *that satisfy (2.49). In the latter case the filter parameters can be extracted using the following equations:*

$$A_c = N^{-T}K_0\bar{X}M^{-1},\ B_c = N^{-T}U,\ C_c = Z\bar{X}M^{-1}, \tag{2.48}$$

where

$$\begin{bmatrix} \Upsilon_1 & \Upsilon_2 & \Upsilon_3 & \Upsilon_4 \\ * & -\epsilon_1 I_n & 0 & 0 \\ * & * & -\epsilon_1 I_n & 0 \\ * & * & * & -\epsilon_2 I_n \end{bmatrix} < 0, \tag{2.49}$$

and where

$$\Upsilon_2^T = \left[E_0^T[\bar{X}\ \ Y]\ \ 0\ \ 0\ \ 0\ \ h\epsilon_f E_0^T[\bar{X}\ \ Y]\ \ 0\ \ 0\ \ 0\ \ 0\right]^T,$$
$$\Upsilon_3^T = \left[E_0^T[\bar{X}\ \ Y]\ \ 0\ \ 0\ \ 0\ \ h\epsilon_f E_0^T[\bar{X}\ \ 0]\ \ 0\ \ 0\ \ 0\ \ 0\right]^T,$$
$$\Upsilon_4^T = \left[E_1^T[\bar{X}\ \ Y]\ \ 0\ \ 0\ \ 0\ \ h\epsilon_f E_1^T[\bar{X}\ \ Y]\ \ 0\ \ 0\ \ 0\ \ 0\right]^T.$$

and where Υ_1 is the matrix in the left side of (2.45) which includes the additional term of $\begin{bmatrix} \bar{H}_0^T \epsilon_1 \bar{H}_0 & 0 \\ 0 & \bar{H}_0^T \epsilon_1 \bar{H}_0 \end{bmatrix}$ in $\bar{\Upsilon}_{11}$ and the additional term of $\begin{bmatrix} \bar{H}_1^T \epsilon_2 \bar{H}_1 & \bar{H}_1^T \epsilon_2 \bar{H}_1 \\ \bar{H}_1^T \epsilon_2 \bar{H}_1 & \bar{H}_1^T \epsilon_2 \bar{H}_1 \end{bmatrix}$ in the $(2,2)$ block of (2.45).

2.6.3 Robust Polytopic Stochastic Filtering

We consider the system of (2.1a–c) with $B_2 = 0$, $D_{12} = 0$, $\bar{C}_2 = 0$ and where the system matrices lie within the polytope $\bar{\Omega}$ of (2.12). Assuming, for simplicity, that $\bar{\alpha} = 0$, we obtain the following theorem:

Theorem 2.12. *Consider the system of (2.1a–c) with $B_2 = 0$ and $D_{12} = 0$, $\bar{C}_2 = 0$, where the system matrices lie within the polytope $\bar{\Omega}$ of (2.12). For a prescribed scalar $\gamma > 0$ and tuning scalar ϵ_f, there exists a filter of the structure (2.4) that achieves (2.6) for all nonzero $w \in \tilde{\mathcal{L}}^2([0,\infty);\mathcal{R}^q)$, $n \in \tilde{\mathcal{L}}^2([0,\infty);\mathcal{R}^p)$, if there exist matrices $\bar{X} > 0$, $Y > 0$, K_0, U, $\bar{R}_1$, $\tilde{M}$, Z, that satisfy (2.50).*

$$\begin{bmatrix} \tilde{\Upsilon}_{i,11} & \tilde{\Upsilon}_{i,12} & \tilde{M} & \tilde{\Upsilon}_{i,14} & \tilde{\Upsilon}_{i,15} & \tilde{\Upsilon}_{i,16} & \tilde{\Upsilon}_{i,17} & \tilde{\Upsilon}_{i,18} & 0 \\ * & -\bar{R}_1 & 0 & 0 & \tilde{\Upsilon}_{i,25} & 0 & 0 & 0 & \tilde{\Upsilon}_{i,29} \\ * & * & -\epsilon_f \bar{X}_Y & 0 & -\epsilon_f h \tilde{M}^T & 0 & 0 & 0 & 0 \\ * & * & * & -\gamma^2 I_{q+p} & \tilde{\Upsilon}_{i,45} & 0 & 0 & 0 & 0 \\ * & * & * & * & -\epsilon_f \bar{X}_Y & 0 & 0 & 0 & 0 \\ * & * & * & * & * & -I_r & 0 & 0 & 0 \\ * & * & * & * & * & * & -\bar{X}_Y & 0 & 0 \\ * & * & * & * & * & * & * & -\bar{X}_Y & 0 \\ * & * & * & * & * & * & * & * & -\bar{X}_Y \end{bmatrix} < 0, \tag{2.50}$$

$\forall i$, $i = 1, 2, \ldots., N$, *where*

$$\tilde{\Upsilon}_{i,11} = \begin{bmatrix} \bar{X}A_0^i + A_0^{i,T}\bar{X} & \bar{X}A_0^i + A_0^{i,T}Y + C_2^{i,T}U^T + K_0^T \\ YA_0^i + UC_2^i + K_0 + A_0^{i,T}\bar{X} & YA_0^i + UC_2^i + A_0^{i,T}Y + C_2^{i,T}U^T \end{bmatrix}$$
$$+\tilde{M} + \tilde{M}^T + \frac{1}{1-d}\bar{R}_1,$$

$$\tilde{\Upsilon}_{i,12} = \begin{bmatrix} \bar{X}A_1^i & \bar{X}A_1^i \\ YA_1^i & YA_1^i \end{bmatrix} - \tilde{M},$$

$$\tilde{\Upsilon}_{i,14} = \begin{bmatrix} \bar{X}B_1^i & 0 \\ YB_1^i & UD_{21}^i \end{bmatrix},$$

$$\tilde{\Upsilon}_{i,15} = \epsilon_f h \begin{bmatrix} A_0^{i,T}\bar{X} & A_0^{i,T}Y + C_2^{i,T}U^T + K_0^T \\ A_0^{i,T}\bar{X} & C_2^{i,T}U^T \end{bmatrix} + \epsilon_f h\tilde{M}^T,$$

$$\tilde{\Upsilon}_{i,16} = \begin{bmatrix} C^{i,T} - Z^T \\ -C^{i,T} \end{bmatrix},$$

$$\tilde{\Upsilon}_{i,17} = \begin{bmatrix} G^{i,T}\bar{X} & G^{i,T}Y \\ G^{i,T}\bar{X} & G^{i,T}Y \end{bmatrix},$$

$$\tilde{\Upsilon}_{i,18} = \begin{bmatrix} 0 & F^{i,T}U^T \\ 0 & F^{i,T}U^T \end{bmatrix},$$

$$\tilde{\Upsilon}_{i,25} = \epsilon_f h \begin{bmatrix} A_1^{i,T}\bar{X} & A_1^{i,T}Y \\ A_1^{i,T}\bar{X} & A_1^{i,T}Y \end{bmatrix} + \epsilon_f h\tilde{M}^T,$$

$$\tilde{\Upsilon}_{i,29} = \begin{bmatrix} H^{i,T}\bar{X} & H^{i,T}Y \\ H^{i,T}\bar{X} & H^{i,T}Y \end{bmatrix}, \tag{2.51}$$
$$\tilde{\Upsilon}_{i,45} = \epsilon_f h\tilde{\Upsilon}_{i,14}.$$

In the latter case the filter parameters can be extracted by using (2.47).

2.6.4 Example – Filtering

We consider the system of [18], where the system (2.1a,c) is described by:

$$A_0 = \begin{bmatrix} -2 & 0 \\ 1 & -1 \end{bmatrix},\ A_1 = \begin{bmatrix} -1 & 0 \\ -0.5 & -1 \end{bmatrix},\ G = H = \sqrt{0.1}I_2,$$

with

$$B_2 = 0,\ D_{12} = 0,\ C_2 = \begin{bmatrix} 1 & 2 \end{bmatrix},\ F = \begin{bmatrix} 0.03 & 0.02 \end{bmatrix},\ D_{21} = 0.01$$

and

$$\bar{C}_2 = [0 \ \ 0],\ C_1 = \begin{bmatrix} -0.5 & 1 \end{bmatrix},\ d = 0.$$

We assume that the stochastic multiplicative noise processes in both the delayed and the nondelayed states are uncorrelated (i.e $\bar{\alpha} = 0$). Taking $h = 2$ and applying the results of Theorem 2.10, for the near minimum attenuation level of $\gamma = 10.073$, the following transfer function for the stationary filter is obtained: $T(s) = \frac{5.075s+10.88}{s^2+13.36s+23.11}$.

2.7 Stochastic Output-Feedback Control for Delayed Systems

2.7.1 Stochastic Output-Feedback Control – The Nominal Case

In this section we address the dynamic output-feedback control problem of the delayed state-multiplicative uncertain noisy system [59]. We consider the system of (2.1a–c) and the general type controller of (2.7).

Denoting $\xi^T(t) \triangleq [x(t)^T \;\; \hat{x}(t)^T]$ and $\bar{w}^T(t) \triangleq [w(t)^T \;\; n(t)^T]$, we obtain the following augmented system:

$$\begin{array}{l} d\xi(t) = [\hat{A}_0\xi(t) + \tilde{B}\bar{w}(t)]dt + \hat{A}_1\xi(t-\tau(t))dt + \tilde{H}\xi(t-\tau(t))d\nu(t) + \\ \tilde{G}\xi(t)d\beta(t) + \tilde{F}\xi(t)d\zeta(t), \quad \xi(\theta) = 0, \; over[-h \; 0], \\ \tilde{z}(t) = \tilde{C}\xi(t), \end{array} \tag{2.52}$$

with the following matrices:

$$\hat{A}_0 = \begin{bmatrix} A_0 & B_2C_c \\ B_cC_2 & A_c \end{bmatrix}, \; \hat{A}_1 = \begin{bmatrix} A_1 & 0 \\ B_c\bar{C}_2 & 0 \end{bmatrix}, \; \tilde{B} = \begin{bmatrix} B_1 & 0 \\ 0 & B_cD_{21} \end{bmatrix},$$

$$\tilde{H} = \begin{bmatrix} H & 0 \\ 0 & 0 \end{bmatrix}, \; \tilde{G} = \begin{bmatrix} G & 0 \\ 0 & 0 \end{bmatrix}, \; \tilde{F} = \begin{bmatrix} 0 & 0 \\ B_cF & 0 \end{bmatrix}, \; \tilde{C} = [C_1 \;\; D_{12}C_c]. \tag{2.53}$$

Using the BRL result of Theorem 2.5, taking for simplicity $\bar{\alpha} = 0$, we obtain that (2.3) is satisfied if there exist $\tilde{Q}$, R_1, R_2, and m that satisfy the following inequality:

$$\begin{bmatrix} \Upsilon_{11} & \tilde{Q}(\hat{A}_1 - m) & \tilde{Q}m & \tilde{Q}\tilde{B} & h\hat{A}_0^T R_2 + hm^T R_2 & \tilde{C}^T & \tilde{G}^T\tilde{Q} & \tilde{F}^T\tilde{Q} & 0 \\ * & -R_1 & 0 & 0 & h\hat{A}_1^T R_2 - hm^T R_2 & 0 & 0 & 0 & \tilde{H}^T\tilde{Q} \\ * & * & -R_2 & 0 & -hm^T R_2 & 0 & 0 & 0 & 0 \\ * & * & * & -\gamma^2 I_{q+p} & h\tilde{B}^T R_2 & 0 & 0 & 0 & 0 \\ * & * & * & * & -R_2 & 0 & 0 & 0 & 0 \\ * & * & * & * & * & -I_r & 0 & 0 & 0 \\ * & * & * & * & * & * & -\tilde{Q} & 0 & 0 \\ * & * & * & * & * & * & * & -\tilde{Q} & 0 \\ * & * & * & * & * & * & * & * & -\tilde{Q} \end{bmatrix} < 0, \tag{2.54}$$

where

$$\Upsilon_{11} = \tilde{Q}(\hat{A}_0 + m) + (\hat{A}_0 + m)^T\tilde{Q} + \frac{1}{1-d}R_1.$$

Denoting the following partition:

$$\tilde{Q} = \begin{bmatrix} Q & \alpha\hat{Q} \\ \alpha\hat{Q} & \hat{Q} \end{bmatrix}, \tag{2.55}$$

where $Q > 0, \hat{Q} > 0 \in \mathcal{R}^{n\times n}$ and α is a tuning scalar parameter, taking $R_2 = \epsilon_f\tilde{Q}$ where $\epsilon_f > 0$ is a tuning scalar parameter and denoting $\tilde{Q}_m = \tilde{Q}m$, we obtain the following inequality:

$$\begin{bmatrix} \bar{\Upsilon}_{11} & \bar{\Upsilon}_{12} & \tilde{Q}_m & \bar{\Upsilon}_{14} & \bar{\Upsilon}_{15} & \begin{bmatrix} C_1^T \\ C_c^T D_{12}^T \end{bmatrix} & \bar{\Upsilon}_{17} & \bar{\Upsilon}_{18} & 0 \\ * & -R_1 & 0 & 0 & \bar{\Upsilon}_{25} & 0 & 0 & 0 & \bar{\Upsilon}_{29} \\ * & * & -\epsilon_f\tilde{Q} & 0 & -h\epsilon_f\tilde{Q}_m^T & 0 & 0 & 0 & 0 \\ * & * & * & -\gamma^2 I_{q+p} & h\epsilon_f\bar{\Upsilon}_{14}^T & 0 & 0 & 0 & 0 \\ * & * & * & * & -\epsilon_f\tilde{Q} & 0 & 0 & 0 & 0 \\ * & * & * & * & * & -I_r & 0 & 0 & 0 \\ * & * & * & * & * & * & -\tilde{Q} & 0 & 0 \\ * & * & * & * & * & * & * & -\tilde{Q} & 0 \\ * & * & * & * & * & * & * & * & -\tilde{Q} \end{bmatrix} < 0, \tag{2.56}$$

where

$$\bar{\Upsilon}_{11} = \begin{bmatrix} QA_0 + A_0^T Q + \alpha B_{f,q} C_2 + \alpha C_2^T B_{f,q}^T & QB_2 C_c + \alpha A_{f,q} + \alpha A_0^T \hat{Q} + C_2^T B_{f,q}^T \\ C_c^T B_2^T Q + \alpha A_{f,q}^T + \alpha \hat{Q} A_0 + B_{f,q} C_2 & \alpha \hat{Q} B_2 C_c + \alpha C_c^T B_2^T \hat{Q} + A_{f,q} + A_{f,q}^T \end{bmatrix}$$

$$+\tilde{Q}_m + \tilde{Q}_m^T + \tfrac{1}{1-d} R_1,$$

$$\bar{\Upsilon}_{12} = \begin{bmatrix} QA_1 + \alpha B_{f,q} \bar{C}_2 & 0 \\ \alpha \hat{Q} A_1 + B_{f,q} \bar{C}_2 & 0 \end{bmatrix} - \tilde{Q}_m,$$

$$\bar{\Upsilon}_{14} = \begin{bmatrix} QB_1 & \alpha B_{f,q} D_{21} \\ \alpha \hat{Q} B_1 & B_{f,q} D_{21} \end{bmatrix},$$

$$\bar{\Upsilon}_{15} = \epsilon_f h \begin{bmatrix} A_0^T Q + \alpha C_2^T B_{f,q}^T & \alpha A_0^T \hat{Q} + C_2^T B_{f,q}^T \\ C_c^T B_2^T Q + \alpha A_{f,q}^T & \alpha C_c^T B_2^T \hat{Q} + A_{f,q}^T \end{bmatrix} + h\epsilon_f \tilde{Q}_m^T,$$

$$\bar{\Upsilon}_{17} = \begin{bmatrix} G^T Q & \alpha G^T \hat{Q} \\ 0 & 0 \end{bmatrix},$$

$$\bar{\Upsilon}_{18} = \begin{bmatrix} \alpha F^T B_{f,q}^T & F^T B_{f,q}^T \\ 0 & 0 \end{bmatrix},$$

$$\bar{\Upsilon}_{25} = \epsilon_f h \begin{bmatrix} A_1^T Q + \alpha \bar{C}_2^T B_{f,q}^T & \alpha A_1^T \hat{Q} + \bar{C}_2^T B_{f,q}^T \\ 0 & 0 \end{bmatrix} - h\epsilon_f \tilde{Q}_m^T,$$

$$\bar{\Upsilon}_{29} = \begin{bmatrix} H^T Q & \alpha H^T \hat{Q} \\ 0 & 0 \end{bmatrix},$$

and where

$$A_{f,q} \triangleq \hat{Q} A_c, \quad B_{f,q} \triangleq \hat{Q} B_c. \tag{2.57}$$

We note that inequality (2.56) is bilinear due to the terms $\alpha C_c^T B_2^T \hat{Q}$ and QB_2C_c in $\bar{\Upsilon}_{11}$. Similar bilinearity is observed in the standard treatment of the H_∞ output-feedback control design of delay-free systems [104], [40]. While in the latter design, a nonlinear transformation on the decision matrices is applied which entangles the biliniarity, in our case the bilinearity involves the special structure assumed for $\tilde{Q}$ in (2.55) and the above transformation will not resolve the problem. There exist few algorithms that may solve bilinear matrix inequalities, however, they do not always converge to a global minimum and they may require considerable computational effort. In order to remain in the linear domain, we pre-choose the matrix C_c at the cost of some overdesign. We arrive at the following theorem:

Theorem 2.13. *Consider the system (2.1a-c). There exists a controller of the structure of (2.7) that achieves negative J_{OF} for all nonzero $w \in \tilde{L}^2_{\mathcal{F}_t}([0,\infty);\mathcal{R}^q)$, $n \in \tilde{L}^2_{\mathcal{F}_t}([0,\infty);\mathcal{R}^p)$, for a prescribed scalar $\gamma > 0$ and tuning parameters α and ϵ_f if there exist matrices $Q > 0$, $\hat{Q} > 0$, $\tilde{Q}_m$, $R_1 > 0$ and $A_{f,q}$, $B_{f,q}$, that satisfy (2.56). In the latter case the controller parameters are:*

$$A_c = \hat{Q}^{-1} A_{f,q} \quad and \quad B_c = \hat{Q}^{-1} B_{f,q}, \tag{2.58}$$

with a pre-chosen structure of C_c.

Remark 2.4. Note that if one pursuits the method of [104] and apply it to the solution of the delayed output-feedback control problem, it will be necessary to incorporate the delay also in the structure of the controller. Since the delay in our problem is not known exactly, a special structure is imposed on $\tilde{Q}$ of (2.54), admittedly involving an overdesign. The above structure of $\tilde{Q}$ has been used in [58] for the solution of the discrete-time stochastic output-feedback control problem and has been shown to yield good results.

A delay-independent result is readily achieved using the results of Remark 2.3 in Section 2.4.1. We arrive at the following:

Corollary 2.7.1 *Consider the system (2.1a–c). There exists a controller of the structure (2.7) that for a prescribed scalar* $\gamma > 0$ *achieves negative* J_{OF} *independently of the delay h, for all nonzero* $w \in \tilde{L}^2_{\mathcal{F}_t}([0,\infty); \mathcal{R}^q)$, $n \in \tilde{L}^2_{\mathcal{F}_t}([0,\infty); \mathcal{R}^p)$, *if there exist matrices* $Q > 0$, $\hat{Q} > 0$, $R_1 > 0$ *and* $A_{f,q}$, $B_{f,q}$ *and a tuning parameter* α, *that satisfy the following LMI:*

$$\begin{bmatrix} \tilde{\Upsilon}_{11} & \begin{bmatrix} QA_1 + \alpha B_{f,q}\bar{C}_2 & 0 \\ \alpha\hat{Q}A_1 + B_{f,q}\bar{C}_2 & 0 \end{bmatrix} & \bar{\Upsilon}_{14} & \begin{bmatrix} C_1^T \\ C_c^T D_{12}^T \end{bmatrix} & \bar{\Upsilon}_{17} & \bar{\Upsilon}_{18} & 0 \\ * & -R_1 & 0 & 0 & 0 & 0 & \bar{\Upsilon}_{29} \\ * & * & -\gamma^2 I_{q+p} & 0 & 0 & 0 & 0 \\ * & * & * & -I_r & 0 & 0 & 0 \\ * & * & * & * & -\tilde{Q} & 0 & 0 \\ * & * & * & * & * & -\tilde{Q} & 0 \\ * & * & * & * & * & * & -\tilde{Q} \end{bmatrix} < 0, \tag{2.59}$$

where $\bar{\Upsilon}_{14}$, $\bar{\Upsilon}_{17}$, $\bar{\Upsilon}_{18}$, *and* $\bar{\Upsilon}_{29}$ *are given following (2.56) and where* $\tilde{\Upsilon}_{11}$ *is the former* $\bar{\Upsilon}_{11}$ *where the terms of* $\bar{Q}_m$ *are omitted. In the latter case the controller parameters are given in (2.58) with a pre-chosen structure of* C_c.

Remark 2.5. We note that a simple strategy is applied for the numerical solution of (2.56). The solution of (2.56) involves a search for two scalar variables: α and ϵ_f. One may start by line searching for α, taking a fixed value for ϵ_f, that leads to a stabilizing controller of minimum γ. Once such a controller is obtained, standard optimization techniques can be used, say Matlab function "fminsearch", which seek the combination of the two scalar parameters that bring γ to a local minimum.

2.7.2 Example – Output-Feedback Control

We bring a stationary modified version of an example which is taken from the field of guidance control ([136], see also [53], Chapter 11). We consider the nominal system obtained from (2.16a–c), with $h = 0.22$ and $d = 0.4$ and with the following system matrices:

$$A = \begin{bmatrix} 0 & 1 & 0 \\ 0 & 0 & 1 \\ 0 & 0 & -30 \end{bmatrix}, \; G = \begin{bmatrix} 0 & 0 & 0 \\ 0 & 0 & 0 \\ 0 & 0 & 4 \end{bmatrix}, \; B_1 = \begin{bmatrix} 0 \\ 1 \\ 0 \end{bmatrix}, \; B_2 = \begin{bmatrix} 0 \\ 0 \\ 30 \end{bmatrix}, \; C_1 = \begin{bmatrix} 0 & 0 & 1 \end{bmatrix},$$

$$C_2 = \begin{bmatrix} 0 & 0 & 0 \\ 0 & 0 & 1 \end{bmatrix}, \; \bar{C}_2 = \begin{bmatrix} 1 & 0 & 0 \\ 0 & 0 & 0 \end{bmatrix}, \; D_{21} = \begin{bmatrix} 0 \\ 0.01 \end{bmatrix},$$

$D_{12} = 0.1$, $A_1 = 0$, $H = 0$ and $F = 0$.

The first two components of the state vector $x(t)$ are the pursuer-evader relative position and velocity, respectively, and the third component is the actual pursuer's acceleration. The control u represents the pursuer's acceleration command and the delay appears in the measurement of the position. It is assumed that the initial conditions effect in the guidance scenario has been first nulled out by an initial phase, e.g., pursuit. Zero initial conditions are therefore assumed as a reasonable approximation to such a situation. We note that the stochastic uncertainty appears only in the dynamics of the acceleration with no delay, where it is originated from a radar noise which is involved, in the finite-horizon version, in the measurement of the time to go (see [53], Chapter 11, page 196 for detailed description). Applying Theorem 2.13 and taking, in line with Remark 2.5, $\alpha = 1.05$, $\epsilon_f = 0.97$, a near minimum attenuation level of $\gamma = 17.30$, was obtained where the controller parameters are:

$$A_c = \begin{bmatrix} -2.050 & 2.457 & -0.000 \\ -0.493 & 0.351 & 0.000 \\ -33.770 & -34.336 & -21.308 \end{bmatrix}, \; B_c = \begin{bmatrix} -2.2221 & 0.1749 \\ -0.539 & -1.009 \\ -1.593 & 8.901 \end{bmatrix}, \; C_c = \begin{bmatrix} 1 & 1 & 0 \end{bmatrix}.$$

Table 2.1 The minimal attenuation levels obtained as a function of the delay derivative bound for $h = 0.22$

Case No.	h=delay length	d= delay derivative	ϵ_f	γ_{min}
1	0	0	–	6.49
2	0.22	0	0.97	14.7
3	0.22	0.1	0.97	15.23
4	0.22	0.3	0.95	16.57
5	0.22	0.5	0.97	18.02

In order to better understand the relation between the two bounds d and h on the solution of the current example, we bring in Table 2.1 the minimum achievable attenuation level that is obtained, via Theorem 2.13 and Remark 2.5, for h=0.22 and various values of d. We note that the attenuation level achieved for the non-delayed case is, obviously, the lowest one. In cases 2–5, it is seen that the attenuations levels increase as the bound on the delay derivative (d) increases. Similar behavior appears for different delay lengths (not shown). Also, the maximum value of the derivative bound is reduced as the delay length is increased.

2.7.3 Robust Stochastic Output-Feedback Control – The Norm-Bounded Case

We consider the system of (2.1a–c) where $A_0,\ A_1,\ B_2$ are replaced by $\tilde{A}_0,\ \tilde{A}_1,\ \tilde{B}_2$ respectively and where the latter matrices are given in (2.10). Noting that

$$\bar{H}_i^T \epsilon_j \bar{H}_i + Q E_i \epsilon_j^{-1} E_i^T Q \geq Q E_i F_i \bar{H}_i + \bar{H}_i^T F_i^T E_i^T Q, \quad \epsilon_j > 0,\ i = 0,1, j = i+1,$$

we arrive at the following theorem [59]:

Theorem 2.14. *Consider the system (2.1a-c) where* $A_0,\ A_1,\ B_2$ *are replaced by* $\tilde{A}_0,\ \tilde{A}_1,\ \tilde{B}_2$ *respectively and where the latter matrices are given in (2.10). There exists a dynamic output-feedback controller of the structure of (2.7) that achieves negative* J_{OF} *for all nonzero* $w \in \tilde{L}^2_{\mathcal{F}_t}([0,\infty);\mathcal{R}^q)$, $n \in \tilde{L}^2_{\mathcal{F}_t}([0,\infty);\mathcal{R}^p)$, *for a prescribed scalar* $\gamma > 0$ *and tuning parameters* α *and* ϵ_f *if there exist matrices* $Q > 0,\ \hat{Q} > 0,\ R_1 > 0,\ \tilde{Q}_m,\ A_{f,q},\ B_{f,q}$ *and scalars* $\epsilon_1,\ \epsilon_2$ *that satisfy the following LMI:*

$$\begin{bmatrix} \Gamma_1 & \Gamma_2 & \Gamma_3 \\ * & -\epsilon_1 I_n & 0 \\ * & * & -\epsilon_2 I_n \end{bmatrix} < 0, \tag{2.60}$$

where

$$\Gamma_2^T = \left[\hat{E}_0^T \tilde{Q}\ 0\ 0\ h\epsilon_f \hat{E}_0^T \tilde{Q}\ 0\ 0\ 0\ 0\ 0 \right]^T,$$

$$\Gamma_3^T = \left[\hat{E}_1^T \tilde{Q}\ 0\ 0\ h\epsilon_f \hat{E}_1^T \tilde{Q}\ 0\ 0\ 0\ 0\ 0 \right]^T,$$

and where Γ_1 *is the matrix in the left side of (2.56) which includes the additional term of* $\hat{H}_0^T \epsilon_1 \hat{H}_0$ *in* $\bar{\Upsilon}_{11}$ *and the additional term of* $\hat{H}_1^T \epsilon_2 \hat{H}_1$ *in the (2,2) block, where*

$$\hat{E}_0 = \begin{bmatrix} E_0 \\ 0 \end{bmatrix}, \quad \hat{E}_1 = \begin{bmatrix} E_1 \\ 0 \end{bmatrix}, \quad \hat{H}_0 = \left[\bar{H}_0\ \ \bar{H}_2 C_c \right], \quad \hat{H}_1 = \left[\bar{H}_1\ \ 0 \right].$$

In this case the controller parameters are given by (2.58).

2.7.4 Robust Polytopic Stochastic Output-Feedback Control

In this subsection we address the dynamic output-feedback control problem of the delayed state-multiplicative polytopic uncertain noisy system. We consider the system of (2.1a–c) where the system matrices lie within the polytope $\bar{\Omega}$ of (2.12) and the general type controller of (2.7).

Denoting $\xi^T(t) \stackrel{\Delta}{=} [x(t)^T \ \ \hat{x}(t)^T]$ and $\bar{w}^T(t) \stackrel{\Delta}{=} [w(t)^T \ \ n(t)^T]$, we obtain the following augmented system:

$$
\begin{aligned}
&d\xi(t) = [\hat{A}_0\xi(t) + \tilde{B}\bar{w}(t)]dt + \hat{A}_1\xi(t-\tau(t))dt + \tilde{H}\xi(t-\tau(t))d\nu(t)\\
&+\tilde{G}\xi(t)d\beta(t) + \tilde{F}\xi(t)d\zeta(t), \quad \xi(\theta) = 0,\ over[-h\ 0],\\
&\tilde{z}(t) = \tilde{C}\xi(t),
\end{aligned}
\tag{2.61}
$$

with the following matrices:

$$
\hat{A}_0 = \begin{bmatrix} A_0 & B_2C_c \\ B_cC_2 & A_c \end{bmatrix}, \quad \hat{A}_1 = \begin{bmatrix} A_1 & 0 \\ B_c\bar{C}_2 & 0 \end{bmatrix}, \quad \tilde{B} = \begin{bmatrix} B_1 & 0 \\ 0 & B_cD_{21} \end{bmatrix}, \quad \tilde{H} = \begin{bmatrix} H & 0 \\ 0 & 0 \end{bmatrix}
$$

$$
\tilde{G} = \begin{bmatrix} G & 0 \\ 0 & 0 \end{bmatrix}, \ \tilde{F} = \begin{bmatrix} 0 & 0 \\ B_cF & 0 \end{bmatrix}, \ \tilde{C} = [C_1 \ \ D_{12}C_c]. \tag{2.62}
$$

Considering the fact that the system matrices in (2.1a–c) lie within the polytope $\bar{\Omega}$ of (2.12), the corresponding matrices of uncertain polytopic counterpart of (2.61) are: $\hat{A}_0^i$, $\hat{A}_1^i$, $\tilde{B}^i$, $\tilde{C}^i$, $\tilde{F}^i$, $\tilde{G}^i$, $\tilde{H}^i$. These matrices are obtained from (2.62) for each index i by taking the ith system matrices of (2.1a–c). Using the BRL result, we obtain that $J_E < 0$ is satisfied if there exist $\tilde{Q}$, R_1, R_2 and m that satisfy the following inequalities:

$$\begin{bmatrix} \Upsilon_{11,i} & \tilde{Q}(\hat{A}_1^i - m) & \tilde{Q}m & \tilde{Q}\tilde{B}^i & h({\hat{A}^i}_0{}^T + m^T)R_2 & \tilde{C^i}^T & \tilde{G^i}^T\tilde{Q} & \tilde{F^i}^T\tilde{Q} & 0 \\ * & -R_1 & 0 & 0 & h({\hat{A}^i}_1{}^T - m^T)R_2 & 0 & 0 & 0 & \tilde{H^i}^T\tilde{Q} \\ * & * & -R_2 & 0 & -hm^TR_2 & 0 & 0 & 0 & 0 \\ * & * & * & -\gamma^2 I_{q+p} & h\tilde{B^i}^T R_2 & 0 & 0 & 0 & 0 \\ * & * & * & * & -R_2 & 0 & 0 & 0 & 0 \\ * & * & * & * & * & -I_r & 0 & 0 & 0 \\ * & * & * & * & * & * & -\tilde{Q} & 0 & 0 \\ * & * & * & * & * & * & * & -\tilde{Q} & 0 \\ * & * & * & * & * & * & * & * & -\tilde{Q} \end{bmatrix} < 0, \quad \forall i, \ i = 1, 2, ...N \tag{2.63}$$

where

$$\Upsilon_{11,i} = \tilde{Q}(\hat{A}_0^i + m) + (\hat{A}_0^i + m)^T\tilde{Q} + \frac{R_1}{1-d} + Qm + m^TQ.$$

Denoting the following partition: $\tilde{Q} = \begin{bmatrix} Q & \alpha\hat{Q} \\ \alpha\hat{Q} & \hat{Q} \end{bmatrix}$, where $Q > 0, \hat{Q} > 0 \in \mathcal{R}^{n\times n}$ and α is a tuning scalar parameter, taking $R_2 = \epsilon_f\tilde{Q}$ where $\epsilon_f > 0$ is a tuning scalar parameter and denoting $\tilde{Q}_m = \tilde{Q}m$, we obtain the following set of inequalities:

$$
\begin{bmatrix}
\bar{\Upsilon}_{11,i} & \begin{bmatrix} QA_1^i + \alpha B_{f,q}\bar{C}_2^i & 0 \\ \alpha\hat{Q}A_1^i + B_{f,q}\bar{C}_2^i & 0 \end{bmatrix} - \tilde{Q}_m & \tilde{Q}_m & \bar{\Upsilon}_{14} & \bar{\Upsilon}_{15,i} & \begin{bmatrix} C^{iT}_1 \\ C_c^T D^{iT}_{12} \end{bmatrix} & \bar{\Upsilon}_{17,i} & \bar{\Upsilon}_{18,i} & 0 \\
* & -R_1 & 0 & 0 & \bar{\Upsilon}_{25,i} & 0 & 0 & 0 & \bar{\Upsilon}_{29,i} \\
* & * & -\epsilon_f\tilde{Q} & 0 & -h\epsilon_f\tilde{Q}_m^T & 0 & 0 & 0 & 0 \\
* & * & * & -\gamma^2 I_{q+p} & h\epsilon_f\bar{\Upsilon}_{14,i}^T & 0 & 0 & 0 & 0 \\
* & * & * & * & -\epsilon_f\tilde{Q} & 0 & 0 & 0 & 0 \\
* & * & * & * & * & -I_r & 0 & 0 & 0 \\
* & * & * & * & * & * & -\tilde{Q} & 0 & 0 \\
* & * & * & * & * & * & * & -\tilde{Q} & 0 \\
* & * & * & * & * & * & * & * & -\tilde{Q}
\end{bmatrix} < 0, \qquad (2.64)
$$

$\forall i,\ i = 1, 2, ...N$, where

$$
\bar{\Upsilon}_{11,i} = \begin{bmatrix} QA_0^i + A^{iT}_0 Q + \alpha B_{f,q}C_2^i + \alpha C^{iT}_2 B_{f,q}^T & QB_2^iC_c + \alpha A_{f,q} + \alpha A^{iT}_0\hat{Q} + C^{iT}_2 B_{f,q}^T \\ C_c^T B^{iT}_2 Q + \alpha A_{f,q}^T + \alpha\hat{Q}A_0^i + B_{f,q}C_2^i & \alpha\hat{Q}B_2^iC_c + \alpha C_c^T B^{iT}_2\hat{Q} + A_{f,q} + A_{f,q}^T \end{bmatrix} + \tilde{Q}_m + \tilde{Q}_m^T + \frac{1}{1-d}R_1
$$

$$
\Upsilon_{14,i} = \begin{bmatrix} QB_1^i & \alpha B_{f,q}D_{21}^i \\ \alpha\hat{Q}B_1^i & B_{f,q}D_{21}^i \end{bmatrix},
$$

$$
\bar{\Upsilon}_{15,i} = \epsilon_f h \begin{bmatrix} A^{iT}_0 Q + \alpha C^{iT}_2 B_{f,q}^T & \alpha A^{iT}_0\hat{Q} + C^{iT}_2 B_{f,q}^T \\ C_c^T B^{iT}_2 Q + \alpha A_{f,q}^T & \alpha C_c^T B^{iT}_2\hat{Q} + A_{f,q}^T \end{bmatrix} + h\epsilon_f\tilde{Q}_m^T,
$$

$$\bar{\Upsilon}_{17,i} = \begin{bmatrix} G^{i^T} Q \ \alpha G^{i^T} \hat{Q} \\ 0 \quad 0 \end{bmatrix},$$

$$\bar{\Upsilon}_{18,i} = \begin{bmatrix} \alpha F^{i^T} B_{f,q}^T \ F^{i^T} B_{f,q}^T \\ 0 \quad 0 \end{bmatrix},$$

$$\bar{\Upsilon}_{25,i} = \epsilon_f h \begin{bmatrix} A^{i}{}_1^T Q + \alpha \bar{C^i}_2^T B_{f,q}^T \ \alpha A^{i}{}_1^T \hat{Q} + \bar{C^i}_2^T B_{f,q}^T \\ 0 \quad 0 \end{bmatrix} - h\epsilon_f \tilde{Q}_m^T,$$

$$\bar{\Upsilon}_{29,i} = \begin{bmatrix} H^{i^T} Q \ \alpha H^{i^T} \hat{Q} \\ 0 \quad 0 \end{bmatrix},$$

where

$$A_{f,q} \stackrel{\Delta}{=} \hat{Q} A_c, \quad B_{f,q} \stackrel{\Delta}{=} \hat{Q} B_c. \tag{2.65}$$

We note that inequalities (2.64) are bilinear because of the terms $\alpha C_c^T B^{i}{}_2^T \hat{Q}$ and $QB_2^i C_c$ in $\bar{\Upsilon}_{11,i}$. In order to remain in the linear domain, we pre-choose the matrix C_c at the cost of some overdesign. We arrive at the following theorem:

Theorem 2.15. *Consider the system (2.1a-c) where the system matrices lie within the polytope $\bar{\Omega}$ of (2.12). There exists a controller of the structure of (2.7) that achieves negative J_E for all nonzero $w \in \tilde{L}^2_{\mathcal{F}_t}([0,\infty);\mathcal{R}^q)$, $n \in \tilde{L}^2_{\mathcal{F}_t}([0,\infty);\mathcal{R}^p)$, for a prescribed $\gamma > 0$ and tuning parameters α and ϵ_f if there exist matrices $Q > 0$, $\hat{Q} > 0$, $\tilde{Q}_m$, $R_1 > 0$ and $A_{f,q}$, $B_{f,q}$, that satisfy (2.64). In the latter case the controller parameters are:*

$$A_c = \hat{Q}^{-1} A_{f,q} \quad B_c = \hat{Q}^{-1} B_{f,q}, \ C_c = [I_{l\times l} \ \ 0_{l\times(n-l)}]. \tag{2.66}$$

2.8 Static Output-Feedback Control

Static output-feedback control is applied in many areas of control engineering including process and flight control. In the latter, designing flight control systems, engineers prefer the simple and physically sound controllers that are recommended as cooked structures [140],[10] (see also [107]). Only gains are included in these simple structures and the closed-loop poles are thus obtained by migration of the open-loop poles that have a clear physical meaning.

In the present solution we adopt a simple design method [54] for deriving the static output feedback gain that satisfies a prescribed H_∞ performance criteria. In [54], it has been shown that since, in designing static output-feedback, a constant gain cannot be achieved in practice and all amplifiers have some finite bandwidth, one can add, in series to the measured output of the system, a simple low-pass component with a very high bandwidth. A parameter dependent Lyapunov function is then described for the augmented system that is obtained by incorporating the states of the additional component into the state space description. This function does not require a specific structure for the matrix that corresponds to the states of the original system and a sufficient condition was then obtained [54].

As explained above, instead of considering the purely constant controller (2.8) we consider the following strictly proper controller

$$d\eta = -\rho\eta dt + \rho y dt, \quad u = K\eta \tag{2.67}$$

where $\eta \in \mathcal{R}^m$ and $1 << \rho$ is a scalar much larger than the open-loop and the desired closed-loop bandwidth. The latter controller is introduced in order to facilitate the convexity of the design method below. It represents however, the actual situation where 'constant' gains are achieved in practice by amplifiers of finite bandwidth.

Augmenting the system (2.1a–c) to include the states of (2.67) we define the augmented state vector $\xi = col\{x, \eta\}$ and obtain the following representation to the closed-loop system, where we take $\bar{C}_2 = 0$:

$$\begin{aligned} d\xi(t) &= [\tilde{A}\xi(t) + \tilde{B}w(t)]dt + \tilde{A}_1\xi(t-\tau)dt + \tilde{G}\xi(t)d\beta(t) \\ &+\tilde{H}\xi(t-\tau)d\nu(t) + \tilde{F}\xi(t)d\zeta(t), \quad \xi(0) = 0, \text{ over } [-h \;\; 0], \\ z &= \tilde{C}\xi(t), \end{aligned} \tag{2.68}$$

where:

$$\tilde{A} \triangleq \begin{bmatrix} A_0 & B_2K \\ \rho C_2 & -\rho I_m \end{bmatrix}, \; \tilde{A}_1 \triangleq \begin{bmatrix} A_1 & 0 \\ 0 & 0 \end{bmatrix}, \; \tilde{G} \triangleq \begin{bmatrix} G & 0 \\ 0 & 0 \end{bmatrix},$$

$$\tilde{H} \triangleq \begin{bmatrix} H & 0 \\ 0 & 0 \end{bmatrix}, \; \tilde{F} \triangleq \begin{bmatrix} 0 & 0 \\ F\rho & 0 \end{bmatrix}, \; \tilde{B} \triangleq \begin{bmatrix} B_1 \\ \rho D_{21} \end{bmatrix}, \; \tilde{C} \triangleq \begin{bmatrix} C_1 & D_{12}K \end{bmatrix}. \tag{2.68}$$

Applying the result of Theorem 2.4 to the system of (2.68) and denoting $Y = Qm$, we obtain, taking $\bar{\alpha} = 0$:

$$\tilde{\Gamma} = \begin{bmatrix} \tilde{\Psi}_{11} & Q\tilde{A}_1 - Y & Y & Q\tilde{B} & h\epsilon\tilde{A}^T Q + h\epsilon Y^T & \tilde{C}^T \\ * & -R_1 + \tilde{H}^T Q\tilde{H} & 0 & 0 & h\epsilon\tilde{A}_1^T Q - h\epsilon Y^T & 0 \\ * & * & -\epsilon Q & 0 & -h\epsilon Y^T & 0 \\ * & * & * & -\gamma^2 I_q & h\epsilon\tilde{B}^T Q & 0 \\ * & * & * & * & -\epsilon Q & 0 \\ * & * & * & * & * & -I_r \end{bmatrix} < 0, \tag{2.69}$$

where

$$\tilde{\Psi}_{11} = Q\tilde{A} + Y + \tilde{A}^T Q + Y^T + \frac{1}{1-d} R_1 + \tilde{G}^T Q\tilde{G} + \tilde{F}^T Q\tilde{F}.$$

Denoting $P = Q^{-1}$ and multiplying (2.69) by $diag\{P, P, P, I_q, P, I_z\}$ from the left and the right we obtain:

$$\begin{bmatrix} \hat{\Psi}_{11} & \tilde{A}_1 P - PYP & PYP & \tilde{B} & h\epsilon P\tilde{A}^T + h\epsilon Pm^T & P\begin{bmatrix} C_1^T \\ K^T D_{12} \end{bmatrix} \\ * & \hat{\Psi}_{22} & 0 & 0 & h\epsilon P\tilde{A}_1^T - h\epsilon Pm^T & 0 \\ * & * & -\epsilon P & 0 & -h\epsilon Pm^T & 0 \\ * & * & * & -\gamma^2 I_q & h\epsilon\tilde{B}^T & 0 \\ * & * & * & * & -\epsilon P & 0 \\ * & * & * & * & * & -I_r \end{bmatrix} \tag{2.70}$$

$$< 0, \tag{2.71}$$

where

$$\begin{aligned}\hat{\Psi}_{11} &= \tilde{A}P + PYP + P\tilde{A}^T + PY^TP + P\tfrac{1}{1-d}R_1P + P\tilde{G}^TP^{-1}\tilde{G}P + \\ &\quad P\tilde{F}^TP^{-1}\tilde{F}P, \\ \hat{\Psi}_{22} &= -PR_1P + P\tilde{H}^TP^{-1}\tilde{H}P,\end{aligned} \tag{2.72}$$

In order to solve the problem, we consider the following structure for P:

$$P = \begin{bmatrix} \bar{P} & C_2^T\hat{P} \\ \hat{P}C_2 & \alpha\hat{P} \end{bmatrix}, \tag{2.73}$$

where $0 < \bar{P} \in \mathcal{R}^{n\times n}$, $0 < \hat{P} \in \mathcal{R}^{m\times m}$ and where α is a positive scalar tuning parameter. Taking $R_2 = \epsilon P^{-1}$ where $\epsilon > 0$ is a tuning scalar, denoting $Y_p = PYP = PQmP = mP$, $R_p = PR_1P$, $\hat{P}_k = K\hat{P}$ and carrying out the various multiplications we obtain the following LMI:

$$\Gamma_n = \begin{bmatrix} \tilde{\Psi}_{11} & \tilde{\Psi}_{12} & 0 & Y_p & \begin{bmatrix} B_1 \\ \rho D_{21} \end{bmatrix} & \tilde{\Psi}_{16} & \tilde{\Psi}_{17} & \tilde{\Psi}_{18} & \tilde{\Psi}_{19} \\ * & -R_p & \tilde{\Psi}_{23} & 0 & 0 & \tilde{\Psi}_{26} & 0 & 0 & 0 \\ * & * & -\epsilon P & 0 & 0 & 0 & 0 & 0 & 0 \\ * & * & * & -\epsilon P & 0 & -h\epsilon Y_p^T & 0 & 0 & 0 \\ * & * & * & * & -\gamma^2 I_q & \epsilon h \begin{bmatrix} B_1 \\ \rho D_{21} \end{bmatrix}^T & 0 & 0 & 0 \\ * & * & * & * & * & -\epsilon P & 0 & 0 & 0 \\ * & * & * & * & * & * & -I_r & 0 & 0 \\ * & * & * & * & * & * & * & -P & 0 \\ * & * & * & * & * & * & * & * & -P \end{bmatrix} < 0, \tag{2.74}$$

where

$$\tilde{\Psi}_{11} = \begin{bmatrix} A_0\bar{P} + \bar{P}A_0^T + B_2\hat{P}_kC_2 + C_2^T\hat{P}_k^TB_2^T \\ \rho[C_2\bar{P} - \hat{P}C_2] + [A_0C_2^T\hat{P} + \alpha B_2\hat{P}_k]^T \end{bmatrix.$$

$$\left. \begin{matrix} A_0C_2^T\hat{P} + \alpha B_2\hat{P}_k + \rho[C_2\bar{P} - \hat{P}C_2]^T \\ \rho[(C_2C_2^T\hat{P} - \alpha\hat{P}) + (C_2C_2^T\hat{P} - \alpha\hat{P})^T] \end{matrix} \right] + Y_p + Y_p^T + \tfrac{R_p}{1-d},$$

$$\tilde{\Psi}_{12} = \begin{bmatrix} A_1\bar{P} & A_1C_2^T\hat{P} \\ 0 & 0 \end{bmatrix} - Y_p,$$

$$\tilde{\Psi}_{16} = h\epsilon \begin{bmatrix} \bar{P}A_0^T + C_2^T\hat{P}_k^TB_2^T & \rho[C_2\bar{P} - \hat{P}C_2]^T \\ \hat{P}C_2A_0^T + \alpha\hat{P}_k^TB_2^T & \rho[C_2C_2^T\hat{P} - \alpha\hat{P}]^T \end{bmatrix} + h\epsilon Y_p^T,$$

$$\tilde{\Psi}_{17} = \begin{bmatrix} \bar{P}C_1^T + C_2^T\hat{P}_k^TD_{12}^T \\ \hat{P}C_2C_1^T + \alpha\hat{P}_k^TD_{12}^T \end{bmatrix},$$

$$\tilde{\Psi}_{18} = \begin{bmatrix} \bar{P}G^T & 0 \\ \hat{P}C_2G^T & 0 \end{bmatrix},$$

$$\tilde{\Psi}_{19} = \begin{bmatrix} 0 & \rho\bar{P}F^T \\ 0 & \rho\hat{P}C_2F^T \end{bmatrix}, \tag{2.75}$$

$$\tilde{\Psi}_{23} = \begin{bmatrix} \bar{P}H^T & 0 \\ \hat{P}C_2H^T & 0 \end{bmatrix},$$

$$\tilde{\Psi}_{26} = h\epsilon \begin{bmatrix} A_1\bar{P} & A_1C_2^T\hat{P} \\ 0 & 0 \end{bmatrix}^T - h\epsilon Y_p^T$$

We thus arrive at the following theorem:

Theorem 2.16. *Consider the system (2.1a–c), with* $\bar{C}_2 = 0$, $\bar{\alpha} = 0$. *There exists a static gain controller* K *that achieves negative* J_{OF} *for all nonzero* $w \in \tilde{L}^2_{\mathcal{F}_t}([0,\infty);\mathcal{R}^q)$, *for a prescribed* $\gamma > 0$ *and positive scalar tuning parameters* α *and* ϵ, *if there exist matrices* $\bar{P} > 0$, $\hat{P} > 0$, $R_p > 0$, Y_p, *and* $\hat{P}_k$, *that satisfy (2.74). In the latter case the controller gain is given by:*

$$K = \hat{P}_k\hat{P}^{-1}. \tag{2.76}$$

2.9 Robust Polytopic Static Output-Feedback Control

In this section we address the problem of static output-feedback control problem of the delayed state-multiplicative polytopic uncertain system. We consider the system of (2.1a–c) where the system matrices, excluding C_2, lie within the polytope $\bar{\Omega}$ of (2.12). We obtain the following result:

Theorem 2.17. *Consider the system (2.1a–c) where the system matrices, excluding C_2, lie within the polytope $\bar{\Omega}$ of (2.12) and where $\bar{C}_2 = 0$, $\bar{\alpha} = 0$. There exists a static gain controller K that achieves negative J_{static} for all nonzero $w \in \tilde{L}^2_{\mathcal{F}_t}([0,\infty);\mathcal{R}^q)$, for a prescribed $\gamma > 0$ and positive scalar tuning parameters α and ϵ, if there exist matrices $\bar{P} > 0$, $\hat{P} > 0$, $R_p > 0$, Y_p, $\hat{P}_k$ that satisfy the LMIs of (2.77). In the latter case the controller gain is given by (2.76).*

$$
\Gamma_{n,i} = \begin{bmatrix}
\tilde{\Psi}_{11,i} & \tilde{\Psi}_{12,i} & 0 & Y_p & \begin{bmatrix} B_{1,i} \\ \rho D_{21,i} \end{bmatrix} & \tilde{\Psi}_{16,i} & \tilde{\Psi}_{17,i} & \tilde{\Psi}_{18,i} & \tilde{\Psi}_{19,i} \\
* & -R_p & \tilde{\Psi}_{23,i} & 0 & 0 & \tilde{\Psi}_{26,i} & 0 & 0 & 0 \\
* & * & -\epsilon P & 0 & 0 & 0 & 0 & 0 & 0 \\
* & * & * & -\epsilon P & 0 & -h\bar{Y}_p^T & 0 & 0 & 0 \\
* & * & * & * & -\gamma^2 I_q & \epsilon h \begin{bmatrix} B_{1,i} \\ \rho D_{21,i} \end{bmatrix}^T & 0 & 0 & 0 \\
* & * & * & * & * & -\epsilon P & 0 & 0 & 0 \\
* & * & * & * & * & * & -I_r & 0 & 0 \\
* & * & * & * & * & * & * & -P & 0 \\
* & * & * & * & * & * & * & * & -P
\end{bmatrix} < 0, \tag{2.77}
$$

where

$$\tilde{\Psi}_{11,i} = \begin{bmatrix} A_{0,i}\bar{P} + \bar{P}A_{0,i}^T + B_{2,i}\hat{P}_k C_2 + C_2^T \hat{P}_k^T B_{2,i}^T & A_{0,i}C_2^T\hat{P} + \alpha B_{2,i}\hat{P}_k + \rho[C_2\bar{P} - \hat{P}C_2]^T \\ \rho[C_2\bar{P} - \hat{P}C_2] + [A_{0,i}C_2^T\hat{P} + \alpha B_{2,i}\hat{P}_k]^T & \rho[(C_2C_2^T\hat{P} - \alpha\hat{P}) + (C_2C_2^T\hat{P} - \alpha\hat{P})^T] \end{bmatrix} + Y_p + Y_p^T + \frac{R_p}{1-d},$$

$$\tilde{\Psi}_{12,i} = \begin{bmatrix} A_{1,i}\bar{P} & A_{1,i}C_2^T\hat{P} \\ 0 & 0 \end{bmatrix} - Y_p,$$

$$\tilde{\Psi}_{16,i} = h\epsilon \begin{bmatrix} \bar{P}A_{0,i}^T + C_2^T\hat{P}_k^T B_{2,i}^T & \rho[C_2\bar{P} - \hat{P}C_2]^T \\ \hat{P}C_2A_{0,i}^T + \alpha\hat{P}_k^T B_{2,i}^T & \rho[C_2C_2^T\hat{P} - \alpha\hat{P}]^T \end{bmatrix} + h\epsilon Y_p^T,$$

$$\tilde{\Psi}_{17,i} = \begin{bmatrix} \bar{P}C_{1,i}^T + C_2^T\hat{P}_k^T D_{12,i}^T \\ \hat{P}C_2C_{1,i}^T + \alpha\hat{P}_k^T D_{12,i}^T \end{bmatrix},$$

$$\tilde{\Psi}_{18,i} = \begin{bmatrix} \bar{P}G_i^T & 0 \\ \hat{P}C_2G_i^T & 0 \end{bmatrix},$$

$$\tilde{\Psi}_{19,i} = \begin{bmatrix} 0 & \rho\bar{P}F_i^T \\ 0 & \rho\hat{P}C_2F_i^T \end{bmatrix},$$

$$\tilde{\Psi}_{23,i} = \begin{bmatrix} \bar{P}H_i & 0 \\ \hat{P}C_2H_i^T & 0 \end{bmatrix}, \tag{2.78}$$

$$\tilde{\Psi}_{26,i} = h\epsilon \begin{bmatrix} A_{1,i}\bar{P} & A_{1,i}C_2^T\hat{P} \\ 0 & 0 \end{bmatrix}^T - h\epsilon Y_p^T,$$

2.10 Conclusions

In this chapter the theory of linear H_∞ control of state-multiplicative noisy systems is extended to time delayed systems where, in addition to deterministic uncertainties (either norm-bounded or polytopic ones), the stochastic uncertainties are encountered in both the delayed and the nondelayed states in the state space model of the system. The delay is assumed to be unknown and time-varying where only the bounds on its size and rate of change are given. Delay dependent analysis and synthesis methods are developed which are based on the input–output approach, in accordance with the approach taken by [55], [59] for the solution of the state-feedback, filtering, dynamic output-feedback and static control problems. This approach transforms the delayed system to a nonretarded system with norm-bounded operators.

Sufficient conditions are thus derived for the stability of the various systems and the existence of solutions to the corresponding BRL. Based on the robust BRL derivation, the robust state-feedback, filtering, and output-feedback control problems are formulated and solved for both: the norm-bounded case and the polytopic case.

Concerning the output-feedback solution, we note that for system with known delay the solution of the output-feedback control problem was traditionally solved by introducing delay in the structure of the controller (even for deterministic systems). In our case, however, the latter approach can not be applied since the delay is not known. Choosing a special structure for the Lyapunov function at the expense of some overdesign, a delay free controller is derived. Some additional overdesign is also admitted in our solution due to the special structure imposed on R_2.

3

Reduced-Order H_∞ Output-Feedback Control

3.1 Introduction

In the this chapter we address the problem of H_∞ reduced-order output-feedback control of state–delayed, continuous-time, state-multiplicative linear systems via the input–output approach, which enables us to transform the retarded systems to uncertain non-retarded ones [64].

The point of view taken in the present chapter **with regard to the time delay** in the system states, is similar to the one taken for the solution of the state-feedback control and filtering problems in Chapter 2. This point of view differs essentially from the one of Lyapunov–Krasovskii (L–K) [130], [18] as we extend the input–output approach of [33] to delay-dependent stochastic solutions of the above problems.

The problem of controlling deterministic linear systems via a reduced-order controller has attracted a lot of interest in the past. In the case where the parameters of the system were all known, some iterative methods have been suggested [74], [82] that achieve, if they converge, a controller of a prescribed order that satisfies prescribed demands on the performance of the closed-loop system. The solutions achieved by these methods, even if they converge to the global minimum of the performance index, cannot guarantee the performance in the case where the parameters of the system are not certain and are only known to reside in a given polytope. To the best of our knowledge, the problem of reduced-order control for retarded stochastic state-multiplicative counterpart of the above deterministic systems has not been addressed in the literature. Here, we extend the simple design method which has been applied for the solution of the robust SOF control of non retarded state-multiplicative systems [54] to the case where a reduced-order output-feedback controller is sought for retarded uncertain systems. However, the approach used for the solution of the nondelayed SOF problem in [54] and in Chapter 2 for the retarded case, is substantially different from the approach applied in this chapter.

E. Gershon & U. Shaked: *Advanced Topics in Control & Estimation*, LNCIS 439, pp. 61–74.
DOI: 10.1007/978-1-4471-5070-1_3

In [54] (see also Chapter 2, Section 2.8), it has been shown that since, in designing static output-feedback, a constant gain cannot be achieved in practice and all amplifiers have some finite bandwidth, one can add, in series to the measured output of the system, a simple low-pass component with a very high bandwidth. A parameter dependent Lyapunov function is then described for the augmented system that is obtained by incorporating the states of the additional component into the state space description.

In this chapter we avoid the use of the above low pass filter. We first transform the retarded closed-loop system to an augmented one, in such a way that a "static" gain is used. We then transform the latter system to an equivalent delay free, norm bounded system. Now, in order to avoid the inclusion of the a low pass filter as in [54], we transform, by further augmentation, the latter system to a descriptor one. We then apply a special parameter dependent Lyapunov function, to guarantee closed-loop stability and to obtain the required controller. In our system we allow for a time-varying delay where the uncertain stochastic parameters multiply both the delayed and the nondelayed states in the state space model of the system. We first derive sufficient condition for obtaining the latter controller for nominal systems and we then extend our theory to the robust polytopic case. To demonstrate the tractability of our solution method, a numerical example taken from the field of aircraft control, is given in Chapter 13 (Example 13.1).

3.2 Problem Formulation

We consider the following linear system

$$\begin{array}{l} dx(t) = [Ax(t) + B_1w(t) + A_1x(t-\tau(t))]dt + B_2u(t)dt + Gx(t)d\beta(t) \\ +Hx(t-\tau(t))d\nu(t), \qquad x(\theta)=0, \;\; \theta \leq 0, \\ y(t) = C_2x(t) + D_{21}w(t), \\ z(t) = C_1x(t) + D_{12}u(t), \end{array} \qquad (3.1)$$

where $x(t) \in \mathcal{R}^n$ is the state vector, $w(t) \in \tilde{L}^2_{\mathcal{F}_t}([0,\infty);\mathcal{R}^q)$ is an exogenous disturbance, $y(t) \in \mathcal{R}^m$ is the measurement vector , $z(t) \in \mathcal{R}^z$ is the objective vector, $u(t) \in \mathcal{R}^\ell$ is the control input signal and A, A_1, B_1, B_2, C_1, C_2 and D_{12}, D_{21}, G, H are constant matrices.

In (3.1a), $\tau(t)$ is an unknown continuous time-delay which satisfies:

$$0 \leq \tau(t) \leq h, \quad \dot{\tau}(t) \leq d < 1, \qquad (3.2)$$

where h is the bound on the delay interval. The zero-mean real scalar Wiener processes $\beta(t)$, $\nu(t)$ satisfy:

$$\mathcal{E}\{\beta(t)\beta(s)\} = min(t,s), \quad \mathcal{E}\{\nu(t)\nu(s)\} = min(t,s), \quad \mathcal{E}\{\beta(t)\nu(s)\} = 0.$$

We seek a controller of order $r \leq n$ with the following dynamics:

$$dx_c(t) = [A_c x_c(t) + B_c y(t)]dt, \quad u(t) = C_c x_c(t) + D_c y(t) \tag{3.3}$$

that, for a prescribed scalar $\gamma > 0$ and for all nonzero $w(t) \in \tilde{L}^2_{\mathcal{F}_t}([0,\infty);\mathcal{R}^q)$, guarantees $J_{OF} < 0$ where

$$J_{OF} \stackrel{\Delta}{=} \mathcal{E}\{\int_0^{\infty} ||z(t)||^2 dt - \gamma^2 \int_0^{\infty} ||w(t)||^2 dt\}. \tag{3.4}$$

Remark 3.1. Obviously $r = n$ results in a full-order dynamic output-feedback control while $r = 0$ yields a static output-feedback gain. We note that in many practical cases, one seeks a simple controller that, on one hand is easily computed and implemented and on the other hand, achieves a disturbance attenuation bound that is smaller than the one obtained by static gains. In these cases, $r < n$ is sought.

In the robust stochastic H_∞control problem, we assume that the system parameters lie within the following polytope

$$\bar{\Omega} \stackrel{\Delta}{=} \left[A\ A_1\ B_1\ B_2\ C_1\ D_{12}\ D_{21}\ H\ G\right], \tag{3.5}$$

which is described by its vertices:

$$\bar{\Omega} = \mathcal{C}o\{\bar{\Omega}_1, \bar{\Omega}_2, ..., \bar{\Omega}_N\}, \tag{3.6}$$

where

$$\bar{\Omega}_i \stackrel{\Delta}{=} \left[A^{(i)}\ A_1^{(i)}\ B_1^{(i)}\ B_2^{(i)}\ C_1^{(i)}\ D_{12}^{(i)}\ D_{21}^{(i)}\ H^{(i)}\ G^{(i)}\right], \tag{3.7}$$

and where N is the number of vertices. In other words:

$$\bar{\Omega} = \sum_{i=1}^{N} \bar{\Omega}_i f_i \quad , \sum_{i=1}^{N} f_i = 1 \quad , f_i \geq 0. \tag{3.8}$$

Similarly to nominal case, a reduced-order controller of the structure of (3.3) is sought that, for a prescribed positive scalar γ and for all nonzero $w(t) \in \tilde{L}^2_{\mathcal{F}_t}([0,\infty);\mathcal{R}^q)$, guarantees $J_{OF} < 0$ for all uncertain parameters in the given polytope, where J_{OF} is given in (3.4).

3.3 The Delayed Stochastic Reduced-Order H_∞ Control

Considering the above problem of reduced-order control for nominal systems and defining the following matrices:

$$\bar{A} \triangleq \begin{bmatrix} A & 0 \\ 0 & 0 \end{bmatrix}, \ \bar{B}_1 \triangleq \begin{bmatrix} B_1 \\ 0 \end{bmatrix}, \ \bar{B}_2 \triangleq \begin{bmatrix} 0 & B_2 \\ I & 0 \end{bmatrix}, \ \bar{C}_1^T \triangleq \begin{bmatrix} C_1^T \\ 0 \end{bmatrix}, \ \bar{C}_2 \triangleq \begin{bmatrix} 0 & I \\ C_2 & 0 \end{bmatrix},$$

$$\bar{D}_{12}^T \triangleq \begin{bmatrix} 0 \\ D_{12}^T \end{bmatrix}, \ \bar{D}_{21} \triangleq \begin{bmatrix} 0 \\ D_{21} \end{bmatrix}, \ \tilde{A}_1 \triangleq \begin{bmatrix} A_1 & 0 \\ 0 & 0 \end{bmatrix}, \ \tilde{G} \triangleq \begin{bmatrix} G & 0 \\ 0 & 0 \end{bmatrix}, \ \tilde{H} \triangleq \begin{bmatrix} H & 0 \\ 0 & 0 \end{bmatrix},$$

$$K \triangleq \begin{bmatrix} A_c & B_c \\ C_c & D_c \end{bmatrix}, \tag{3.9}$$

the following theorem provides a sufficient condition for the existence of a solution to the problem:

Theorem 3.1. *Consider the system (3.1a–c). There exists a stabilizing reduced-order controller K that achieves negative J_{OF} for all nonzero $w \in \tilde{L}^2_{\mathcal{F}_t}([0,\infty);\mathcal{R}^q)$, for prescribed $\alpha > 0$, $\gamma > 0$ and positive scalar tuning parameters ϵ and $\bar{\epsilon}$, if there exist matrices $Q > 0$, $\hat{Q} > 0$, $R_Q > 0$ and $\hat{K}$ that satisfy the following LMI:*

$$\begin{bmatrix} \hat{\Upsilon}_{11} & \hat{\Upsilon}_{12} & Q_{\bar{\epsilon}} & \begin{bmatrix} \bar{B}_1 \\ \bar{D}_{21} \end{bmatrix} & \hat{\Upsilon}_{15} & \hat{\Psi}_{16} & 0 & \hat{\Upsilon}_{18} \\ * & -R_Q & 0 & 0 & \hat{\Upsilon}_{25} & 0 & \hat{\Upsilon}_{27} & 0 \\ * & * & -\hat{\Upsilon}_{33} & 0 & \hat{\Upsilon}_{35} & 0 & 0 & 0 \\ * & * & * & -\gamma^2 I_q & \hat{\Upsilon}_{45} & 0 & 0 & 0 \\ * & * & * & * & -\hat{\Upsilon}_{55} & 0 & 0 & 0 \\ * & * & * & * & * & -I_z & 0 & 0 \\ * & * & * & * & * & * & -Q & 0 \\ * & * & * & * & * & * & * & -Q \end{bmatrix} < 0, \tag{3.10}$$

where

$$\hat{\Psi}_{11,a} = \begin{bmatrix} \bar{A}Q + Q\bar{A}^T + \bar{B}_2\hat{K}\bar{C}_2 + \bar{C}_2^T\hat{K}^T\bar{B}_2^T & \alpha\bar{B}_2\hat{K} + Q\bar{C}_2^T - \bar{C}_2^T\hat{Q} \\ \alpha\hat{K}^T\bar{B}_2^T + \bar{C}_2Q - \hat{Q}\bar{C}_2 & -2\alpha\hat{Q} \end{bmatrix},$$

$$Q_{\bar{\epsilon}} = \bar{\epsilon}\begin{bmatrix} Q & 0 \\ 0 & 0 \end{bmatrix},$$

$$\hat{\Upsilon}_{11} = \hat{\Psi}_{11,a} + 2Q_{\bar{\epsilon}} + \tfrac{R_Q}{1-d},$$

$$\hat{\Upsilon}_{12} = \begin{bmatrix} (\tilde{A}_1 - \bar{\epsilon}I)Q & 0 \\ 0 & 0 \end{bmatrix},$$

$$\hat{\Upsilon}_{15} = h\begin{bmatrix} \epsilon Q\bar{A}^T + \epsilon\bar{C}_2^T\hat{K}^T\bar{B}_2^T + Q\bar{C}_2^T\bar{C}_2 - \bar{C}_2^T\hat{Q}\bar{C}_2 & \alpha Q\bar{C}_2^T - \alpha\bar{C}_2^T\hat{Q} \\ \epsilon\alpha\hat{K}^T\bar{B}_2^T - h\alpha\hat{Q}\bar{C}_2 & -\alpha^2\hat{Q} \end{bmatrix} + h\epsilon Q_{\bar{\epsilon}},$$

$$\hat{\Upsilon}_{16} = \begin{bmatrix} Q\bar{C}_1^T + \bar{C}_2^T\hat{K}^T\bar{D}_{12}^T \\ \alpha\hat{K}^T\bar{D}_{12}^T \end{bmatrix},$$

$$\hat{\Upsilon}_{18} = \begin{bmatrix} Q\tilde{G}^T \\ 0 \end{bmatrix},$$

$$\hat{\Upsilon}_{25} = h\epsilon\begin{bmatrix} Q\tilde{A}_1^T & 0 \\ 0 & 0 \end{bmatrix} - h\epsilon Q_{\bar{\epsilon}},$$

$$\hat{\Upsilon}_{27} = \begin{bmatrix} Q\tilde{H}^T \\ 0 \end{bmatrix},$$

$$\hat{\Upsilon}_{35} = -h\epsilon Q_{\bar{\epsilon}},$$

$$\hat{\Upsilon}_{45} = h\begin{bmatrix} \epsilon\bar{B}_1^T + \bar{D}_{21}^T\bar{C}_2 & \alpha\bar{D}_{21}^T \end{bmatrix},$$

$$\hat{\Upsilon}_{33} = \hat{\Upsilon}_{55} = \begin{bmatrix} \epsilon Q + \bar{C}_2^T\hat{Q}\bar{C}_2 & \alpha\bar{C}_2^T\hat{Q} \\ \alpha\hat{Q}\bar{C}_2 & \alpha^2\hat{Q} \end{bmatrix}. \tag{3.11}$$

In the latter case the controller gain is given by:

$$K = \hat{K}\hat{Q}^{-1}. \tag{3.12}$$

Proof: Augmenting the system (3.1a) to include the states of (3.3a), we define the augmented state vector $\xi(t) = col\{x(t),\ x_c(t)\}$ and obtain the following representation:

$$\begin{aligned} d\xi(t) &= [\tilde{A}\xi(t) + \tilde{B}w(t)]dt + \tilde{A}_1\xi(t-\tau(t))dt + \tilde{G}\xi(t)d\beta(t) + \tilde{H}\xi(t-\tau(t))d\nu(t), \\ \xi(0) &= 0, \quad \text{over } [-h \ \ 0], \\ z(t) &= \tilde{C}\xi(t) + \tilde{D}w(t), \end{aligned} \tag{3.13}$$

where

$$\begin{aligned}\tilde{A} &\stackrel{\Delta}{=} \bar{A} + \bar{B}_2 K \bar{C}_2,\\ \tilde{B} &\stackrel{\Delta}{=} \bar{B}_1 + \bar{B}_2 K \bar{D}_{21},\\ \tilde{C} &\stackrel{\Delta}{=} \bar{C}_1 + \bar{D}_{12} K \bar{C}_2,\\ \tilde{D} &= \bar{D}_{12} K \bar{D}_{21}.\end{aligned} \tag{3.14}$$

Recalling the definition of (9k), the system (3.13a,b) can be described as a closed-loop system with

$$\tilde{u}(t) = K\tilde{y}(t),$$

that is applied to the following system:

$$\begin{aligned}&d\tilde{\xi}(t) = [\bar{A}\tilde{\xi}(t) + \bar{B}_1 w(t) + \bar{B}_2\tilde{u}(t)]dt + \tilde{A}_1\tilde{\xi}(t-\tau(t))dt + \tilde{G}\tilde{\xi}(t)d\beta(t)\\ &+\tilde{H}\tilde{\xi}(t-\tau(t))d\nu(t),\\ &\tilde{\xi}(0) = 0, \quad \text{over } [-h \;\; 0],\\ &\tilde{y}(t) = \bar{C}_2\tilde{\xi}(t) + \bar{D}_{21}w(t),\\ &z = \bar{C}_1\tilde{\xi}(t) + \bar{D}_{12}\tilde{u}(t),\end{aligned} \tag{3.15}$$

where $\tilde{\xi}(t) \in \mathcal{R}^{\mathbf{n+r}}$, $\tilde{y}(t) \in \mathcal{R}^{\mathbf{m+r}}$, $\tilde{u}(t) \in \mathcal{R}^{\mathbf{l+r}}$, and where $\bar{A}$, $\bar{B}_1$, $\bar{B}_2$, $\bar{C}_1$, $\bar{C}_2$ and $\bar{D}_{12}$, $\tilde{A}_1$, $\tilde{G}$, $\tilde{H}$, $\bar{D}_{21}$ are given in (3.9). Note that in the latter formulation, the problem of finding the dynamic controller parameters (i.e A_c, B_c, C_c, and D_c) becomes one of seeking a static output-feedback gain that achieves the required attenuation level.

In order to solve the above problem, we introduce the following operators:

$$(\mathcal{D}_0 g)(t) \stackrel{\Delta}{=} g(t-\tau(t)), \quad (\mathcal{D}_1 g)(t) \stackrel{\Delta}{=} \int_{t-\tau(t)}^{t} g(s)ds, \tag{3.16}$$

where $g(t)$ is a piecewise continuous-time function. In what follows we use the fact that the induced L_2-norm of $\mathcal{D}_0$ is bounded by $\frac{1}{\sqrt{1-d}}$, and the fact that the induced L_2-norm of $\mathcal{D}_1$ is bounded by h (see [77], Lemma 1 for the latter norm). Using the operator notations of (3.16), defining

$$w_1(t) = (\mathcal{D}_0\tilde{\xi})(t) = \tilde{\xi}(t-\tau(t)), \quad \text{and} \quad w_2(t) = (\mathcal{D}_1\bar{y})(t) = \{\int_{t-\tau}^{t} \bar{y}(s)ds\},$$

where $\bar{y}(t)$ is given in (3.17b), and introducing a matrix $M \in \mathcal{R}^{(n+r)\times(n+r)}$, we obtain the following system:

$$\begin{aligned}&d\tilde{\xi}(t) = [\bar{A}+M]\tilde{\xi}(t)dt + [\bar{B}_1 w(t) + \bar{B}_2\tilde{u}(t)]dt + (\tilde{A}_1 - M)w_1(t)dt\\ &-Mw_2(t)dt + \tilde{G}\tilde{\xi}(t)d\beta(t) + \tilde{H}w_1(t)d\nu - \Gamma_\beta dt - \Gamma_\nu dt,\\ &\bar{y}(t) = [\bar{A}+M]\tilde{\xi}(t) + (\tilde{A}_1 - M)w_1(t) - Mw_2(t) + \bar{B}_1 w(t) + \bar{B}_2\tilde{u}(t) - \Gamma_\beta - \Gamma_\nu\end{aligned} \tag{3.17}$$

where

$$\Gamma_\beta = M\int_{t-\tau}^{t} \tilde{G}\tilde{\xi}(s)d\beta(s), \text{ and } \Gamma_\nu = M\int_{t-\tau}^{t} \tilde{H}w_1(s)d\nu(s).$$

We note that the definitions of $\mathcal{D}_1 g$ in (3.16b) is different from the one used in [77]. The reason is that in the stochastic case the latter definition does not lead to the bound of h. We also note that the matrix M is an unknown constant matrix which is introduced into the dynamics of (3.17a,b) in order to achieve additional degree of freedom in the design of the reduced-order controller.

Remark 3.2. The system (3.15a) is a special case of the dynamics of (3.17a), similarly to what is shown in Remark 1.1, Chapter 1. This follows by replacing $Mw_2(t)dt$ in (3.17a) by $M\{\int_{t-\tau}^{t}\bar{y}(t')dt'\}dt$, using the above definition of $w_2(t)$. Noting that $\bar{y}(t')dt' = d\tilde{\xi}(t') - \tilde{G}\tilde{\xi}(t')d\beta(t') - \tilde{H}w_1(t')d\nu(t')$, one obtains that

$$-Mw_2(t) = -M\int_{t-\tau}^{t}\bar{y}(t')dt' = -M\int_{t-\tau}^{t}\{d\tilde{\xi}(t') - \tilde{G}\tilde{\xi}(t')d\beta(t') - \tilde{H}w_1(t')d\nu(t')\}$$

$$= -M\tilde{\xi}(t) + M\tilde{\xi}(t-\tau(t)) + \Gamma_\beta + \Gamma_\nu = -M\tilde{\xi}(t) + Mw_1(t) + \Gamma_\beta + \Gamma_\nu,$$

where Γ_β and Γ_ν are defined following (3.17). Substituting the right hand side of the latter equation for $-Mw_2(t)$ in (3.17a), the dynamics of (3.15a) is recovered.

Using the fact that $||\mathcal{D}_0||_\infty \le \frac{1}{\sqrt{1-d}}$ and $||\mathcal{D}_1||_\infty \le h$ [77], (3.17) may be cast into what is entitled: the norm-bounded uncertain model, by introducing into (3.17) the following new variables:

$$w_1(t) = \Delta_1\tilde{\xi}(t), \quad and \quad w_2(t) = \Delta_2\bar{y}(t), \tag{3.18}$$

where $||\Delta_1||_\infty \le \frac{1}{\sqrt{1-d}}$ and $||\Delta_2||_\infty \le h$ are diagonal operators having identical scalar operators on the main diagonal. In order to avoid the use of a low pass filter for the solution of the "static" control problem of (17a,b) [54], we define $\bar{\xi}(t) \triangleq col\{\tilde{\xi}(t), \tilde{y}(t)\}$ and obtain the following equivalent descriptor system representation to the closed-loop system:

$$\begin{array}{l} Ed\bar{\xi}(t) = (\hat{A} + \tilde{M})\bar{\xi}(t)dt + \hat{B}w(t) + (\hat{A}_1 - \tilde{M})\bar{w}_1(t)dt + \hat{G}\bar{\xi}(t)d\beta(t) \\ -\tilde{M}\bar{w}_2(t) + \hat{H}\bar{w}_1(t)d\nu(t) + \Gamma_{\beta,\nu}dt, \quad \bar{\xi}(0) = 0, \text{over } [-h \;\; 0], \\ z = [\bar{C}_1 \;\; \bar{D}_{12}K]\bar{\xi}(t), \end{array} \tag{3.19}$$

where $\bar{\xi}(t) \in \mathcal{R}^{\mathbf{n+m+2r}}$ and where

$$E \triangleq \begin{bmatrix} I & 0 \\ 0 & 0 \end{bmatrix}, \; \bar{w}_1(t) = \begin{bmatrix} w_1(t) \\ 0 \end{bmatrix}, \; \bar{w}_2(t) = \begin{bmatrix} w_2(t) \\ 0 \end{bmatrix}, \; \hat{A} \triangleq \begin{bmatrix} \bar{A} & \bar{B}_2K \\ \bar{C}_2 & -I \end{bmatrix},$$

$$\hat{A}_1 \triangleq \begin{bmatrix} \tilde{A}_1 & 0 \\ 0 & 0 \end{bmatrix}, \; \hat{B} \triangleq \begin{bmatrix} \bar{B}_1 \\ \bar{D}_{21} \end{bmatrix}, \; \tilde{M} = \begin{bmatrix} M & 0 \\ 0 & 0 \end{bmatrix}, \; \hat{G} \triangleq \begin{bmatrix} \tilde{G} & 0 \\ 0 & 0 \end{bmatrix}, \; \hat{H} \triangleq \begin{bmatrix} \tilde{H} & 0 \\ 0 & 0 \end{bmatrix},$$

$$\Gamma_{\beta,\nu} \triangleq \begin{bmatrix} -M[\Gamma_\beta + \Gamma_\nu] \\ 0 \end{bmatrix}. \tag{3.20}$$

In order to derive conditions for the stability of the closed-loop system, we consider the system of (3.19), with $\hat{B} = 0$ and define the following Lyapunov function:

$$V(t, \bar{\xi}(t)) = \bar{\xi}^T E \bar{P} \bar{\xi}, \quad \bar{P} \triangleq \begin{bmatrix} P & 0 \\ -\alpha^{-1}\bar{C}_2 P & \alpha^{-1}\hat{P} \end{bmatrix} \tag{3.21}$$

where $P \in \mathcal{R}^{(\mathbf{n+r})\times(\mathbf{n+r})}$ is a positive definite matrix, $\hat{P} > 0$ is a matrix in $\mathcal{R}^{(\mathbf{m+r})\times(\mathbf{m+r})}$, and α is chosen for simplicity to be a positive scalar.

Noting that $\mathcal{E}\{\Gamma_{\beta,\nu}\} = 0$, we apply the *Itô* lemma [9], and taking expectation we obtain:

$$\begin{aligned} \mathcal{E}\{(\mathcal{L}V)(t)\} &= \mathcal{E}\{2\langle \bar{P}\xi(t), [(\hat{A} + \tilde{M})\bar{\xi}(t) + (\hat{A}_1 - \tilde{M})\bar{w}_1(t) - \tilde{M}\bar{w}_2(t)]\rangle\} \\ &+ \mathcal{E}\{Tr\{\bar{P}[\hat{G}\xi(t) \ \ \hat{H}\bar{w}_1(t)]\tilde{Q}[\hat{G}\xi(t) \ \ \hat{H}\bar{w}_1(t)]^T\}\}, \end{aligned}$$

where $\mathcal{L}$ is the infinitesimal generator associated with the differential equation of (3.19) [9], $\tilde{Q} \triangleq \begin{bmatrix} 1 & 0 \\ 0 & 1 \end{bmatrix}$ is the covariance matrix of the augmented Wiener process vector $col\{\beta(t) \ \ v(t)\}$. We also have the following:

$$\begin{aligned} &Tr\{\bar{P}[\hat{G}\bar{\xi}(t) \ \ \hat{H}\bar{w}_1(t)]\tilde{Q}[\hat{G}\bar{\xi}(t) \ \ \hat{H}\bar{w}_1(t)]^T\} \\ &= Tr\{\begin{bmatrix} \bar{\xi}^T(t)\hat{G}^T \\ \bar{w}_1^T(t)\hat{H}^T \end{bmatrix} \bar{P}[\hat{G}\bar{\xi}(t) \ \ \hat{H}\bar{w}_1(t)]\tilde{Q}\} \\ &= Tr\{\begin{bmatrix} \bar{\xi}^T(t)\hat{G}^T\bar{P}\hat{G}\bar{\xi}(t) & \bar{\xi}^T(t)\hat{G}^T\bar{P}\hat{H}\bar{w}_1(t) \\ \bar{w}_1^T(t)\hat{H}^T\bar{P}\hat{G}\bar{\xi}(t) & \bar{w}_1^T(t)\hat{H}^T\bar{P}\hat{H}\bar{w}_1(t) \end{bmatrix} \begin{bmatrix} 1 & 0 \\ 0 & 1 \end{bmatrix}\} \\ &= \bar{\xi}^T(t)\hat{G}^T\bar{P}\hat{G}\bar{\xi}(t) + \bar{w}_1^T(t)\hat{H}^T\bar{P}\hat{H}\bar{w}_1(t). \end{aligned}$$

In the attempt to establish $\mathcal{E}\{(\mathcal{L}V)(t)\} \leq -k||\bar{\xi}(t)||^2$, for some $k > 0$ (uniformly in t), we have the following requirement:

$$\begin{aligned} &2\bar{\xi}^T(t)\bar{P}^T[(\hat{A} + \tilde{M})\bar{\xi}(t) + (\hat{A}_1 - \tilde{M})\bar{w}_1(t) - \tilde{M}\bar{w}_2(t)] + k\bar{\xi}^T(t)\bar{\xi}(t) \\ &\quad + \bar{\xi}^T(t)\hat{G}^T\bar{P}\hat{G}\bar{\xi}(t) + \bar{w}_1^T(t)\hat{H}^T\bar{P}\hat{H}\bar{w}_1(t) < 0. \end{aligned} \tag{3.22}$$

To the latter we add the following term which is nonnegative due to the diagonality of Δ_1

$$\bar{\xi}^T(t)(\frac{1}{1-d}R_1 - \bar{\Delta}_1^T R_1 \bar{\Delta}_1)\bar{\xi}(t) = \bar{\xi}^T(t)\frac{1}{1-d}R_1\bar{\xi}(t) - \bar{w}_1^T(t)R_1\bar{w}_1(t).$$

Using the facts:

$$-2\bar{\xi}^T(t)\bar{P}^T\tilde{M}\bar{w}_2(t) \leq h^2\bar{y}^T(t)R_2\bar{y}(t) + \bar{\xi}^T(t)\bar{P}^T\tilde{M}R_2^{-1}\tilde{M}^T\bar{P}\bar{\xi}(t) \tag{3.23}$$

where R_1 and R_2 are constant positive definite matrices, and

$$\bar{w}_2^T(t)\bar{w}_2(t) = [\bar{y}^T(t)\bar{\Delta}_2^T \ \ 0] \begin{bmatrix} \bar{\Delta}_2\bar{y}(t) \\ 0 \end{bmatrix} \leq h^2\bar{y}^T(t)R_2\bar{y}(t),$$

we find that (3.22) holds if the following inequality is satisfied:

$$\begin{gathered} 2\bar{\xi}^T\bar{P}^T[(\hat{A}+\tilde{M})\bar{\xi}(t) + (\hat{A}_1 - \tilde{M})\bar{w}_1(t)] + k\bar{\xi}^T(t)\bar{\xi}(t) + \bar{\xi}^T(t)\hat{G}^T\bar{P}\hat{G}\bar{\xi}(t) \\ +\bar{w}_1^T(t)\hat{H}^T\bar{P}\hat{H}\bar{w}_1(t) + \bar{\xi}^T(t)\tfrac{1}{1-d}R_1\bar{\xi}(t) - \bar{w}_1^T(t)R_1\bar{w}_1(t) + h^2\bar{y}^T(t)R_2\bar{y}(t) \\ +\bar{\xi}^T(t)\bar{P}^T\tilde{M}R_2^{-1}\tilde{M}^T\bar{P}\bar{\xi}(t) < 0. \end{gathered} \tag{3.24}$$

Denoting $\zeta(t) = col\{\bar{\xi}(t), \bar{w}_1(t)\}$, (3.24) is equivalent to the following:

$$\zeta^T(t)\begin{bmatrix} \Psi_{11} & \bar{P}^T(\hat{A}_1 - \tilde{M})+ \\ * & -R_1 + \hat{H}^T\bar{P}\hat{H} \end{bmatrix}\zeta(t) + h^2\bar{y}^T(t)R_2\bar{y}(t) < 0,$$

where
$\Psi_{11} = \bar{P}^T(\hat{A}+\tilde{M}) + (\hat{A}^T + \tilde{M}^T)\bar{P} + \frac{1}{1-d}R_1 + \hat{G}^T\bar{P}\hat{G} + \bar{P}^T\tilde{M}R_2^{-1}\tilde{M}^T\bar{P}$.
Substituting for $\bar{y}$ from (3.17b), the following inequality is then obtained:

$$\bar{\Psi} = \begin{bmatrix} \bar{\Psi}_{11} & \bar{P}^T\hat{A}_1 - Y & Y & h(\hat{A}^TR_2 + \bar{Y}^T) \\ * & -R_1 + \hat{H}^T\bar{P}\hat{H} & 0 & h(\hat{A}_1^TR_2 - \bar{Y}^T) \\ * & * & -R_2 & -h\bar{Y}^T \\ * & * & * & -R_2 \end{bmatrix} < 0,$$

where

$$\bar{\Psi}_{11} = \bar{P}^T\hat{A} + \hat{A}^T\bar{P} + Y + Y^T + \frac{1}{1-d}R_1 + \hat{G}^T\bar{P}\hat{G}, \quad Y = \bar{P}^T\tilde{M}, \quad \bar{Y} = R_2\tilde{M}. \tag{3.25}$$

In order to avoid using BMI for the solution of (3.25a) (since $\tilde{M}$ appears in both Y and $\bar{Y}$), we choose $M = \bar{\epsilon}I$ in $\tilde{M}$ of (3.20g) where $\bar{\epsilon}$ is a positive tuning scalar. We thus obtain:

$$Y = \bar{P}^T\tilde{M} = \bar{P}^T\begin{bmatrix} \bar{\epsilon}I & 0 \\ 0 & 0 \end{bmatrix}, \quad \bar{Y} = R_2\tilde{M} = R_2\begin{bmatrix} \bar{\epsilon}I & 0 \\ 0 & 0 \end{bmatrix}. \tag{3.26}$$

This choice transforms (3.25a) into a LMI, for a given $\bar{\epsilon}$. The stability of the closed-loop system is therefore guaranteed if there exist $\bar{P} > 0$, K, $R_1 > 0$, $R_2 > 0$, and a positive tuning scalar $\bar{\epsilon}$ that satisfy (3.25a). To satisfy $J_{OF} < 0$ where J_{OF} is given in (3.4) it is required that:

$$\mathcal{E}\{\int_0^\infty [\mathcal{L}V + z^T(t)z(t) - \gamma^2 w^T(t)w(t)]dt\} < 0, \tag{3.27}$$

where in the expression for $\mathcal{E}\{\mathcal{L}V\}$ the operators Δ_1 and Δ_2 are used. Using arguments similar to those that guarantee stability, the latter requirement is described by the following inequality:

$$\Gamma = \begin{bmatrix} \bar{\Psi}_{11} & \bar{P}^T\hat{A}_1 - Y & Y & \bar{P}^T\hat{B} & h\hat{A}^T R_2 + h\bar{Y}^T & [\bar{C}_1 \;\; \bar{D}_{12}K]^T \\ * & -R_1 + \hat{H}^T\bar{P}\hat{H} & 0 & 0 & h\hat{A}_1^T R_2 - h\bar{Y}^T & 0 \\ * & * & -R_2 & 0 & -h\bar{Y}^T & 0 \\ * & * & * & -\gamma^2 I_q & h\hat{B}^T R_2 & 0 \\ * & * & * & * & -R_2 & 0 \\ * & * & * & * & * & -I_z \end{bmatrix} < 0, \tag{3.28}$$

where $\bar{\Psi}_{11}$ is given in (3.25b).

The latter inequality is not affine in the decision variables $\bar{P}$ and K. Denoting,

$$\bar{Q} = \bar{P}^{-1} \triangleq \begin{bmatrix} Q & 0 \\ \hat{Q}\bar{C}_2 & \alpha\hat{Q} \end{bmatrix},$$

we multiply (3.28) by $diag\{\bar{Q}^T, \bar{Q}^T, \bar{Q}^T, I, \bar{Q}^T, I\}$ and $diag\{\bar{Q}, \bar{Q}, \bar{Q}, I, \bar{Q}, I\}$ from the left and the right, respectively. Carrying out the various multiplications we obtain the following inequality: $\hat{\Psi} =$

$$\begin{bmatrix} \hat{\Psi}_{11} & \hat{\Psi}_{12} & \bar{Q}^T Y \bar{Q} & \hat{B} & h\bar{Q}^T[\hat{A}^T R_2 + \bar{Y}^T]\bar{Q} & \hat{\Psi}_{16} & 0 & \hat{\Psi}_{18} \\ * & -\bar{Q}^T R_1 \bar{Q} & 0 & 0 & h\bar{Q}^T[\hat{A}_1^T R_2 - \bar{Y}^T]\bar{Q} & 0 & \hat{\Psi}_{27} & 0 \\ * & * & -\bar{Q}^T R_2 \bar{Q} & 0 & -h\bar{Q}^T\bar{Y}^T\bar{Q} & 0 & 0 & 0 \\ * & * & * & -\gamma^2 I_q & h\hat{B}^T R_2 \bar{Q} & 0 & 0 & 0 \\ * & * & * & * & -\bar{Q}^T R_2 \bar{Q} & 0 & 0 & 0 \\ * & * & * & * & * & -I_z & 0 & 0 \\ * & * & * & * & * & * & -Q & 0 \\ * & * & * & * & * & * & * & -Q \end{bmatrix} < 0, \tag{3.29}$$

where we define $\hat{K} \triangleq K\hat{Q}$ and where

$$
\begin{aligned}
\hat{\Psi}_{11} &= \hat{\Psi}_{11,a} + \bar{Q}^T[Y + Y^T]\bar{Q} + \tfrac{\bar{Q}^T R_1 \bar{Q}}{1-d},\\
\hat{\Psi}_{11,a} &= \begin{bmatrix} \bar{A}Q + Q\bar{A}^T + \bar{B}_2\hat{K}\bar{C}_2 + \bar{C}_2^T\hat{K}^T\bar{B}_2^T & \alpha\bar{B}_2\hat{K} + Q\bar{C}_2^T - \bar{C}_2^T\hat{Q} \\ \alpha\hat{K}^T\bar{B}_2^T + \bar{C}_2 Q - \hat{Q}\bar{C}_2 & -\alpha[\hat{Q} + \hat{Q}] \end{bmatrix},\\
\hat{\Psi}_{12} &= \hat{A}_1\bar{Q} - \bar{Q}^T Y \bar{Q},\\
\hat{\Psi}_{16} &= \bar{Q}^T[\bar{C}_1 \;\; \bar{D}_{12}K]^T = \begin{bmatrix} Q^T\bar{C}_1^T + \bar{C}_2^T\hat{K}^T\bar{D}_{12}^T \\ \alpha\hat{K}^T\bar{D}_{12}^T \end{bmatrix},\\
\hat{\Psi}_{18} &= \bar{Q}^T\hat{G}^T \begin{bmatrix} I \\ 0 \end{bmatrix},\\
\hat{\Psi}_{2,7} &= \bar{Q}^T\hat{H}^T \begin{bmatrix} I \\ 0 \end{bmatrix}.
\end{aligned}
\tag{3.30}
$$

Taking $R_2 = \begin{bmatrix} \epsilon P & 0 \\ 0 & \hat{Q}^{-1} \end{bmatrix}$, where ϵ is a positive scalar tuning parameter, denoting $R_Q = \bar{Q}^T R_1 \bar{Q}$, $\bar{R} = R_2\bar{Q} = \begin{bmatrix} \epsilon I & 0 \\ \bar{C}_2 & \alpha I \end{bmatrix}$ and carrying further the various multiplications, we obtain the following LMI:

$$
\hat{\Upsilon} = \begin{bmatrix}
\hat{\Upsilon}_{11} & \hat{A}_1\bar{Q} - Q_{\bar{\epsilon}} & Q_{\bar{\epsilon}} & \hat{B} & h[\bar{Q}^T\hat{A}^T\bar{R} + \epsilon Q_{\bar{\epsilon}}] & \hat{\Psi}_{16} & 0 & \bar{Q}^T\hat{G}^T\begin{bmatrix} I \\ 0 \end{bmatrix} \\
* & -R_Q & 0 & 0 & h[\bar{Q}^T\hat{A}_1^T\bar{R} - \epsilon Q_{\bar{\epsilon}}] & 0 & \hat{\Psi}_{2,7} & 0 \\
* & * & -\hat{\Upsilon}_{33} & 0 & -h\epsilon Q_{\bar{\epsilon}} & 0 & 0 & 0 \\
* & * & * & -\gamma^2 I_q & h\hat{B}^T\bar{R} & 0 & 0 & 0 \\
* & * & * & * & -\hat{\Upsilon}_{55} & 0 & 0 & 0 \\
* & * & * & * & * & -I_z & 0 & 0 \\
* & * & * & * & * & * & -Q & 0 \\
* & * & * & * & * & * & * & -Q
\end{bmatrix} < 0,
\tag{3.31}
$$

where $Q_{\bar{\epsilon}}$ is defined in (3.11b),

$$
\hat{\Upsilon}_{11} = \hat{\Psi}_{11,a} + Q_{\bar{\epsilon}} + Q_{\bar{\epsilon}}^T + \frac{R_Q}{1-d},
$$

$$
\hat{\Upsilon}_{33} = \hat{\Upsilon}_{55} = \bar{Q}^T R_2 \bar{Q} = \begin{bmatrix} \epsilon Q + \bar{C}_2^T\hat{Q}\bar{C}_2 & \alpha\bar{C}_2^T\hat{Q} \\ \alpha\hat{Q}\bar{C}_2 & \alpha^2\hat{Q} \end{bmatrix}.
$$

Solving for $\hat{A}$, $\hat{A}_1$, $\hat{B}$ and $\hat{H}$, $\hat{G}$ and carrying out the various multiplications, the LMI of (3.10) is obtained.

Remark 3.3. Considering the inequality of (3.29) for the case where h tends to zero, one would expect a condition that is equivalent to the one obtained for the stochastic nonretarded case [53] . Indeed, letting h and R_1 tend to zero and R_2 tend to infinity in (3.29) and choosing in (3.26a,b) $\tilde{M} = \hat{A}_1$ in Y and $\bar{Y}$, the result of [53] is recovered.

Theorem 3.1 provides a sufficient condition for the existence of the gain matrix K that guarantees the required performance for a single plant (3.15) in the polytope $\bar{\Omega}$. As such, this gain matrix provides by (3.9k) the reduced-order controller for this single plant that satisfies (3.4). Since the elements in (3.31) are affine in the system matrices, the condition of Theorem 3.1 can be extended to cope with all the parameters in $\bar{\Omega}$.

Denoting the matrices of (3.15) that correspond to the i-th vertex of $\bar{\Omega}_i$ (and thus also to the vertices of $\bar{\Omega}$) by: $\bar{A}^{(i)}$, $\tilde{A}_1^{(i)}$,$\bar{B}_1^{(i)}$, $\bar{B}_2^{(i)}$, $\bar{C}_1^{(i)}$, $\bar{D}_{12}^{(i)}$ and $\bar{D}_{21}^{(i)}$, $\tilde{H}^{(i)}$, $\tilde{G}^{(i)}$, $i = 1, 2, ..., N$, we obtain the following result:

Corollary 3.3.1 *There exists a reduced-order controller K that achieves negative J_{OF} of (3.4) over the entire polytope $\bar{\Omega}$, for all nonzero $w \in \tilde{L}^2_{\mathcal{F}_t}([0,\infty);\mathcal{R}^q)$, for prescribed $\gamma > 0$, $\alpha > 0$ and positive scalar tuning parameters ϵ and $\bar{\epsilon}$, if there exist matrices $Q > 0$, $\hat{Q} > 0$, $R_Q > 0$ and $\hat{K}$ that satisfy the following set of LMIs:*

$$\begin{bmatrix} \hat{\Upsilon}_{11}^{(i)} & \hat{\Upsilon}_{12}^{(i)} & Q_{\bar{\epsilon}} & \begin{bmatrix} \bar{B}_1^{(i)} \\ \bar{D}_{21}^{(i)} \end{bmatrix} & \hat{\Upsilon}_{15}^{(i)} & \hat{\Upsilon}_{16}^{(i)} & 0 & \hat{\Upsilon}_{18}^{(i)} \\ * & -R_Q & 0 & 0 & \hat{\Upsilon}_{25}^{(i)} & 0 & \hat{\Upsilon}_{27}^{(i)} & 0 \\ * & * & -\hat{\Upsilon}_{33}^{(i)} & 0 & \hat{\Upsilon}_{35}^{(i)} & 0 & 0 & 0 \\ * & * & * & -\gamma^2 I_q & \hat{\Upsilon}_{45}^{(i)} & 0 & 0 & 0 \\ * & * & * & * & -\hat{\Upsilon}_{33}^{(i)} & 0 & 0 & 0 \\ * & * & * & * & * & -I_z & 0 & 0 \\ * & * & * & * & * & * & -Q & 0 \\ * & * & * & * & * & * & * & -Q \end{bmatrix} < 0, \quad i = 1, 2, ..., N \tag{3.32}$$

where $\hat{\Upsilon}_{11}^{(i)}$, $\hat{\Upsilon}_{12}^{(i)}$, $\hat{\Upsilon}_{15}^{(i)}$, $\hat{\Upsilon}_{16}^{(i)}$, $\hat{\Upsilon}_{18}^{(i)}$, $\hat{\Upsilon}_{25}^{(i)}$, $\hat{\Upsilon}_{27}^{(i)}$, $\hat{\Upsilon}_{33}^{(i)}$, $\hat{\Upsilon}_{35}^{(i)}$ and $\hat{\Upsilon}_{45}^{(i)}$ are given in (3.11) and where the nominal matrices are replaced by those corresponding to the vertex i of the polytope.

If a solution to the latter set of LMIs exists, the matrices of the controller (3.3) that stabilizes the system (3.1) and achieves the required performance over the polytope $\bar{\Omega}$ is given by

$$K \triangleq \begin{bmatrix} A_c & B_c \\ C_c & D_c \end{bmatrix} = \hat{K}\hat{Q}^{-1}. \tag{3.33}$$

Remark 3.4. We note that a simple strategy is applied for the numerical solution of (3.32), which also applies to the solution of the LMI of (3.10) for the nominal case. The solution of (3.32) involves a search for two scalar variables: ϵ and $\bar{\epsilon}$. One may start by taking $\epsilon = \bar{\epsilon}$ and seek, using line searching, values for both tuning parameters that leads to a stabilizing controller of minimum γ. Once such a controller is obtained, standard optimization techniques can be used, say Matlab function "fminsearch", which seeks the combination of the two scalar parameters that bring γ to a local minimum.

The above theorem seeks a single set of Q and $\hat{Q}$ that will satisfy the LMIs of (3.32). As such it provides what is called the quadratic stabilizing solution. This solution which may be sometimes quite conservative can be replaced by another, less conservative, still quadratic, solution as follows: It turns out, however, that the special structure of $\bar{P}$ may sometimes lead to zero B_c and C_c, thus resulting in suboptimal values of γ and will lead to static, rather than dynamic, output-feedback controllers. To circumvent this difficulty, we have the following remark:

Remark 3.5. An efficient way for computing the reduced-order controller which avoids the occurrence of $B_c = 0$ and $C_c = 0$, on the expense of restricting the structure of the controller, is one where we fix D_c a priori (say $D_c = 0$) and take $C_c = B_2^T F$ where $F \in \mathcal{R}^{n \times r}$ is a free tuning matrix. The matrices A_c and B_c are then sought which lead to the required performance. In this case, the closed-loop system is described by:

$$\begin{aligned} d\xi(t) &= \mathcal{A}\xi(t)dt + \mathcal{B}w(t)dt + \tilde{A}_1\xi(t-\tau(t))dt + \tilde{G}\xi(t)d\beta(t) + \tilde{H}\xi(t-\tau(t))d\nu(t), \\ z(t) &= \mathcal{C}\xi(t)dt + \mathcal{D}w(t)dt \end{aligned} \tag{3.34}$$

where $\xi \stackrel{\Delta}{=} col\{x,\ x_c\}$ and where

$$\mathcal{A} = \begin{bmatrix} A + B_2D_cC_2 & B_2B_2^T F \\ B_cC_2 & A_c \end{bmatrix},$$
$$\mathcal{B} = \begin{bmatrix} B_1 + B_2D_cD_{21} \\ B_cD_{21} \end{bmatrix},$$
$$\mathcal{C} = \begin{bmatrix} C_1 + D_{12}D_cC_2 & D_{12}B_2^T F \end{bmatrix},$$

and $\mathcal{D} = D_{12}D_cD_{21}$. Defining $K \stackrel{\Delta}{=} \begin{bmatrix} A_c & B_c \end{bmatrix}$ we obtain the following:

$$\begin{aligned} \mathcal{A} &= \bar{A} + \bar{B}_2K\bar{C}_2, \\ \mathcal{B} &= \bar{B}_1 + K\bar{D}_{21}, \\ \mathcal{C} &= \bar{C}_1 + \bar{D}_{12}K\bar{C}_2, \\ \mathcal{D} &= \bar{D}_{11} + \bar{D}_{12}K\bar{D}_{21} \end{aligned}$$

where

$$\begin{aligned}
\bar{A} &= \begin{bmatrix} A + B_2 D_c C_2 & B_2 B_2^T F \\ 0 & 0 \end{bmatrix}, \\
\bar{D}_{11} &= D_{12} D_c D_{21}, \\
\bar{B}_1 &= \begin{bmatrix} B_1 + B_2 D_c D_{21} \\ 0 \end{bmatrix}, \\
\bar{B}_2 &= \begin{bmatrix} 0 \\ I \end{bmatrix}, \\
\bar{C}_1 &= \begin{bmatrix} C_1 + D_{12} D_c C_2 & D_{12} B_2^T F \end{bmatrix}, \\
\bar{C}_2 &= \begin{bmatrix} 0 & I \\ C_2 & 0 \end{bmatrix}, \\
\bar{D}_{21} &= \begin{bmatrix} 0 \\ D_{21} \end{bmatrix}, \\
\bar{D}_{12} &= 0.
\end{aligned}$$

Obviously, this approach is suitable for the case where B_2 is without uncertainty. A similar approach can be applied for the case where C_2 is without uncertainty, in which case we assign similar structure to B_c.

3.4 Conclusions

In this chapter the problem of linear H_∞ reduced-order output-feedback control of state-multiplicative retarded systems has been presented and solved for both nominal and uncertain polytopic systems. The stochastic uncertainties have been encountered in both the delayed and the nondelayed states in the state space model of the system. The delay has been assumed to be unknown and time-varying where only the bounds on its length and rate of change are given. Delay dependent synthesis method has been applied which is based on the input–output approach. The resulting nonretarded system is then augmented to obtain a descriptor system which avoids the inclusion of an artificial low pass filter, in series to the output. A stability condition is derived for the closed-loop system, which subsequently leads to the solution of the reduced-order output-feedback control problem.

Some over-design is entailed in our solution due to the special structure chosen for the Lyapunov function and due to the special structure imposed on R_2 in the resulting LMIs. In the uncertain case, an additional over-design stems from the fact that a quadratically stable solution is sought. The efficiency of the proposed results is demonstrated via example 13.1 in Chapter 13.

4

Tracking Control with Preview

4.1 Introduction

In this chapter we address the problem of H_∞ state-feedback preview tracking control of state-delayed, continuous-time, state-multiplicative, linear systems, as formulated in Section 4.2 (for a similar deterministic setup see [78], [106], [112]). Similarly to [55],[59], the point of view taken here differs from the "traditional" L-K point of view as we apply the input–output approach.

The problem of tracking control with preview may be treated by the standard theory of non-retarded systems with state-multiplicative noise (see [53] and the references therein). The approach there leads to over-conservative conditions which, in the case of retarded systems, also require too many tunings of scalar parameters. The main source of conservatism stems there from the way stochastic stability is guaranteed via Linear Matrix Inequalities (LMIs) in the standard approach [53].

To reduce this conservatism, a new approach is adopted here where we first derive, in Section 4.3, Lemma 4.3.1, a matrix inequality condition for the stability of the closed-loop system that involves the yet undetermined control input. Applying the resulting stability condition to the control design problem in Section 4.4, a matrix inequality condition is obtained that guarantees a prescribed H_∞ performance. This inequality involves the undetermined control input too. Realizing that application of Schur complement is equivalent to completion to squares, the optimal control strategy is then derived, in Theorem 4.1.

In our system, we allow for a time-varying delay where the uncertain stochastic parameters multiply both the delayed and the non delayed states in the state space model of the system. We obtain sufficient conditions for stability and H_∞ state-feedback tracking control design, in terms of LMIs in Corollaries 4.4.1, 4.4.2 of Section 4.4. We demonstrate our theory, in Section 4.5, via an example that demonstrates the impact of the delay length, delay derivative and the preview length on the system performance. An additional example is given in Chapter 13 (Example 13.2).

E. Gershon & U. Shaked: *Advanced Topics in Control & Estimation*, LNCIS 439, pp. 75–93.
DOI: 10.1007/978-1-4471-5070-1_4

4.2 Problem Formulation

Given the following linear continuous time-invariant system:

$$\begin{array}{l} dx(t) = [A_0x(t) + A_1x(t-\tau(t)) + B_1w(t) + B_2u(t) + B_3r(t)]dt \\ +Gx(t-\tau(t))d\zeta(t) + Fx(t)d\beta(t), \quad x(\theta) = 0,\ over[-\bar{h}\ 0], \\ z(t) = C_1x(t) + D_{12}u(t) + D_{13}r(t) \end{array} \tag{4.1}$$

where $x(t) \in \mathcal{R}^n$ is the system state vector, $u(t) \in \mathcal{R}^l$ is the input vector, $r(t) \in L_2$ is the signal to be tracked, $w(t) \in \tilde{L}^2([0,\infty);\mathcal{R}^p)$ is the exogenous disturbance signal, $z(t) \in \mathcal{R}^q$ is the signal to be controlled, A, A_1, B_1, B_2, B_3, C_1, D_{12}, D_{13}, F, and G are constant matrices of the appropriate dimensions and $\tau(t)$ is the unknown time-delay that satisfies:

$$0 \le \tau(t) \le \bar{h}, \quad \dot{\tau}(t) \le d < 1. \tag{4.2}$$

The variables $\beta(t)$ and $\zeta(t)$ are zero-mean real scalar Wiener processes that satisfy:

$$\mathcal{E}\{d\beta(t)\} = 0,\ \mathcal{E}\{d\zeta(t)\}=0,\ \mathcal{E}\{d\beta(t)^2\}=dt,\ \mathcal{E}\{d\zeta(t)^2\}=dt,$$
$$\mathcal{E}\{d\beta(t)d\zeta(t)\}= 0.$$

We seek a state-feedback control law $u(t)$ that minimizes, for the worst-case of the process disturbance $w(t)$, the expected value of the energy of $z(t)$ with respect to the uncertain parameters, by using the available knowledge on the reference signal. We, therefore, consider, for a given scalar $\gamma > 0$, the following performance index:

$$J_E \stackrel{\Delta}{=} \mathcal{E}\{\int_0^\infty (||z(t)||^2 - \gamma^2||w(t)||^2)dt\}. \tag{4.3}$$

We consider two different tracking problems differing on the information pattern over $r(t)$:

1) Stochastic H_∞ finite-fixed preview tracking of $r(t)$: The tracking signal $r(t)$ is previewed in a known fixed interval, i.e., $r(\tilde{\tau})$ is known for $\tilde{\tau} \le t + h$ where h is a known preview length.
2) Stochastic H_∞-tracking with zero preview of $r(t)$: The tracking signal is measured on line, i.e., at time t, $r(\bar{\tau})$ is known for $\bar{\tau} \le t$.

In these two cases we seek a control law $u(t)$ of the form $u(t) = H_x x(t) + H_r r(t)$, where H_x is a causal operator and where the causality of the operator H_r depends on the information pattern of the reference signal.

For these tracking problems we consider a related linear quadratic min-max game in which the controller plays against nature. We, thus, consider the following game:

Find r(t)-dependent strategies $w^* \in \tilde{L}^2([0,\infty);\mathcal{R}^p)$ and $u^*(t) \in R^l$ that satisfy, $\forall r(t) \in L_2$:

$$J_E(r,u^*,w) \le J_E(r,u^*,w^*) \le J_E(r,u,w^*), \quad \forall w(t) \in \tilde{L}^2([0,\infty);\mathcal{R}^p)$$

$$\text{and } \forall u(t) \in R^l.$$

Also find the minimum upper-bound $\bar{J}(r)$ to the saddle point value $J_E(r,u^*,w^*)$.

In order to solve the above problems, we introduce the following operators:

$$(\Delta_1 g)(t) \stackrel{\Delta}{=} g(t-\tau(t)),\ (\Delta_2 g)(t) \stackrel{\Delta}{=} \int_{t-\tau(t)}^{t} g(s)ds. \tag{4.4}$$

Considering the above two operators the system (4.1a) can be written as:

$$\begin{aligned}
&dx(t) = [A_0 + M]x(t)dt + [A_1 - M]w_1(t)dt - Mw_2(t)dt + B_1 w(t)dt \\
&+B_2 u(t)dt + B_3 r(t)dt + Fx(t)d\beta(t) + Gw_1(t)d\zeta(t) - \Gamma_\beta dt - \Gamma_\zeta dt, \\
&\bar{y}(t) = [A_0 + M]x(t) + [A_1 - M]w_1(t) - Mw_2(t) + B_1 w(t) + B_2 u(t) \\
&+B_3 r(t) - \Gamma_\beta - \Gamma_\zeta,
\end{aligned} \tag{4.5}$$

where

$$\Gamma_\beta = M\int_{t-\tau}^{t} Fx(s)d\beta(s), \quad \Gamma_\zeta = M\int_{t-\tau}^{t} Gw_1(s)d\zeta(s),$$

$$w_1(t) = \Delta_1 x(t) \text{ and } w_2(t) = \Delta_2 \bar{y}(t), \tag{4.6}$$

where the matrix M is an unknown constant matrix to be determined.

Remark 4.1. Similarly to what is shown in Remark 1.1 in Chapter 1, the dynamics of (4.1a) is a special case of that of (4.5a) as follows:

Noting (4.6c,d) and applying the operators of (4.4a,b), Equation (4.5a) can be written as:

$$dx(t) = [A_0 + M]x(t)dt + [A_1 - M]w_1(t)dt - M\{\int_{t-\tau}^{t} \bar{y}(t')dt'\}dt + B_1 w(t)dt$$

$$+B_2 u(t)dt + B_3 r(t)dt + Fx(t)d\beta(t) + Gw_1(t)d\zeta(t) - \Gamma_\beta dt - \Gamma_\zeta dt,$$

$$w_1(t) = x(t-\tau(t)).$$

Now, recalling $\bar{y}(t)$ of (4.5b) one can write:

$$dx(t) = \bar{y}(t)dt + Fx(t)d\beta(t) + Gw_1(t)d\zeta(t)$$

and therefore $\bar{y}(t')dt' = dx(t') - Fx(t')d\beta(t') - Gw_1(t')d\zeta(t')$. Hence,

$$-Mw_2(t) = -M\int_{t-\tau}^{t} \bar{y}(t')dt' = -M\int_{t-\tau}^{t} \{dx(t') - Fx(t')d\beta(t') - Gw_1(t')d\zeta(t')\}$$

$$= -Mx(t) + Mx(t-\tau) + \Gamma_\beta + \Gamma_\zeta = -Mx(t) + Mw_1(t) + \Gamma_\beta + \Gamma_\zeta,$$

where Γ_β and Γ_ζ are defined in (4.6a,b) respectively. Replacing the right hand side of the latter equation for $-Mw_2(t)$ in (4.5a), the dynamics of (4.1a) is recovered.

Using the facts that $||\Delta_1||_\infty \leq \frac{1}{\sqrt{1-d}}$ and $||\Delta_2||_\infty \leq \bar{h}$ (this follows from arguments similar to those in [77], see Lemma 1), the system of (5) can be described in the frame work of norm-bounded systems.

4.3 Stability of the Delayed Tracking System

In the sequel, we derive a stability result that is based on the assertion that the conditions for the stability of the system (4.5a,b) with the 'feedback' that is described by (4.6c,d), are usually derived using the small gain theorem. It is well known [15], however, that these conditions are identical to those obtained by applying the Lyapunov approach to the same system with feedback that is described by the diagonal 'feedback' operators $\bar{\Delta}_1$ and $\bar{\Delta}_2$ that, for all x and $\bar{y}$ in $\mathcal{R}^n$, satisfy the following:

$$x^T(t)\bar{\Delta}_1^T\bar{\Delta}_1 x(t) \leq (1-d)^{-1}\|x(t)\|^2$$

and

$$\bar{y}^T(t)\bar{\Delta}_2^T\bar{\Delta}_2\bar{y}(t) \leq \bar{h}^2\|\bar{y}(t)\|^2.$$

In order to derive our stability result, we replace Δ_i by $\bar{\Delta}_i,\ i = 1,2$. Noting that $\mathcal{E}\{\Gamma_\beta\} = 0$ and $\mathcal{E}\{\Gamma_\zeta\} = 0$, we consider $V(t,x(t)) = \langle x(t), Qx(t)\rangle$. Applying Ito formula to $V(t,x(t))$, defining

$$\tilde{A} = A_0 + M, \quad \tilde{A}_1 = A_1 - M$$

and taking expectation we obtain, defining the infinitesimal generator associated with the differential equation of (5a) by $\mathcal{L}$:

$$\mathcal{E}\{\mathcal{L}V(t,x(t))\} = 2\mathcal{E}\langle Qx(t), \tilde{A}x(t) + \tilde{A}_1\bar{\Delta}_1 x(t) - M\bar{\Delta}_2\bar{y}(t) + B_1 w(t)$$

$$+ B_2 u(t) + B_3 r(t)\rangle + \mathcal{E}Tr\{Q[Fx(t)\ \ Gw_1(t)]\bar{P}[Fx(t)\ \ Gw_1(t)]^T\}, \qquad (4.7)$$

where $\bar{P} \triangleq \begin{bmatrix} 1 & 0 \\ 0 & 1 \end{bmatrix}$ is the covariance matrix of the stationary augmented Wiener process vector $col\{\beta(t),\ \ \zeta(t)\}$ and where the condition for the stability of the system is $\mathcal{E}\{\mathcal{L}V(t,x(t))\} < 0$. Now,

$$Tr\{Q[Fx(t)\ \ Gw_1(t)]\bar{P}[Fx(t)\ \ Gw_1(t)]^T\} = Tr\{\begin{bmatrix} x^T(t)F^T \\ w_1^T(t)G^T \end{bmatrix} Q[Fx(t)\ \ Gw_1(t)]\bar{P}\}$$

$$= x^T(t)F^T QFx(t) + w_1^T(t)G^T QGw_1(t).$$

Adding the following term, which is nonnegative due to the diagonality of $\bar{\Delta}_1$, to (4.7):

$$x^T(t)(\frac{1}{1-d}R_1 - \bar{\Delta}_1^T R_1 \bar{\Delta}_1)x(t) = x^T(t)\frac{1}{1-d}R_1 x(t) - w_1^T(t)R_1 w_1(t),$$

and noting that:

$$w_2^T(t)w_2(t) = \bar{y}^T(t)\bar{\Delta}_2^T \bar{\Delta}_2 \bar{y}(t) \leq h^2 \bar{y}^T(t)R_2 \bar{y}(t),$$

where R_1 and R_2 are constant positive definite matrices we obtain:

$$\mathcal{E}\{\mathcal{L}V(t,x(t))\} \leq 2\mathcal{E}\langle Qx(t), \tilde{A}x(t) + \tilde{A}_1 w_1 - Mw_2 + B_1 w(t) + B_2 u(t) + B_3 r(t)\rangle$$
$$-w_1^T(t)R_1 w_1(t) + \bar{h}^2 \bar{y}^T(t)R_2 \bar{y}(t) - w_2^T R_2 w_2 + x^T(t)F^T QFx(t)$$
$$+ x^T(t)\frac{1}{1-d}R_1 x(t) + w_1^T(t)G^T QGw_1(t). \tag{4.8}$$

Taking $w(t) = 0$ and $r(t) = 0$ in (4.8) and in $\bar{y}$ of (4.5b), defining $\xi(t) = col\{x(t), w_1(t), w_2(t), u(t)\}$, we readily obtain from (4.8), that stability is guaranteed for a feedback control signal $u(t)$ that satisfies the following inequality $\mathcal{E}\{\xi^T(t)\bar{\phi}\xi(t) + \bar{y}^T(t)\bar{h}^2 R_2 \bar{y}(t)\} =$

$$\mathcal{E}\{\xi^T(t)\{\begin{bmatrix} \Upsilon_{11} & Q\tilde{A}_1 & -QM & QB_2 \\ * & -R_1 + G^T QG & 0 & 0 \\ * & * & -R_2 & 0 \\ * & * & * & 0 \end{bmatrix}$$

$$+ \begin{bmatrix} \tilde{A}_0^T \\ \tilde{A}_1^T \\ -M^T \\ B_2^T \end{bmatrix} \bar{h}^2 R_2 \begin{bmatrix} \tilde{A}_0 & \tilde{A}_1 & -M & B_2 \end{bmatrix}\}\xi(t)\} < 0,$$

$$\text{where} \qquad \Upsilon_{11} = Q\tilde{A} + \tilde{A}^T Q + \tfrac{1}{1-d}R_1 + F^T QF, \tag{4.9}$$

and where $\bar{\phi}$ is the leftmost block matrix in the last inequality. The later inequality leads, in turn, to the following inequality:

$$\mathcal{E}\{\bar{\xi}^T(t)\phi\bar{\xi}(t)\} < 0, \text{ where } \phi = \begin{bmatrix} \Upsilon_{11} & Q\tilde{A}_1 & -QM & QB_2 & A_0^T + M^T \\ * & -R_1 + G^TQG & 0 & 0 & A_1^T - M^T \\ * & * & -R_2 & 0 & -M^T \\ * & * & * & 0 & B_2^T \\ * & * & * & * & -R_2^{-1}\bar{h}^{-2} \end{bmatrix}, \tag{4.10}$$

and where $\bar{\xi}(t) \triangleq \begin{bmatrix} \xi(t) \\ \bar{y}(t) \end{bmatrix}$. We note that (4.10a) should be satisfied along the trajectories $\bar{\xi}$. This does not mean that $\phi < 0$. Defining $Q_M = QM$, taking $R_2 = \epsilon_1 Q$ where ϵ_1 is a positive tuning scalar, and multiplying (4.10a) from the left and the right by $diag\{I, I, I, I, \bar{h}R_2\}$, we obtain the following requirement for $u(t)$ to achieve closed-loop stability.

$$\mathcal{E}\{\xi_y^T(t)\{\begin{bmatrix} \Upsilon_{11} & \Upsilon_{12} & -Q_M & QB_2 & \bar{h}\epsilon_1(A_0^TQ + Q_M^T) \\ * & -R_1 + G^TQG & 0 & 0 & \bar{h}\epsilon_1(A_1^TQ - Q_M^T) \\ * & * & -\epsilon_1 Q & 0 & -\bar{h}\epsilon_1 Q_M^T \\ * & * & * & 0 & \bar{h}\epsilon_1 Q \\ * & * & * & * & -\epsilon_1 Q \end{bmatrix}\}\xi_y(t)\} < 0, \tag{4.11}$$

where $\xi_y(t) \triangleq \begin{bmatrix} \xi(t) \\ \bar{h}^{-1}R_2^{-1}\bar{y}(t) \end{bmatrix}$ and where $\Upsilon_{11} = QA_0 + Q_M + A_0^TQ + Q_M^T + \frac{1}{1-d}R_1 + F^TQF$, $\Upsilon_{12} = QA_1 - Q_M$. We thus obtain the following lemma:

Lemma 4.3.1 *The system of (4.1a) with $w = 0$ and $r = 0$ is exponentially mean square stable for a given feedback control signal $u(t)$ and a given positive scalar ϵ_1, if there exist positive definite matrices $0 < Q \in \mathcal{R}^{n\times n}$, $0 < R_1 \in \mathcal{R}^{n\times n}$ and a matrix $Q_M \in \mathcal{R}^{n\times n}$ that satisfy (4.11).*

4.4 The State-Feedback Tracking

In the sequel, the solution to the stochastic state-feedback tracking control problem for retarded systems is obtained by applying a game theory approach which enables us to obtain the strategies of the signals involved, alongside the two preview tracking patterns that are applied. We note that the solution to the non-retarded case in ([53], pages 57–65) has first been obtained for the finite-horizon case using the game theoretic approach. It has been extended there to infinite-horizon where mean square stability of the close-loop system is obtained in the limit where $t \to \infty$.

The approach taken in the present manuscript, is somewhat different. Here, we restrict the theory to the infinite-horizon case. We assume, first, that the closed-loop system is exponentially mean square stable and then derive a LMI condition that assures the H_∞ control performance if the 'minimizing control' input indeed stabilizes the closed-loop system. We then show, in the Appendix, that the condition of Lemma 4.3.1, for the stability of the closed-loop system, is satisfied as a part of the resulting LMI.

We consider the following inequality:

$$\hat{\Psi} = \begin{bmatrix} \hat{\Psi}_{11} & \hat{\Psi}_{12} & A_1 - M & 0 & M & \frac{PR_1}{\sqrt{1-d}} & PC_1^T & PF^T \\ * & \hat{\Psi}_{22} & \bar{h}R_2(A_1 - M) & 0 & \bar{h}R_2 M & 0 & 0 & 0 \\ * & * & -R_1 & G^T & 0 & 0 & 0 & 0 \\ * & * & * & -P & 0 & 0 & 0 & 0 \\ * & * & * & * & -R_2 & 0 & 0 & 0 \\ * & * & * & * & * & -R_1 & 0 & 0 \\ * & * & * & * & * & * & -I_q & 0 \\ * & * & * & * & * & * & * & -P \end{bmatrix} < 0, \quad (4.12)$$

$\forall t \in [0\ \infty)$, where

$$\begin{aligned} \hat{\Psi}_{11} &= (A_0 + M)P + P(A_0 + M)^T + \gamma^{-2}B_1B_1^T - B_2\hat{R}^{-1}B_2^T, \\ \hat{\Psi}_{12} &= \bar{h}PA_0^T R_2 + \bar{h}P\bar{R}_2^T + \bar{h}\gamma^{-2}B_1B_1^T - \bar{h}B_2\hat{R}^{-1}B_2^T, \\ \hat{\Psi}_{22} &= \bar{h}^2[\gamma^{-2}B_1B_1^T - B_2\hat{R}^{-1}B_2^T] - R_2, \\ \hat{R} &= D_{12}^T D_{12}. \end{aligned} \quad (4.13)$$

The cost functional J_E of (4.3) cannot be possibly made negative for all $w \in \tilde{L}^2([0,\infty);\mathcal{R}^p)$ and $r(t) \in L_2$. Denoting, therefore, by $f(r,t)$, $f(0,t) = 0$ the yet unknown positive scalar function that describes the upper-bound on the integrand of (4.3) and by $\bar{J}(r) = \int_0^\infty f(r,t)dt$, assuming for the moment that $\int_0^\infty f(r,t)dt$ exists, the following main result of the paper is obtained for the case of fixed-finite preview (i.e. $h \neq 0$):

Theorem 4.1. *Consider the system of (4.1a,b) and J_E of (4.3). Given $\gamma > 0$, there exists a control signal $u(t)$ that possesses a solution that minimizes $\bar{J}(r)$ if there exist a positive definite matrix $P \in \mathcal{R}^{n\times n}$, a matrix $M \in \mathcal{R}^{n\times n}$ and positive definite matrices $R_1 \in \mathcal{R}^{n\times n}$, $R_2 \in \mathcal{R}^{n\times n}$ that solve (4.12). When a solution exists, the maximizing and minimizing strategies of Nature and the controller are respectively given by:*

$$w^*(t) = -\gamma^{-2}[B_1^T P^{-1}x(t) + \bar{h}B_1^T R_2^{-1}\bar{y}(t)], \quad u^*(t) = [\bar{K}_x x(t) + \bar{K}_r r(t) + \bar{K}_\theta \theta(t)],$$

where

$$
\begin{aligned}
\bar{K}_x &= -[\bar{h}^2 B_2^T \phi^{-1} B_2 + D_{12}^T D_{12}]^{-1}[\bar{h} g \phi^{-1} B_2 + C_1^T D_{12} + P^{-1} B_2]^T, \\
\bar{K}_r &= -[\bar{h}^2 B_2^T \phi^{-1} B_2 + D_{12}^T D_{12}]^{-1}(D_{12}^T D_{13} + \bar{h}^2 B_2^T \phi^{-1} B_3), \\
\bar{K}_\theta &= -[\bar{h}^2 B_2^T \phi^{-1} B_2 + D_{12}^T D_{12}]^{-1} \bar{B}_2^T
\end{aligned}
\tag{4.14}
$$

and where

$$
\begin{aligned}
& Q = P^{-1}, \\
& \bar{R}_2 = R_2 M, \\
& \phi = -(\bar{h}^2[(R_2 A_1 - \bar{R}_2)[R_1 - G^T Q G]^{-1}(R_2 A_1 - \bar{R}_2)^T + \bar{R}_2 R_2^{-1} \bar{R}_2^T] \\
& + \bar{h}^2 \gamma^{-2} B_1 B_1^T - R_2), \\
& g = \bar{h}(A_0^T R_2 + \bar{R}_2^T) + \bar{h}[Q(A_1 - M)[R_1 - G^T Q G]^{-1}(A_1^T R_2 - \bar{R}_2^T) \\
& + Q M R_2^{-1} \bar{R}_2^T] + \bar{h}\gamma^{-2} Q B_1 B_1^T, \\
& \bar{B}_2 = B_2 + [\hat{A}_w \tilde{R}_{\tilde{w},\bar{y}} - \bar{h}\gamma^{-2} \bar{B}_1 B_1^T] B_2, \\
& \hat{A}_w = ([\tilde{A}_1 \;\; - M] - \bar{h}\gamma^{-2} \bar{B}_1 B_1^T [\tilde{A}_1 \;\; - M])(I - \tilde{R}_{\tilde{w},\bar{y}}[\tilde{A}_1 \;\; - M])^{-1}, \\
& \tilde{R}_{\tilde{w}} = \begin{bmatrix} R_1 - G^T Q G & 0 \\ 0 & R_2 \end{bmatrix}, \\
& \tilde{R}_{\tilde{w},x} = \tilde{R}_{\tilde{w}}^{-1} \begin{bmatrix} \tilde{A}_1^T Q \\ -M^T Q \end{bmatrix}, \\
& \tilde{R}_{\tilde{w},\bar{y}} = \tilde{R}_{\tilde{w}}^{-1} \begin{bmatrix} \bar{h}\tilde{A}_1^T \\ -\bar{h} M^T \end{bmatrix} R_2^{-1}, \\
& \bar{B}_1 = (I + \hat{A}_w \tilde{R}_{\tilde{w},\bar{y}}) B_1 [I + \bar{h}\gamma^{-2} B_1^T B_1]^{-1}.
\end{aligned}
\tag{4.15}
$$

The signal $\theta(\tau)$ is described by

$$
\dot{\theta}(\tau) = -\bar{A}^T \theta(\tau) + \bar{B}_r r(\tau), \quad \tau \in [t \;\; t+h], \; \theta(t+s) \equiv 0, \; s \geq h, \tag{4.16}
$$

with

$$
\begin{aligned}
& \bar{A} = A - \gamma^{-2} \bar{B}_1 [B_1^T Q + \bar{h} B_1^T \tilde{A}] - B_2 [\bar{h}^2 B_2^T \phi^{-1} B_2 + D_{12}^T D_{12}]^{-1} [\bar{h} g \phi^{-1} B_2 \\
& + C_1^T D_{12} + Q B_2] + \hat{A}_w \tilde{R}_{\tilde{w},x}, \\
& \bar{B}_r = B_3 - \bar{B}_2 \bar{K}_r - \gamma^{-2} \bar{h} \bar{B}_1 B_1^T B_3 + \hat{A}_w \tilde{R}_{\tilde{w},\bar{y}} B_3,
\end{aligned}
\tag{4.17}
$$

where h is the preview length. The bound on the performance index is then given by:

$$
J_E(r, u^*, w^*) \leq \bar{J}(r) \tag{4.18}
$$

where

$$\bar{J}(r) = \mathcal{E}\int_0^{\infty}\{r^T(\alpha_6 + \alpha_5^T\alpha_4^{-1}\alpha_5)r\}dt + \mathcal{E}\int_0^{\infty}||\bar{B}_1^T\theta||^2_{\gamma^{-2}}dt$$

$$+ 2\mathcal{E}\int_0^{\infty}\theta^T(t)\{\bar{B}_r r(t)\}dt - \mathcal{E}\int_0^{\infty}||\hat{A}_w^T\theta||^2_{\tilde{R}_{\bar{w}}^{-1}}dt - \mathcal{E}\int_0^{\infty}||\alpha_4^{-1/2}\bar{B}_2^T\theta||^2 dt \tag{4.19}$$

and where α_4, α_5, α_6 *are given in the following:*

$$\tilde{\Upsilon}_3 = \begin{bmatrix} \tilde{\Upsilon}_{3,1} & \bar{h}g\phi^{-1}B_2 + C_1^T D_{12} + QB_2 & QB_3 + \bar{h}g\phi^{-1}B_3 + C_1^T D_{13} \\ * & D_{12}^T D_{12} + \bar{h}^2 B_2^T\phi^{-1}B_2 & D_{12}^T D_{13} + \bar{h}^2 B_2^T\phi^{-1}B_3 \\ * & * & D_{13}^T D_{13} + \bar{h}^2 B_3^T\phi^{-1}B_3 \end{bmatrix}$$

$$\triangleq \begin{bmatrix} \alpha_1 & \alpha_2 & \alpha_3 \\ \alpha_2^T & \alpha_4 & \alpha_5 \\ \alpha_3^T & \alpha_5^T & \alpha_6 \end{bmatrix}, \quad \tilde{\Upsilon}_{3,1} = \bar{\Upsilon}_{11} + g\phi^{-1}g^T + C_1^T C_1. \tag{4.20}$$

The proof of the theorem is given in the Appendix where the well-posedness of $\bar{J}(r)$ is also discussed.

Noting that, unfortunately, the inequality (4.12) is trilinear in the decision variables, two possible solution methods can be applied: the first involves P, $(M,\ R_1)$ iterations, where we take $R_2 = \epsilon I_n$ and choose ϵ to be a tuning parameter, similar to the $P-K$ iterations method. This method, if it converges, can only achieve a local minimum. The second method, which we apply in the statement of Lemma 4.4.1 below, is to assign predetermined values for M, R_1 and R_2 which entails an over-design while rendering a simple tractable solution. It should be noted here that M was **an additional degree of freedom** in our solution method. Substituting in (4.12) $M = A_1 + \epsilon_m I_n$, $R_1 = \epsilon_r I_n$ and $R_2 = \epsilon_2 I_n$, we obtain the following condition:

$$\tilde{\Upsilon} = \begin{bmatrix} \tilde{\Upsilon}_{11} & \tilde{\Upsilon}_{12} & -\epsilon_m I_n & 0 & A_1 + \epsilon_m I_n & \frac{P\epsilon_r}{\sqrt{1-d}} & PC_1^T & PF^T \\ * & \tilde{\Upsilon}_{22} & -\bar{h}\epsilon_2\epsilon_m I_n & 0 & \bar{h}\epsilon_2(A_1 + \epsilon_m I_n) & 0 & 0 & 0 \\ * & * & -\epsilon_r I_n & G^T & 0 & 0 & 0 & 0 \\ * & * & * & -P & 0 & 0 & 0 & 0 \\ * & * & * & * & -\epsilon_2 I_n & 0 & 0 & 0 \\ * & * & * & * & * & -\epsilon_r I_n & 0 & 0 \\ * & * & * & * & * & * & -I_q & 0 \\ * & * & * & * & * & * & * & -P \end{bmatrix} < 0, \tag{4.21}$$

$\forall t \in [0\ \infty)$, where

$$\begin{aligned}\tilde{\Upsilon}_{11} &= (A_0 + A_1 + \epsilon_m I_n)P + P(A_0 + A_1 + \epsilon_m I_n)^T + \gamma^{-2}B_1B_1^T - B_2\hat{R}^{-1}B_2^T,\\ \tilde{\Upsilon}_{12} &= \bar{h}PA_0^T\epsilon_2 + \bar{h}P(A_1 + \epsilon_m I_n)^T\epsilon_2 + \bar{h}\gamma^{-2}B_1B_1^T - \bar{h}B_2\hat{R}^{-1}B_2^T,\\ \tilde{\Upsilon}_{22} &= \bar{h}^2[\gamma^{-2}B_1B_1^T - B_2\hat{R}^{-1}B_2^T] - \epsilon_2 I_n.\end{aligned} \tag{4.22}$$

We thus have the following result for the case of finite fixed-preview tracking control:

Lemma 4.4.1 *Consider the system of (4.1a,b) and J_E of (4.3). Given $\gamma > 0$, and three tuning scalars: ϵ_m, $\epsilon_2 > 0$ and $\epsilon_r > 0$, the state-feedback tracking controller minimizes $\bar{J}(r)$ if there exists a positive definite matrix $P \in \mathcal{R}^{n\times n}$ that solve (4.21).*

The following preview strategies are obtained:

Corollary 4.4.1 Stochastic H_∞-tracking with finite fixed-preview : *Since at time t, $r(\tilde{\tau})$ is known for $\tilde{\tau} \leq t + h$ the following control law is obtained:*

$$u(t) = -[\bar{K}_x x(t) + \bar{K}_r r(t) + \bar{K}_\theta \theta(t)], \tag{4.23}$$

where K_x, K_r and K_θ are defined in (4.14). The latter controller achieves the bound $\bar{J}(r)$ of (4.19).

Corollary 4.4.2 Stochastic H_∞-tracking with no preview : *In this case the control law is given by $u(t) = -[\bar{K}_x x(t) + \bar{K}_r r(t)]$, where $\bar{K}_x$ and $\bar{K}_r$ are the same as in Corollary 4.4.1. The bound of the performance index is $\bar{J}(r)$ of (4.19) where all the terms that contain the signal $\theta(t)$ are zero.*

Remark 4.2. Considering (4.12), deleting the second to the fifth column and row blocks of $\hat{\Psi}$ and taking $M = A_1$, $R_1 = O(\bar{h})$ and $R_2 = O(\frac{1}{\bar{h}})$, one obtains the result of Theorem 4.2, page 63 of [53] for the nondelayed stochastic state-feedback tracking case. We note also that (4.12) and therefore (4.21), do not depend on B_3 and D_{13}, similarly to the result in Theorem 4.2 of [53].

Remark 4.3. We note that a simple strategy is applied for the numerical solution of (4.21). The solution of (4.21) involves a search for three scalar variables: ϵ_m, ϵ_2 and ϵ_r. One may start by taking $\epsilon_m = \epsilon_2 = \epsilon_r$ and seek, using line searching, values for these tuning parameters that leads to a stabilizing controller of minimum γ. Once such a controller is obtained, standard optimization techniques can be used, say Matlab function "fminsearch", which seek the combination of the three scalar parameters that bring γ to a local minimum.

4.5 Example

We consider the system of (4.1a,b) taken from [53] with the objective function of (3) where :

$$A = \begin{bmatrix} 0 & 1 \\ -1 & -0.4 \end{bmatrix},\ F = \begin{bmatrix} 0 & 0 \\ 0 & -0.1 \end{bmatrix},\ B_1 = \begin{bmatrix} 1 \\ -1 \end{bmatrix},\ B_2 = \begin{bmatrix} 0 \\ 1 \end{bmatrix},\ B_3 = \begin{bmatrix} 1 \\ 0 \end{bmatrix}$$

$C_1 = \begin{bmatrix} -0.5 0.4 \end{bmatrix}$, $D_{12} = 0.1,\ D_{13} = -1$ and $G = 0$ in (4.1a).

The time varying delay $\tau(t)$ is assumed to satisfy $|\tau(t)| \le \bar{h} = 0.20sec$ and $|\dot{\tau}(t)| \le 0.1$ and we consider a preview length of $h = 4sec$. Seeking a state-feedback controller, we apply the result of Corollary 4.4.1 and obtain a near minimum attenuation level of $\gamma = 1.77$, which is found for $\epsilon_2 = 40,\ \epsilon_m = 0.013$ and $\epsilon_r = 1.6$. The control law is given by (4.14a-c) where:

$$\bar{K}_x = \begin{bmatrix} -249.4425 & -48.2705 \end{bmatrix},$$
$$\bar{K}_r = \begin{bmatrix} 1.6735e-004 \end{bmatrix},$$
$$\bar{K}_\theta = \begin{bmatrix} -5.8024 & -85.0952 \end{bmatrix}.$$

The corresponding closed-loop poles are $[-5.8484,\ -42.8221]$.

In order to demonstrate the effect of the delay and its derivative bounds on the attenuation level in this example we consider, for simplicity, the case of zero preview and obtain, applying Corollary 4.4.2 and taking $\epsilon_2 = 40$, the following near minimum attenuation levels in Table 4.1.

Table 4.1 The minimal attenuation levels obtained as a function of the delay length and the delay derivative bound

$\bar{h}$	d	ϵ_r	ϵ_m	γ_{min}
0	0	1	0.012	1.06
0.25	0	1.6	0.035	1.97
0.4	0	1.4	0.087	2.78
0.6	0	1.45	0.066	3.92
0.6	0.4	1.5	0.067	4.15
0.6	0.9	1.14	0.095	4.60

We note that the result for $\bar{h} = 0$ (i.e., zero delay) (given in the first line of Table 4.1) is in line with Remark 4.2: the minimal attenuation level of $\gamma = 1.06$ was also obtained in [53] for the non-delayed stochastic state-feedback tracking control.

4.6 Conclusions

In this chapter the problem of tracking signals with preview in presence of Wiener-type stochastic parameter uncertainties is solved for retarded LTI systems. The retarded systems have been transformed to norm-bounded uncertain delay-free systems via the input-output approach. The state-feedback tracking control problem has then been solved for the infinite horizon case. Applying a min–max approach to the equivalent norm-bounded system, minimizing control and maximizing disturbance strategies have been obtained, which are based on the measurement of the system states and the previewed reference signal. The performance index in the tracking problem includes averaging over the statistics of the stochastic parameters in the system state-space model. An upper-bound on this index of performance has been obtained for the two patterns of preview control tracking problems treated in the paper.

The solution method used in the present chapter is different from the one applied to the non delayed case. In the latter case, a finite-horizon solution has been obtained at first and then it has been extended to the infinite-horizon case where stability was guaranteed in the limit of $t \to \infty$. An attempt to "combine" the already solved preview tracking result with the input–output approach along the approach used for the non-delayed case yields, unfortunately, a complicated and a most conservative solution to the delayed tracking problem. This is why an alternative approach has been adopted in this problem. Inequality conditions for stability and performance are first derived in terms of the yet undetermined control signal $u(t)$ and only in the last step of the solution the optimal control signal is derived. This approach yields a less conservative result that is most amenable for calculating a stabilizing controller.

It is shown that when the delay in the system tends to zero, the results of our theory coincide with those of the non delayed stochastic state-feedback tracking control design. The tractability and solvability of our theory is demonstrated via two examples, where the effect of the time delay and it's derivative on the performance index is shown in Example 4.5 and where the applicability of our theory to real practical control engineering is demonstrated in Chapter 13, Section 13.2 . Extension of these results to the case where there is no access to the system states and the controller has to rely on noisy measurements of the output, is not treated here – it is left for a future work.

4.7 Appendix

Proof of Theorem 4.1: The proof is based on minimizing a positive upper-bound on the index of performance. By applying several completions to square operations and using Schur complement formula, the optimal strategy for $w(t)$ is first obtained. We then introduce a new signal $\theta(t)$, which allows the derivation of the control strategy for two preview patterns. We first treat the fixed-finite preview case and then the zero-preview control case.

We start by seeking a finite positive scalar $\bar{J}(r) = \int_0^\infty f(r,t)dt$ such that

$$\mathcal{E}\int_0^\infty [\mathcal{L}V + z^T(t)z(t) - \gamma^2 w^T(t)w(t)]dt < \bar{J}(r),\ \forall r(t) \in L_2,$$

$$\forall w(t) \in \tilde{L}^2([0,\infty);\mathcal{R}^p), \tag{4.24}$$

while minimizing $\bar{J}(r)$ and requiring that the inequality 4.24 will also guarantee that $\mathcal{E}\{\mathcal{L}V\} < 0$. We note that since $x_0 \equiv 0$ and V is a Lyapunov function, 4.24 will guarantee that $J_E < \bar{J}(r)$. Considering (4.8), we define $\bar{\xi}(t) = col\{x(t), w_1(t), w_2(t), w(t), u(t), r(t)\}$, and require $f(r,t)$ to satisfy the following inequality:

$$\mathcal{E}\{\bar{\xi}^T(t)\{\begin{bmatrix} \Upsilon_{11} & \Upsilon_{12} & -QM & QB_1 & QB_2 & QB_3 \\ * & -R_1 + G^T QG & 0 & 0 & 0 & 0 \\ * & * & -R_2 & 0 & 0 & 0 \\ * & * & * & -\gamma^2 I & 0 & 0 \\ * & * & * & * & 0 & 0 \\ * & * & * & * & * & 0 \end{bmatrix}$$

$$+\tilde{\Gamma}_1^T \bar{h}^2 R_2 \tilde{\Gamma}_1 + \tilde{\Gamma}_2^T \tilde{\Gamma}_2\}\bar{\xi}(t)\} < f(r,t),$$

where

$$\tilde{\Gamma}_1^T = col\{(A_0 + M)^T,\ (A_1 - M)^T,\ -M^T,\ B_1^T,\ B_2^T,\ B_3^T\ \}$$

$$\tilde{\Gamma}_2^T = col\{C_1^T,\ 0,\ 0,\ 0,\ D_{12}^T,\ D_{13}^T\},$$

and where

$$\Upsilon_{11} = Q\tilde{A} + \tilde{A}^T Q + \frac{1}{1-d}R_1 + F^T QF, \quad \Upsilon_{12} = Q\tilde{A}_1.$$

We thus obtain the requirement of $\mathcal{E}\{\bar{\xi}_y^T(t)\bar{\Gamma}_1\bar{\xi}_y(t)\} < f(r,t)$, where $\bar{\xi}_y^T(t) = [\bar{\xi}^T(t)\ \ \bar{h}\bar{\xi}^T(t)\tilde{\Gamma}_1^T\ \ \bar{\xi}^T(t)\tilde{\Gamma}_2^T]$, and where

$$\bar{\Gamma}_1 = \begin{bmatrix} \Upsilon_{11} & \Upsilon_{12} & -QM & QB_1 & QB_2 & QB_3 & \bar{h}(A_0^T + M^T) & C_1^T \\ * & -R_1 + G^T QG & 0 & 0 & 0 & 0 & \bar{h}(A_1^T - M^T) & 0 \\ * & * & -R_2 & 0 & 0 & 0 & -\bar{h}m^T & 0 \\ * & * & * & -\gamma^2 I & 0 & 0 & \bar{h}B_1^T & 0 \\ * & * & * & * & 0 & 0 & \bar{h}B_2^T & D_{12}^T \\ * & * & * & * & * & 0 & \bar{h}B_3^T & D_{13}^T \\ * & * & * & * & * & * & -R_2^{-1} & 0 \\ * & * & * & * & * & * & * & -I \end{bmatrix}.$$

Multiplying the above $\bar{\Gamma}_1$ from both sides by $diag\{I, I, I, I, I, I, R_2, I\}$ and denoting $\bar{R}_2 = R_2 M$, the following requirement is obtained:

$$\mathcal{E}\{\tilde{\xi}_y^T(t) \begin{bmatrix} \Upsilon_{11} & \Upsilon_{12} & -QM & QB_1 & QB_2 & QB_3 & \bar{h}(A_0^T R_2 + \bar{R}_2^T) & C_1^T \\ * & -R_1 + G^T QG & 0 & 0 & 0 & 0 & \bar{h}(A_1^T R_2 - \bar{R}_2^T) & 0 \\ * & * & -R_2 & 0 & 0 & 0 & -\bar{h}\bar{R}_2^T & 0 \\ * & * & * & -\gamma^2 I & 0 & 0 & \bar{h}B_1^T R_2 & 0 \\ * & * & * & * & 0 & 0 & \bar{h}B_2^T R_2 & D_{12}^T \\ * & * & * & * & * & 0 & \bar{h}B_3^T R_2 & D_{13}^T \\ * & * & * & * & * & * & -R_2 & 0 \\ * & * & * & * & * & * & * & -I \end{bmatrix} \tilde{\xi}_y(t)\} \tag{4.25}$$

$$< f(r,t),$$

where $\tilde{\xi}_y^T(t) = [\tilde{\xi}^T(t) \;\; \bar{h}\xi^T(t)\tilde{\Gamma}_1^T R_2^{-1} \;\; \xi^T(t)\tilde{\Gamma}_2^T]$.

Remark 4.4. The requirement of (4.25), if satisfied for a given $u(t)$, guarantees the stability of the closed-loop system by Lemma 4.3.1. The latter stems from the fact that for $r \equiv 0$, $f(r,t) = 0$ (by definition) and the five columns and rows of (12) are the first, second, third, fifth, and seventh columns and rows of (4.25), respectively. The latter fact leads to $\mathcal{E}\{\mathcal{L}V\} < 0$.

Next, we apply Schur complement to the second and third columns and rows of (4.25) and we obtain, defining $\tilde{w}(t) \triangleq col\{w_1(t),\; w_2(t)\}$ the following requirement:

$$\mathcal{E}\{\tilde{\xi}^T(t) \begin{bmatrix} \hat{\Upsilon}_{11} & QB_1 & QB_2 & QB_3 & \hat{\Upsilon}_{15} & C_1^T \\ * & -\gamma^2 I & 0 & 0 & \bar{h}B_1^T & 0 \\ * & * & 0 & 0 & \bar{h}B_2^T & D_{12}^T \\ * & * & * & 0 & \bar{h}B_3^T & D_{13}^T \\ * & * & * & * & \hat{\Upsilon}_{55} & 0 \\ * & * & * & * & * & -I \end{bmatrix} \tilde{\xi}(t)\} - \mathcal{E}\{(\tilde{w}-\tilde{w}^*)^T \tilde{R}_{\tilde{w}}(\tilde{w}-\tilde{w}^*)\} < f(r,t), \tag{4.26}$$

where $\tilde{\xi}(t) = col\{x(t), w(t), u(t), r(t), \bar{h}R_2^{-1}\tilde{\Gamma}_1\tilde{\xi}(t), \tilde{\Gamma}_2\tilde{\xi}(t)\}$,

$\hat{\Upsilon}_{11} = \Upsilon_{11} + Q\tilde{A}_1[R_1 - G^T QG]^{-1}\tilde{A}_1^T Q + QMR_2^{-1}M^T Q$,
$\hat{\Upsilon}_{15} = \bar{h}(A_0^T R_2 + \bar{R}_2^T) + \bar{h}[Q(A_1 - M)[R_1 - G^T QG]^{-1}(A_1^T R_2 - \bar{R}_2^T)$
$+QMR_2^{-1}\bar{R}_2^T]$,

$\hat{\Upsilon}_{55} = -R_2 + \bar{h}^2[(R_2 A_1 - \bar{R}_2)[R_1 - G^T QG]^{-1}(R_2 A_1 - \bar{R}_2)^T + \bar{R}_2 R_2^{-1}\bar{R}_2^T]$,

$$\tilde{w}^* = \tilde{R}_{\tilde{w}}^{-1}[\begin{bmatrix} [A_1^T - M^T]Q \\ -M^T Q \end{bmatrix} x(t) + \begin{bmatrix} \bar{h}(A_1 - M)^T \\ -\bar{h}M^T \end{bmatrix} R_2^{-1}\bar{y}(t)],$$

and where $\tilde{R}_{\tilde{w}}$ is given in (4.15g). The latter operation is equivalent to completion to squares for $\tilde{w}$ that makes the $\tilde{w}^*$ the optimal strategy. We note that the second term of the left side of (4.26) is not omitted (by choosing the optimal strategy) but is rather left for the latter part of the proof. Completing to squares for w, we apply Schur complement to the second column and row blocks of (4.26) and we obtain:

$$\mathcal{E}\{\hat{\xi}^T(t) \begin{bmatrix} \bar{\Upsilon}_{11} & QB_2 & QB_3 & \bar{\Upsilon}_{14} & C_1^T \\ * & 0 & 0 & \bar{h}B_2^T & D_{12}^T \\ * & * & 0 & \bar{h}B_3^T & D_{13}^T \\ * & * & * & \bar{\Upsilon}_{44} & 0 \\ * & * & * & * & -I \end{bmatrix} \hat{\xi}(t)\} - \mathcal{E}\{(\tilde{w}-\tilde{w}^*)^T \tilde{R}_{\tilde{w}}(\tilde{w}-\tilde{w}^*)\}$$

$$- \mathcal{E}\{(w-w^*)^T\gamma^2(w-w^*)\} < f(r,t), \tag{4.27}$$

where $\hat{\xi}(t) = col\{x(t),\ u(t),\ r(t), \bar{h}R_2^{-1}\tilde{\Gamma}_1\bar{\xi}(t), \tilde{\Gamma}_2\bar{\xi}(t)\}$.

The signal $w^*(t)$ is now the optimal strategy for w where,

$$\begin{aligned} \bar{\Upsilon}_{11} &= \hat{\Upsilon}_{11} + \gamma^{-2}QB_1B_1^TQ, \\ \bar{\Upsilon}_{14} &= \hat{\Upsilon}_{15} + \bar{h}\gamma^{-2}QB_1B_1^T, \\ \bar{\Upsilon}_{44} &= \hat{\Upsilon}_{55} + \bar{h}^2\gamma^{-2}B_1B_1^T, \\ w^* &= -\gamma^{-2}[B_1^TQx + \bar{h}B_1^TR_2^{-1}\bar{y}]. \end{aligned}$$

Applying Schur complement to the fourth and fifth columns and rows of the leftmost block matrix of (4.27) the following requirement is obtained:

$$\bar{J}_E(t) \triangleq \mathcal{E}\{\xi_3^T\tilde{\Upsilon}_3\xi_3\} - \mathcal{E}\{(\tilde{w}-\tilde{w}^*)^T\tilde{R}_{\tilde{w}}(\tilde{w}-\tilde{w}^*)\} - \mathcal{E}\{(w-w^*)^T\gamma^2(w-w^*)\} < f(r,t), \tag{4.28}$$

where $\tilde{\Upsilon}_3$ is defined in (4.20), $\xi_3(t) = col\{x(t),\ u(t),\ r(t)\}$ and where ϕ and g of (4.15c,d) are obtained by:

$$\begin{aligned} &\phi = -\bar{\Upsilon}_{44} = -(\hat{\Upsilon}_{55} + \bar{h}^2\gamma^{-2}B_1B_1^T) = \\ &-(\bar{h}^2[(R_2A_1 - \bar{R}_2)[R_1 - G^TQG]^{-1}(R_2A_1 - \bar{R}_2)^T + \bar{R}_2R_2^{-1}\bar{R}_2^T] \\ &+\bar{h}^2\gamma^{-2}B_1B_1^T - R_2), \quad g = \bar{\Upsilon}_{14} = \hat{\Upsilon}_{15} + \bar{h}\gamma^{-2}QB_1B_1^T = \\ &\bar{h}(A_0^TR_2 + \bar{R}_2^T) + \bar{h}[Q(A_1 - M)[R_1 - G^TQG]^{-1}(A_1^TR_2 - \bar{R}_2^T) + QMR_2^{-1}\bar{R}_2^T] \\ &+\bar{h}\gamma^{-2}QB_1B_1^T. \end{aligned}$$

In order to find the minimizing strategy for the controller, we seek to minimize the expression of $\mathcal{E}\{\Phi(r)\} \triangleq \mathcal{E}\{\xi_3^T(t)\tilde{\Upsilon}_3\xi_3(t)\}$ in $\bar{J}_E(t)$ of (4.28). The latter minimization will also hold for the integral form of (4.28). We obtain:

$$\mathcal{E}\{\Phi(r)\} \triangleq \mathcal{E}\{x^T(t)\alpha_1x(t) + 2x^T(t)\alpha_2u(t) + 2x^T(t)\alpha_3r(t) + u^T(t)\alpha_4u(t)$$

$$+2u^T(t)\alpha_5r(t) + r^T(t)\alpha_6r(t)\},$$

where $\alpha_1, \alpha_2, ..\alpha_6$ are defined in (4.20). Completing to squares for u(t), we find that

$$\mathcal{E}\{\Phi(r)\} = \mathcal{E}\{x^T(t)(\alpha_1 - \alpha_2\alpha_4^{-1}\alpha_2^T)x(t) + (u^T(t) + (r^T(t)\alpha_5^T + x^T(t)\alpha_2)\alpha_4^{-1})\alpha_4$$

$$(u(t) + \alpha_4^{-1}(\alpha_2^T x(t) + \alpha_5 r(t))) + 2x^T(t)\alpha_2\alpha_4^{-1}\alpha_5 r(t) + r^T(t)(\alpha_6 + \alpha_5^T\alpha_4^{-1}\alpha_5)r(t)\}.$$

In order to allow for the different preview tracking patterns, we add the following identically zero-term to $\bar{J}_E(t)$ of (4.28) following integration, where we apply the *Itô* lemma [9]:

$$0 = 2\mathcal{E}\int_0^\infty d\{\theta^T(t)x(t)\} + 2\theta^T(0)x(0) = 2\mathcal{E}\theta^T(\infty)x(\infty) - 2\theta^T(0)x(0) + 2\theta^T(0)x(0)$$

$$= 2\mathcal{E}\int_0^\infty \dot{\theta}^T(t)x(t)dt + 2\theta^T(0)x(0)$$

$$+2\mathcal{E}\int_0^\infty \theta^T(t)\{[\bar{A}x(t) + \bar{B}_1\bar{w}(t) + \bar{B}_2\hat{u}(t) + \bar{B}_r r(t) + \hat{A}_w\tilde{w}_0(t)]dt$$

$$+ Fx(t)d\beta(t) + Gw_1(t)d\zeta - \Gamma_\beta dt - \Gamma_\zeta dt\} \qquad (4.29)$$

where $\theta(t)$ is defined in (4.16) and where we denote:

$$\tilde{w}_0(t) = \tilde{w} - \tilde{w}^*, \ \ \bar{w} = w - w^*.$$

The result of (4.29) makes use of the fact

$$dx(t) = [\bar{A}x(t) + \bar{B}_1\bar{w}(t) + \bar{B}_2\hat{u}(t) + \bar{B}_r r(t) + \hat{A}_w\tilde{w}_0(t)]dt + Fx(t)d\beta(t)$$

$$+Gw_1(t)d\zeta - \Gamma_\beta dt - \Gamma_\zeta dt$$

$$= [Ax(t) + B_1 w(t) + B_2 u(t) + B_3 r(t) + \tilde{A}_1 w_1(t) - Mw_2(t)]dt + Fx(t)d\beta(t)$$

$$+Gw_1(t)d\zeta - \Gamma_\beta dt - \Gamma_\zeta dt,$$

where $\bar{B}_1$, $\bar{B}_2$, $\hat{A}_w$ and $\bar{A}$, $\bar{B}_r$ are given in (4.15) and (4.17) respectively.

Remark 4.5. The signal $\theta(t)$ plays a central roll only in the case of fixed-finite preview pattern of Corollary 4.4.1. In this case, the signal $r(\tau)$ is known for $t < \tau < t + h$. Note that, following (4.16), $\theta(t)$ is retrieved by backwards integration from $t + h$ towards t. In the zero preview case (Corollary 4.4.2), $\theta(\tau) \equiv 0$.

We note that due to the expectation operator in (4.29), the noise terms in $d\beta$ and $d\zeta$ vanish. Completing to squares for $\bar{w}(t)$ and then for $\tilde{w}_0(t)$ we obtain, for $\tilde{J}_E(t) = \int_0^\infty \bar{J}_E dt$, noting that $x(0) = 0$:

$$\tilde{J}_E(t) = \mathcal{E}\int_0^\infty \{x^T\hat{\Upsilon}_x x\}dt + \mathcal{E}\int_0^\infty \{\hat{u}^T\alpha_4\hat{u}\}dt + 2\mathcal{E}\int_0^\infty \{x^T\alpha_2\alpha_4^{-1}\alpha_5 r\}dt$$

$$+\mathcal{E}\int_0^\infty \{r^T(\alpha_6+\alpha_5^T\alpha_4^{-1}\alpha_5)r\}dt-\mathcal{E}\int_0^\infty ||[\bar{w}-\gamma^{-2}\bar{B}_1^T\theta]||^2_{\gamma^2}$$

$$+\mathcal{E}\int_0^\infty ||[\tilde{w}_0+\tilde{R}_{\tilde{w}}^{-1}\hat{A}_w^T\theta]||^2_{\tilde{R}_{\tilde{w}}}-\mathcal{E}\int_0^\infty ||\hat{A}_w^T\theta||^2_{\tilde{R}_{\tilde{w}}^{-1}}dt$$

$$+2\mathcal{E}\int_0^\infty \dot{\theta}^T(t)x(t)dt+2\mathcal{E}\int_0^\infty \theta^T(t)\{\bar{A}x(t)+\bar{B}_2\hat{u}(t)+\bar{B}_r r(t)\}dt$$

$$+\mathcal{E}\int_0^\infty ||\bar{B}_1^T\theta||^2_{\gamma^{-2}}dt$$

where

$$\begin{aligned}\hat{\Upsilon}_x&=\alpha_1-\alpha_2\alpha_4^{-1}\alpha_2^T,\\ \hat{u}(t)&=u(t)+\alpha_4^{-1}[\alpha_2x(t)+\alpha_5r(t)].\end{aligned}\tag{4.30}$$

Next, completing to squares for $\hat{u}(t)$ we obtain:

$$\tilde{J}_E(t)=\mathcal{E}\int_0^\infty \{x^T\hat{\Upsilon}_x x\}dt+\mathcal{E}\int_0^\infty \{(\hat{u}+\alpha_4^{-1}\bar{B}_2^T\theta)^T\alpha_4(\hat{u}+\alpha_4^{-1}\bar{B}_2^T\theta)\}dt$$

$$+\mathcal{E}\int_0^\infty \{r^T(\alpha_6+\alpha_5^T\alpha_4^{-1}\alpha_5)r\}dt-\mathcal{E}\int_0^\infty ||[\bar{w}-\gamma^{-2}\bar{B}_1^T\theta]||^2_{\gamma^2}+\mathcal{E}\int_0^\infty ||\bar{B}_1^T\theta||^2_{\gamma^{-2}}dt$$

$$-\mathcal{E}\int_0^\infty ||\hat{A}_w^T\theta||^2_{\tilde{R}_{\tilde{w}}^{-1}}dt-\mathcal{E}\int_0^\infty ||\alpha_4^{-1/2}\bar{B}_2^T\theta||^2dt+2\mathcal{E}\int_0^\infty \theta^T(t)\{B_r r(t)\}dt$$

$$+\mathcal{E}\int_0^\infty ||[\tilde{w}_0+\tilde{R}_{\tilde{w}}^{-1}\hat{A}_w^T\theta]||^2_{\tilde{R}_{\tilde{w}}}+2\mathcal{E}\int_0^\infty [\dot{\theta}+\bar{A}^T\theta+[\alpha_2\alpha_4^{-1}\alpha_5]r]^T x dt.$$

Taking $\tilde{w}_0=-\tilde{R}_{\tilde{w}}^{-1}\hat{A}_w^T\theta$, $\dot{\theta}=-\bar{A}^T\theta-[\alpha_2\alpha_4^{-1}\alpha_5]r$, the above equation can be expressed by:

$$\tilde{J}_E(t)=\mathcal{E}\int_0^\infty \{x^T\hat{\Upsilon}_x x\}dt+\mathcal{E}\int_0^\infty \{(\hat{u}+\alpha_4^{-1}\bar{B}_2^T\theta)^T\alpha_4(\hat{u}+\alpha_4^{-1}\bar{B}_2^T\theta)\}dt$$

$$-\mathcal{E}\int_0^\infty ||[\bar{w}-\gamma^{-2}\bar{B}_1^T\theta]||^2_{\gamma^2}+\bar{J}(r),$$

where

$$\bar{J}(r)=\mathcal{E}\int_0^\infty \{r^T(t)(\alpha_6+\alpha_5^T\alpha_4^{-1}\alpha_5)r(t)\}dt+\mathcal{E}\int_0^\infty ||\bar{B}_1^T\theta(t)||^2_{\gamma^{-2}}dt$$

$$-\mathcal{E}\int_0^\infty ||\hat{A}_w^T\theta(t)||^2_{\tilde{R}_{\tilde{w}(t)}^{-1}}dt-\mathcal{E}\int_0^\infty ||\alpha_4^{-1/2}\bar{B}_2^T\theta(t)||^2dt$$

$$+2\mathcal{E}\int_0^\infty \theta^T(t)\{B_r r(t)\}dt,$$

is independent of $u(t)$ and $w(t)$ and is given in explicit form in (4.19).

In order to minimize $\tilde{J}_E(t)$ for all possible r, namely to achieve a minimum $\bar{J}(r)$ and thus find the minimal possible value of $f(r,t)$, we seek a condition which will assure $\hat{\Upsilon}_x < 0$. Considering (4.30), assuming for simplicity that $C_1^T D_{12} = 0$ and recalling that $\hat{R} = D_{12}^T D_{12}$, we require:

$$\bar{\Upsilon}_{11} + g\phi^{-1}g^T + C_1^T C_1 - (\bar{h}g\phi^{-1}B_2 + QB_2)\Psi_3^{-1}(\bar{h}g\phi^{-1}B_2 + QB_2)^T < 0,$$

where $\Psi_3 = \hat{R} + \bar{h}^2 B_2^T \phi^{-1} B_2$. Defining then:

$$\Gamma_1 = g\phi^{-1}g^T - \bar{h}^2 g\phi^{-1}B_2\Psi_3^{-1}B_2^T\phi^{-1}g^T,$$
$$\Gamma_2 = -2\bar{h}g\phi^{-1}B_2\Psi_3^{-1}B_2^T Q,$$
$$\Gamma_3 = -QB_2\Psi_3^{-1}B_2^T Q,$$

it is thus required that:

$$\bar{\Upsilon}_{11} + C_1^T C_1 + \Gamma_1 + \Gamma_2 + \Gamma_3 < 0.$$

Using the matrix inversion lemma we find:

$$\begin{aligned}
\Gamma_1 &= g(\phi + \bar{h}^2 B_2\hat{R}^{-1}B_2^T)^{-1}g^T,\\
\Gamma_2 &= -2\bar{h}g(\phi + \bar{h}^2 B_2\hat{R}^{-1}B_2^T)^{-1}B_2\hat{R}^{-1}B_2^T Q,\\
\Gamma_3 &= -Q\phi(\phi + \bar{h}^2 B_2\hat{R}^{-1}B_2^T)^{-1}B_2\hat{R}^{-1}B_2^T Q,
\end{aligned}$$

and have

$$\begin{aligned}
\Gamma_1 + \Gamma_2 + \Gamma_3 = (g - \bar{h}QB_2\hat{R}^{-1}B_2^T)(\phi + \bar{h}^2 B_2\hat{R}^{-1}B_2^T)^{-1}(g^T - \bar{h}B_2\hat{R}^{-1}B_2^T Q)\\
-QB_2\hat{R}^{-1}B_2^T Q,
\end{aligned}$$

where we note that

$$\begin{aligned}
-\bar{h}^2 QB_2\hat{R}^{-1}B_2^T(\phi + \bar{h}^2 B_2\hat{R}^{-1}B_2^T)^{-1}B_2\hat{R}^{-1}B_2^T Q - Q\phi(\phi + \bar{h}^2 B_2\hat{R}^{-1}B_2^T)^{-1}\\
B_2\hat{R}^{-1}B_2^T Q = -QB_2\hat{R}^{-1}B_2^T Q.
\end{aligned}$$

Namely, we require:

$$\begin{aligned}
\hat{\Upsilon}_x = \bar{\Upsilon}_{11} + C_1^T C_1 - QB_2\hat{R}^{-1}B_2^T Q + (g - \bar{h}QB_2\hat{R}^{-1}B_2^T)(\phi + \bar{h}^2 B_2\hat{R}^{-1}B_2^T)^{-1}\\
(g^T - \bar{h}B_2\hat{R}^{-1}B_2^T Q) < 0.
\end{aligned}$$

The latter inequality can be written as :

$$\tilde{\Gamma} = \begin{bmatrix} \bar{\Upsilon}_{11} + C_1^T C_1 - QB_2\hat{R}^{-1}B_2^T Q & g - \bar{h}QB_2\hat{R}^{-1}B_2^T \\ * & -(\phi + \bar{h}^2 B_2\hat{R}^{-1}B_2^T) \end{bmatrix} < 0.$$

Multiplying the above $\tilde{\Gamma}$ by $diag\{Q^{-1},\ I\}$ from the left and the right, replacing for ϕ and g and applying Schur complement, we readily obtain the following inequality:

$$\begin{bmatrix} \tilde{\Psi}_{11} & \tilde{\Psi}_{12} & A_1 - M & M \\ * & \tilde{\Psi}_{22} & \bar{h}R_2(A_1 - M) & \bar{h}R_2M \\ * & * & -[R_1 - G^T P^{-1} G] & 0 \\ * & * & * & -R_2 \end{bmatrix} < 0,$$

where we denote $P = Q^{-1}$ and where

$$\begin{aligned} &\tilde{\Psi}_{11} = (A_0 + M)P + P(A_0 + M)^T + \gamma^{-2}B_1B_1^T - B_2\hat{R}^{-1}B_2^T + PR_1P\tfrac{1}{1-d} \\ &+PC_1^TC_1P + PF^TP^{-1}FP, \\ &\tilde{\Psi}_{12} = \bar{h}PA_0^TR_2 + \bar{h}P\bar{R}_2^T + \bar{h}\gamma^{-2}B_1B_1^T - \bar{h}B_2\hat{R}^{-1}B_2^T, \\ &\tilde{\Psi}_{22} = \bar{h}^2[\gamma^{-2}B_1B_1^T - B_2\hat{R}^{-1}B_2^T] - R_2, \quad \hat{R} = D_{12}^TD_{12}. \end{aligned}$$

Applying Schur complement again, the inequality of (4.12) then follows.

Remark 4.6. In order to prove that $\int_0^\infty f(r,t)dt$ exists (and therefore that $\bar{J}(r)$ of (4.19) is well defined) we consider (4.16) and obtain that

$$\theta(t) = \int_t^{t+h} e^{-\bar{A}^T(t-\tau)}\bar{B}_r r(\tau)d\tau = \int_{-h}^0 e^{-\bar{A}^T\bar{\tau}}\bar{B}_r r(t-\bar{\tau})d\bar{\tau}.$$

Since $\bar{J}(r)$ includes integral terms that contain the signal $\theta(t)$ multiplied by some constants, the L_2 norm of $\theta(t)$ should be derived. We readily find that

$$\int_0^\infty \theta^T(t)\theta(t)dt = Tr\{\int_0^\infty \theta(t)\theta^T(t)dt\}$$

$$= Tr\{\int_0^\infty \{[\int_{-h}^0 e^{-\bar{A}^T\bar{\tau}}\bar{B}_r r(t-\bar{\tau})d\bar{\tau}][\int_{-h}^0 e^{-\bar{A}^T\tau}\bar{B}_r r(t-\tau)d\tau]^T\}dt\}$$

$$= Tr\{\int_{-h}^0\int_{-h}^0 e^{-\bar{A}^T\bar{\tau}}\bar{B}_r\Psi(r,\tau,\bar{\tau})\bar{B}_r^Te^{-\bar{A}\tau}d\tau d\bar{\tau}\}$$

where $\Psi(r,\tau,\bar{\tau}) \triangleq \int_0^\infty r(t-\bar{\tau})r^T(t-\tau)dt$ and where by the Cauchy-Shwartz inequality $\Psi(r,\tau,\bar{\tau})$ is finite for all $r(t) \in L_2$. Obviously the additional quadratic term of $r(t)$ in $\bar{J}(r)$ of (4.19) is finite too.

5

H_∞ Control and Estimation of Retarded Linear Discrete-Time Systems

5.1 Introduction

In this chapter we address the problems of H_∞ state-feedback control and filtering of state-delayed, discrete-time, state-multiplicative linear systems via the *input–output* approach based on the stability and Bounded Real Lemma (BRL) of these systems, which are developed here. The multiplicative noise appears in the system model in both the delayed and the non delayed states of the system.

A major part of the stochastic H_∞ control and estimation research for LTI delayed systems, has centered around continuous-time systems. As reviewed in the Introduction, this includes the stability and control of stochastic delayed systems of various types (i.e constant time-delay, slow and fast varying delay etc.). The field of discrete-time control of delayed systems in the stochastic state-multiplicative context has not been fully investigated. Over the last decade, the mean square exponential stability and the control and filtering problems of these systems were treated by several groups [131], [135]. In [131], the state-feedback control problem solution is solved for norm-bounded uncertain systems, for the restrictive case where the same multiplicative noise sequence multiplies both the states and the input of the system. The solution in [131] is delay-dependent.

The point of view taken in the present chapter is similar to the one taken for the solution of both the continuous-time state-feedback control and filtering problems in Chapter 2 (see also [55],[59]). Here, we adopt the input–output approach of Chapter 2, to delay-dependent solutions of the above discrete-time counter part stochastic problems. This approach is based on the representation of the system's delay action by linear operators, with no delay, which in turn allows one to replace the underlying system with an equivalent one which possesses a norm-bounded uncertainty, and therefore may be treated by the theory of norm bounded uncertain, non-retarded systems with state-multiplicative noise [53].

E. Gershon & U. Shaked: *Advanced Topics in Control & Estimation,* LNCIS 439, pp. 95–113.
DOI: 10.1007/978-1-4471-5070-1_5

Similarly to Chapter 2, in our system we allow for a time-varying delay where the uncertain stochastic parameters multiply both the delayed and the non delayed states in the state space model of the system. Based on the input–output approach, the solution of the stability issue is achieved in Section 5.3, followed by the solution of the BRL in Section 5.4. The state-feedback control problem is treated in Section 5.5. The filtering problem solution is achieved in Section 5.6, for a general-type filter, resulting in a single LMI that yields the filter parameters. Each of the above sections is ended with a numerical example that demonstrates the tractability of our solution method.

Notation: For convenience we bring the necessary notations for the discrete-time setting. We denote, by $L^2(\Omega, \mathcal{R}^n)$ the space of square-integrable $\mathcal{R}^n-$ valued functions on the probability space $(\Omega, \mathcal{F}, \mathcal{P})$, where Ω is the sample space, $\mathcal{F}$ is a σ algebra of a subset of Ω called events and $\mathcal{P}$ is the probability measure on $\mathcal{F}$. By $(\mathcal{F}_k)_{k\in\mathcal{N}}$ we denote an increasing family of σ-algebras $\mathcal{F}_k \subset \mathcal{F}$. We also denote by $\tilde{l}^2(\mathcal{N}; \mathcal{R}^n)$ the n-dimensional space of nonanticipative stochastic processes $\{f_k\}_{k\in\mathcal{N}}$ with respect to $(\mathcal{F}_k)_{k\in\mathcal{N}}$ where $f_k \in L^2(\Omega, \mathcal{R}^n)$. On the latter space the following l^2-norm is defined:

$$||\{f_k\}||^2_{\tilde{l}_2} = E\{\textstyle\sum_0^\infty ||f_k||^2\} = \sum_0^\infty E\{||f_k||^2\} < \infty, \quad \{f_k\} \in \tilde{l}_2(\mathcal{N}; \mathcal{R}^n), \tag{5.1}$$

where $||\cdot||$ is the standard Euclidean norm. Throughout this chapter we refer to the notation of exponential l^2 stability, or internal stability, in the sense of [13] (see Definition 2.1, page 927, there).

5.2 Problem Formulation

We consider the following linear retarded system:

$$\begin{aligned} x_{k+1} &= (A_0 + D\nu_k)x_k + (A_1 + F\mu_k)x_{k-\tau(k)} + B_1 w_k + (B_2 + G\zeta_k)u_k, \\ x_l &= 0,\ l \le 0, \\ y_k &= C_2 x_k + D_{21} n_k \end{aligned} \tag{5.2}$$

with the objective vector

$$z_k = C_1 x_k + D_{12} u_k, \tag{5.3}$$

where $x_k \in \mathcal{R}^n$ is the system state vector, $w_k \in \mathcal{R}^q$ is the exogenous disturbance signal, $n_k \in \mathcal{R}^p$ is the the measurement noise signal, $u_k \in \mathcal{R}^\ell$ is the control input, $y_k \in \mathcal{R}^m$ is the measured output and $z_k \in \mathcal{R}^r$ is the state combination (objective function signal) to be regulated and where the time delay is denoted by the integer τ_k and it is assumed that $0 \le \tau_k \le h, \forall k$. The variables $\{\zeta_k\}$ $\{\mu_k\}$ and $\{\nu_k\}$ are zero-mean real scalar white-noise sequences that satisfy:

$$E\{\nu_k \nu_j\} = \delta_{kj},\ E\{\zeta_k \zeta_j\} = \delta_{kj},\ E\{\mu_k \mu_j\} = \delta_{kj},$$

$$E\{\zeta_k\mu_j\} = E\{\zeta_k\nu_j\} = E\{\mu_k\nu_j\} = 0, \ \ \forall k, j \quad \geq 0.$$

The matrices in (5.2a,b) and (5.3) are constant matrices of appropriate dimensions.

We treat the following two problems:

i) H_∞ state-feedback control:
We consider the system of (5.2a) and (5.3) and the following performance index:

$$J_E \stackrel{\Delta}{=} ||z_k||^2_{\tilde{l}_2} - \gamma^2 ||w_k||^2_{\tilde{l}_2}. \tag{5.4}$$

Our objective is to find a state-feedback control law $u_k = Kx_k$ that achieves $J_E < 0$, for the worst-case of the process disturbance $w_k \in \tilde{l}^2_{\mathcal{F}_k}([0,\infty);\mathcal{R}^q)$ and for the prescribed scalar $\gamma > 0$.

ii) H_∞ filtering:
We consider the system of (5.2a,b) and (5.3), where $B_2 = 0, \ G = 0$ and $D_{12} = 0$ and the following general form estimator:

$$\begin{array}{l} \hat{x}_{k+1} = A_c\hat{x}_k + B_c y_k, \quad \hat{x}_i = 0, \forall i \leq 0 \\ \hat{z}_k = C_c\hat{x}_k. \end{array} \tag{5.5}$$

We denote

$$e_k = x_k - \hat{x}_k, \quad \text{and} \quad \bar{z}_k = z_k - \hat{z}_k, \tag{5.6}$$

and for a given scalar parameter γ, we consider the following cost function:

$$J_F \stackrel{\Delta}{=} ||\bar{z}_k||^2_{\tilde{l}_2} - \gamma^2[||w_k||^2_{\tilde{l}_2} + ||n_{k+1}||^2_{\tilde{l}_2}]. \tag{5.7}$$

Given γ , we seek an estimate $C_c\hat{x}_k$ of C_1x_k over the infinite time horizon $[0,\infty)$ such that J_F given by (5.7) is negative for all nonzero w_k, n_k where $w_k \in \tilde{l}^2_{\mathcal{F}_k}([0,\infty);\mathcal{R}^q)$, $n_k \in \tilde{l}^2_{\mathcal{F}_k}([0,\infty];\mathcal{R}^p)$.

In the robust stochastic H_∞control and estimation problems treated here, we assume that the system parameters lie within the following polytope

$$\bar{\Omega} \stackrel{\Delta}{=} \left[A_0 \, A_1 \, B_1 \, B_2 \, C_1 \, C_2 \, D_{12} \, D_{21} \, D \, F \, G \right], \tag{5.8}$$

which is described by the vertices:

$$\bar{\Omega} = \mathcal{C}o\{\bar{\Omega}_1, \bar{\Omega}_2, ..., \bar{\Omega}_N\}, \tag{5.9}$$

where

$$\bar{\Omega}_i \stackrel{\Delta}{=} \left[A_0^{(i)} \, A_1^{(i)} \, B_1^{(i)} \, B_2^{(i)} \, C_1^{(i)} \, C_2^{(i)} \, D_{12}^{(i)} \, D_{21}^{(i)} \, D^{(i)} \, F^{(i)} \, G^{(i)} \right] \tag{5.10}$$

and where N is the number of vertices. In other words:

$$\bar{\Omega} = \sum_{i=1}^{N} \bar{\Omega}_i f_i, \quad \sum_{i=1}^{N} f_i = 1 \quad, f_i \geq 0. \tag{5.11}$$

5.3 Mean-Square Exponential Stability

In order to solve the above two problems, we investigate first the stability of the retarded discrete-time system. We introduce the following scalar operators which are needed, in the sequel, for transforming the delayed system to an equivalent norm-bounded nominal system:

$$\Delta_1(g_k) = g_{k-h}, \quad \Delta_2(g_k) = \sum_{j=k-h}^{k-1} g_j. \tag{5.12}$$

Denoting $\bar{y}_k = x_{k+1} - x_k$ and using the fact that $\Delta_2(\bar{y}_k) = x_k - x_{k-h}$, the following state space description of the system is obtained:

$$\begin{aligned} x_{k+1} = (A_0 + D\nu_k + M)x_k + (A_1 - M + F\mu_k)\Delta_1(x_k) - M\Delta_2(\bar{y}_k) + B_1 w_k + \\ (B_2 + G\zeta_k)u_k, \quad x_l = 0,\ l \le 0, \\ y_k = C_2 x_k + D_{21} n_k, \\ z_k = C_1 x_k + D_{12} u_k, \end{aligned} \tag{5.13}$$

where the matrix M is a free decision variable that will be determined later.

We consider then the following auxiliary system where we take $B_2 = 0$ and $G = 0$:

$$x_{k+1} = (A_0 + D\nu_k + M)x_k + (A_1 - M + F\mu_k)w_{1,k} - Mw_{2,k} + B_1 w_k, \tag{5.14}$$

with the feedback

$$w_{1,k} = \Delta_1(x_k), \quad w_{2,k} = \Delta_2(\bar{y}_k). \tag{5.15}$$

We consider the system of (5.14) where $B_1 = 0$ and the following Lyapunov function:

$$V_k \triangleq x_k^T Q x_k. \tag{5.16}$$

Taking expectation with respect to v_k and μ_k and solving for (5.14) we obtain:

$$\begin{aligned} &E\{V_{k+1}\} - V_k = \\ &E\{[x_k^T(A_0^T + M^T + D^T\nu_k) + w_{1,k}^T(A_1^T - M^T + F^T\mu_k) - w_{2,k}^T M^T]Q \\ &[(A_0 + M + D\nu_k)x_k + (A_1 - M + F\mu_k)w_{1,k} - Mw_{2,k}]\} - x_k^T Q x_k. \end{aligned} \tag{5.17}$$

We thus arrive at the following condition for $E\{V_{k+1}\} - V_k < 0$.

Theorem 5.1. *The exponential stability in the mean square sense of the system (5.2a) where $B_1 = 0$, $B_2 = 0$, and $G = 0$, is guaranteed if there exist $n \times n$ matrices $Q > 0$, $R_1 > 0$, $R_2 > 0$, and M that satisfy the following inequality:*

$$\Gamma \stackrel{\Delta}{=} \begin{bmatrix} \Gamma_{11} & (A_0+M)^T Q & 0 & 0 & h(A_0^T+M^T)R_2 - R_2 h \\ * & -Q & Q(A_1-M) & QM & 0 \\ * & * & -R_1+F^T(Q+h^2R_2)F & 0 & h(A_1^T-M^T)R_2 \\ * & * & * & -R_2 & -hM^T R_2 \\ * & * & * & * & -R_2 \end{bmatrix} < 0, \tag{5.18}$$

where $\Gamma_{11} = -Q + D^T(Q+h^2R_2)D + R_1$.

Proof: Define $\tilde{x}_{k+1} = x_{k+1} - D\nu_k x_k - F\mu_k w_{1,k}$ and $\tilde{y}_k = \bar{y}_k - D\nu_k x_k - F\mu_k w_{1,k}$ and denote

$$\eta_k = col\{x_k,\ \tilde{x}_{k+1},\ w_{1,k},\ w_{2,k},\ h\tilde{y}_k\}.$$

If (5.18) is satisfied for the appropriate $Q,\ R_1,\ R_2$, and M, then the following holds:

$$\theta_k \stackrel{\Delta}{=} \eta_k^T \Gamma \eta_k < 0, \quad \eta_k \neq 0. \tag{5.19}$$

Carrying out the multiplications in (5.19) we find that

$$\begin{aligned}\theta_k = & -x_k^T Q x_k + x_k^T D^T(Q+h^2R_2)Dx_k + x_k^T R_1 x_k - \tilde{x}_{k+1}^T Q \tilde{x}_{k+1} + \bar{\eta}_k \\ & -w_{1,k}^T(R_1 - F^T(Q+R_2)F)w_{1,k} + h^2\tilde{y}_k^T R_2 \tilde{y}_k - w_{2,k}^T R_2 w_{2,k},\end{aligned}$$

where $\bar{\eta}_k = 2x_k^T(A_0^T+M^T)Q\tilde{x}_{k+1} - 2w_2^T M^T Q\tilde{x}_{k+1} + 2\tilde{x}_{k+1}^T Q(A_1-M)w_1 = 2\tilde{x}_{k+1}^T Q\tilde{x}_{k+1}$.

It thus follows that

$$\begin{aligned}\theta_k = \tilde{x}_{k+1}^T Q\tilde{x}_{k+1} - x_k^T Q x_k + x_k^T D^T(Q+h^2R_2)Dx_k + [x_k^T R_1 x_k - w_{1,k}^T R_1 w_{1,k}] + \\ [h^2\tilde{y}_k^T R_2\tilde{y}_k - w_{2,k}^T R_2 w_{2,k}] + w_{1,k}^T F^T(Q+h^2R_2)Fw_{1,k}.\end{aligned}$$

Since

$$E\{\bar{y}_k^T R_2 \bar{y}_k\} = E\{\tilde{y}_k^T R_2 \tilde{y}_k\} + E\{x_k^T D^T R_2 D x_k\} + E\{w_{1,k}^T F^T R_2 F w_{1,k}\}$$

and

$$E\{x_{k+1}^T Q x_{k+1}\} = E\{\tilde{x}_{k+1}^T Q\tilde{x}_{k+1}\} + E\{x_k^T D^T Q D^T x_k\} + E\{w_{1,k}^T F^T Q F w_{1,k}\},$$

we find that if (5.18) is satisfied then:

$$\begin{aligned}E\{\theta_k\} = E\{x_{k+1}^T Q x_{k+1} - x_k^T Q x_k\} + E\{\bar{y}_k^T h^2 R_2 \bar{y}_k - w_{2,k}^T R_2 w_{2,k}\} + \\ E\{x_k^T R_1 x_k - w_{1,k}^T R_1 w_{1,k}\} < 0.\end{aligned} \tag{5.20}$$

Since for all sequences $\{r_k\}$ in R^n

$$\|\Delta_1 r_k\|^2 \leq \|r_k\|^2, \quad \text{and} \quad \|\Delta_2 r_k\|^2 \leq h^2\|r_k\|^2,$$

we have that $E\{h^2\bar{y}_k^T R_2\bar{y}_k - w_{2,k}^T R_2 w_{2,k}\} > 0$ and $E\{x_k^T R_1 x_k - w_{1,k}^T R_1 w_{1,k}\} > 0$.

Thus, (5.20) implies that $E\{V_{k+1}\} - V_k = E\{x_{k+1}^T Q x_{k+1}\} - x_k^T Q x_k < 0$ and the stability is guaranteed.

We note that inequality (5.18) is bilinear due to the terms QM and R_2M. Similar bilinearity is observed in the continuous-time counterpart of the stability problem (see Chapter 2, Section 2.3.1). There exist few algorithms that may solve bilinear matrix inequalities, however they do not always converge to a global minimum and they may require considerable computational effort. In order to remain in the linear domain, we choose $R_2 = \epsilon Q$ where ϵ is a positive tuning scalar. Defining $Q_M = QM$ we obtain the following result:

Corollary 5.3.1 *The exponential stability in the mean square sense of the system (5.2a) where $B_1 = 0$, $B_2 = 0$, and $G = 0$, is guaranteed if there exist $n \times n$ matrices $Q > 0$, $R_1 > 0$ and Q_M, and a tuning scalar $\epsilon > 0$ that satisfy the following inequality:*

$$\bar{\Gamma} \stackrel{\Delta}{=} \begin{bmatrix} \bar{\Gamma}_{11} & A_0^T Q + Q_M^T & 0 & 0 & \epsilon h[A_0^T Q + Q_M^T] - \epsilon h Q \\ * & -Q & QA_1 - Q_M & Q_M & 0 \\ * & * & -R_1 + (1+\epsilon h^2) F^T Q F & 0 & \epsilon h[A_1^T Q - Q_M^T] \\ * & * & * & -\epsilon Q & -h\epsilon Q_M^T \\ * & * & * & * & -\epsilon Q \end{bmatrix} < 0, \tag{5.21}$$

where $\bar{\Gamma}_{11} = -Q + D^T Q[1 + \epsilon h^2] D + R_1$.

In the uncertain case we obtain the following result:

Corollary 5.3.2 *The exponential stability in the mean square sense of the system (5.2a) where $B_1 = 0$, $B_2 = 0$, and $G = 0$ and where the system matrices lie within the polytope $\bar{\Omega}$ of (5.8) is guaranteed if there exist $n \times n$ matrices $Q > 0$, $R_1 > 0$ and Q_M, and tuning scalar $\epsilon > 0$ that satisfy the following set of LMIs:*

$$\bar{\Gamma}^i \stackrel{\Delta}{=} \begin{bmatrix} -Q + R_1 & \bar{\Gamma}_{12}^i & 0 & 0 & \bar{\Gamma}_{15}^i & D^{i,T} Q \bar{h} & 0 \\ * & -Q & QA_1^i - Q_M & Q_M & 0 & 0 & 0 \\ * & * & -R_1 & 0 & \epsilon h[A_1^{i,T} Q - Q_M^T] & 0 & \bar{\epsilon} F^{i,T} Q \\ * & * & * & -\epsilon Q & -h\epsilon Q_M^T & 0 & 0 \\ * & * & * & * & -\epsilon Q & 0 & 0 \\ * & * & * & * & * & -Q & 0 \\ * & * & * & * & * & * & -Q \end{bmatrix} < 0, \tag{5.22}$$

$\forall i$, $i = 1, 2, ..., N$, where

$$\bar{\Gamma}_{12}^i = A_0^{i,T} Q + Q_M^T,$$
$$\bar{\Gamma}_{15}^i = \epsilon h[A_0^{i,T} Q + Q_M^T] - \epsilon h Q,$$
$$\bar{\epsilon}^2 = 1 + \epsilon h^2.$$

5.3.1 *Example – Stability*

We consider the system of (5.2a), with the following system matrices:

$$A = \begin{bmatrix} 0.8 & 0 \\ 0 & 0.97 \end{bmatrix}, \quad D = \begin{bmatrix} 0.1 & 0 \\ 0 & 0 \end{bmatrix}, \quad A_1 = \begin{bmatrix} -0.1 & 0 \\ -0.1 & -0.1 \end{bmatrix}, \quad F = \begin{bmatrix} 0.05 & 0 \\ 0 & 0 \end{bmatrix},$$

where $B_1 = 0,\ B_2 = 0$, and $G = 0$. Applying the result of Corollary 5.3.1, we obtain a maximum delay bound of $h = 7$ for $\epsilon = 0.08$.

5.4 The Bounded Real Lemma

In the case where $w_k \neq 0$, (5.19) implies that $E\{[\tilde{x}_{k+1} - B_1 w_k]^T Q [\tilde{x}_{k+1} - B_1 w_k]\} - x_k^T Q x_k < 0$. Denoting

$$J_B = E\{x_{k+1}^T Q x_{k+1}\} - x_k^T Q x_k + z_k^T z_k - \gamma^2 w_k^T w_k, \tag{5.23}$$

where $z_k = C_1 x_k$, we add $B_1 w_k$ to the previously defined $\tilde{x}_{k+1}$ and readily find that $J_B < 0,\ \forall\ w_k \in \tilde{l}^2_{\mathcal{F}_k}([0,\infty);\mathcal{R}^q)$ if

$$\hat{\eta}_k^T \hat{\Gamma} \hat{\eta}_k < 0, \tag{5.24}$$

where $\hat{\eta}_k = col\{x_k,\ \tilde{x}_{k+1},\ w_{1,k},\ w_{2,k},\ h\tilde{y}_k,\ w_k,\ z_k\}$ and where

$$\hat{\Gamma} \triangleq \begin{bmatrix} \hat{\Gamma}_{11} & (A_0+M)^T Q & 0 & 0 & h(A_0^T + M^T)R_2 - R_2 h & 0 & C_1^T \\ * & -Q & Q(A_1 - M) & QM & 0 & QB_1 & 0 \\ * & * & \hat{\Gamma}_{33} & 0 & h(A_1^T - M^T)R_2 & 0 & 0 \\ * & * & * & -R_2 & -hM^T R_2 & 0 & 0 \\ * & * & * & * & -R_2 & hR_2B_1 & 0 \\ * & * & * & * & * & -\gamma^2 I_q & 0 \\ * & * & * & * & * & * & -I_r \end{bmatrix}, \tag{5.25}$$

with

$$\hat{\Gamma}_{11} = -Q + D^T(Q + h^2 R_2)D + R_1,$$
$$\hat{\Gamma}_{33} = -R_1 + F^T(Q + h^2 R_2)F.$$

Note that when carrying out the multiplications in (5.24) the product $\tilde{x}_{k+1}^T \hat{\Gamma}_2 \hat{\eta}_k$ is zero, where $\hat{\Gamma}_2$ denotes the second row block of $\hat{\Gamma}$. We also note that the matrix in the 5th row and the 6th column blocks in the latter inequality stems from the fact that the expression for $\tilde{y}_k$ includes now the additional term $B_1 w_k$.

Similarly to the stability result of Corollary 5.3.1, the following result is readily obtained:

Theorem 5.2. *Consider the system (5.2a) and (5.3) with $B_2 = 0,\ G = 0$, and $D_{12} = 0$. The system is exponentially stable in the mean square sense and, for a prescribed scalar $\gamma > 0$ and a given scalar tuning parameter $\epsilon_b > 0$, the requirement of $J_B < 0$ is achieves for all nonzero $w \in \tilde{l}^2_{\mathcal{F}_k}([0,\infty);\mathcal{R}^q)$, if there exist $n \times n$ matrices $Q > 0,\ R_1 > 0$ and a $n \times n$ matrix Q_m that satisfy the following LMI:*

$$\begin{bmatrix} \hat{\Gamma}_{11} & A_0^T Q + Q_m^T & 0 & 0 & \hat{\Gamma}_{15} & 0 & C_1^T \\ * & -Q & QA_1 - Q_m & Q_m & 0 & QB_1 & 0 \\ * & * & \hat{\Gamma}_{33} & 0 & \epsilon_b h[A_1^T Q - Q_m^T] & 0 & 0 \\ * & * & * & -\epsilon_b Q & -h\epsilon_b Q_m^T & 0 & 0 \\ * & * & * & * & -\epsilon_b Q & \epsilon_b h Q B_1 & 0 \\ * & * & * & * & * & -\gamma^2 I_q & 0 \\ * & * & * & * & * & * & -I_r \end{bmatrix} < 0, \tag{5.26}$$

where

$$\hat{\Gamma}_{11} = -Q + D^T Q[1 + \epsilon_b h^2]D + R_1,$$
$$\hat{\Gamma}_{33} = -R_1 + (1 + \epsilon_b h^2)F^T QF.$$

In the uncertain case we obtain the following result:

Corollary 5.4.1 *Consider the system (5.2a) and (5.3) with $B_2 = 0,\ G = 0$, and $D_{12} = 0$ and where the system matrices lie within the polytope $\bar{\Omega}$ of (5.8). The system is exponentially stable in the mean square sense and, for a prescribed $\gamma > 0$ and a given tuning parameter $\epsilon_b > 0$, the requirement of $J_B < 0$ is achieves for all nonzero $w \in \tilde{l}^2_{\mathcal{F}_k}([0,\infty);\mathcal{R}^q)$, if there exist $n \times n$ matrices $Q > 0,\ R_1 > 0$ and a $n \times n$ matrix Q_m that satisfy the following set of LMIs:*

$$\hat{\Gamma}^i \triangleq \begin{bmatrix} \hat{\Gamma}^i_{11} & \hat{\Gamma}^i_{12} & 0 & 0 & \hat{\Gamma}^i_{15} & 0 & C_1^{i,T} & \hat{\Gamma}^i_{18} & 0 \\ * & -Q & \hat{\Gamma}^i_{23} & Q_m & 0 & QB_1^i & 0 & 0 & 0 \\ * & * & -R_1 & 0 & \hat{\Gamma}^i_{35} & 0 & 0 & 0 & \bar{\epsilon}F^{i,T}Q \\ * & * & * & -\epsilon_b Q & -h\epsilon_b Q_m^T & 0 & 0 & 0 & 0 \\ * & * & * & * & -\epsilon_b Q & \epsilon_b h Q B_1^i & 0 & 0 & 0 \\ * & * & * & * & * & -\gamma^2 I_q & 0 & 0 & 0 \\ * & * & * & * & * & * & -I_r & 0 & 0 \\ * & * & * & * & * & * & * & -Q & 0 \\ * & * & * & * & * & * & * & * & -Q \end{bmatrix} < 0, \tag{5.27}$$

$\forall i,\ i = 1, 2, ..., N$, where

$$\begin{aligned}
\hat{\Gamma}^i_{11} &= -Q + R_1,\\
\hat{\Gamma}^i_{12} &= A_0^{i,T} Q + Q_m^T\\
\hat{\Gamma}^i_{15} &= \epsilon_b h[A_0^{i,T} Q + Q_m^T] - \epsilon_b h Q,\\
\hat{\Gamma}^i_{18} &= \bar{\epsilon} D^{i,T} Q,\\
\hat{\Gamma}^i_{23} &= Q A_1^i - Q_m\\
\hat{\Gamma}^i_{35} &= \epsilon_b h[A_1^{i,T} Q - Q_m^T]\\
\bar{\epsilon}^2 &= 1 + \epsilon_b h^2.
\end{aligned}$$

5.4.1 Example – BRL

We consider the system of (5.2a) and (5.3) with the following system matrices:

$$A = \begin{bmatrix} 0.1 & 0.6 \\ -1 & -0.5 \end{bmatrix}, \quad D = \begin{bmatrix} 0 & 0.63 \\ 0 & 0 \end{bmatrix}, \quad A_1 = \begin{bmatrix} 0 & 0.1 \\ 0 & 0 \end{bmatrix}, \quad F = \begin{bmatrix} 0 & 0.02 \\ 0 & 0 \end{bmatrix},$$
$$B1 = \begin{bmatrix} -0.225 \\ 0.45 \end{bmatrix}, \quad C1 = \begin{bmatrix} -0.5 & 04 \\ 0 & 0 \end{bmatrix},$$

where $B_2 = 0$, $G = 0$, and $D_{12} = 0$. Applying the result of Theorem 5.2, we obtain for a delay interval of $h = 120$, a near minimum attenuation level of $\gamma = 44.97$ for $\epsilon_b = 10^{-6}$.

5.5 State-Feedback Control

In this section we address the problem of finding the following state-feedback control law

$$u_k = K x_k, \tag{5.28}$$

that stabilizes the system and achieves a prescribed level of attenuation. We consider the system of (5.2a) and (5.3) and we apply the control law of (5.28), where A_0 is replaced by $(A_0 + B_2 K)$ and C_1 is replaced by $C_1 + D_{12}K$. We obtain the following inequality:

$$\begin{bmatrix}
-Q + R_1 + \Upsilon_{11,a} & \Upsilon_{12} & 0 & 0 & \Upsilon_{15} & 0 & \Upsilon_{17}\\
* & -Q & QA_1 - Q_M & Q_M & 0 & QB_1 & 0\\
* & * & \Upsilon_{33} & 0 & \epsilon_b h[A_1^T Q - Q_M^T] & 0 & 0\\
* & * & * & -\epsilon_b Q & -h\epsilon_b Q_M^T & 0 & 0\\
* & * & * & * & -\epsilon_b Q & h\epsilon_b Q B_1 & 0\\
* & * & * & * & * & -\gamma^2 I_q & 0\\
* & * & * & * & * & * & -I_r
\end{bmatrix} < 0, \tag{5.29}$$

where

$$
\begin{aligned}
\Upsilon_{11,a} &= D^T Q[1+\epsilon_b h^2]D + K^T G^T Q[1+\epsilon_b h^2]GK,\\
\Upsilon_{12} &= [A_0 + B_2 K]^T Q + Q_M^T,\\
\Upsilon_{15} &= \epsilon_b h[(A_0 + B_2 K)^T Q + Q_M^T] - \epsilon_b h Q,\\
\Upsilon_{17} &= [C_1 + D_{12} K]^T,\\
\Upsilon_{33} &= -R_1 + (1+\epsilon_b h^2) F^T Q F.
\end{aligned}
$$

Multiplying the above inequality by $diag\{Q^{-1}, Q^{-1}, Q^{-1}, Q^{-1}, Q^{-1}, I_q, I_r\}$, from the left and the right and denoting , $\bar{R}_1 = Q^{-1} R_1 Q^{-1}$, $P \stackrel{\Delta}{=} Q^{-1}$, $M_P = MP$, $K_P = KP$, we obtain the following LMI:

$$
\bar{\Upsilon} \stackrel{\Delta}{=} \begin{bmatrix}
-P+\bar{R}_1 & \bar{\Upsilon}_{12} & 0 & 0 & \bar{\Upsilon}_{15} & 0 & \bar{\Upsilon}_{17} & \bar{\epsilon} P D^T & 0 & \bar{\epsilon} K_P^T G^T\\
* & -P & A_1 P - M_P & M_P & 0 & B_1 & 0 & 0 & 0 & 0\\
* & * & -\bar{R}_1 & 0 & \bar{\Upsilon}_{35} & 0 & 0 & 0 & \bar{\epsilon} P F^T & 0\\
* & * & * & -\epsilon_b P & -h\epsilon_b M_P^T & 0 & 0 & 0 & 0 & 0\\
* & * & * & * & -\epsilon_b P & h\epsilon_b B_1 & 0 & 0 & 0 & 0\\
* & * & * & * & * & -\gamma^2 I_q & 0 & 0 & 0 & 0\\
* & * & * & * & * & * & -I_r & 0 & 0 & 0\\
* & * & * & * & * & * & * & -P & 0 & 0\\
* & * & * & * & * & * & * & * & -P & 0\\
* & * & * & * & * & * & * & * & * & -P
\end{bmatrix} < 0, \tag{5.30}
$$

where

$$
\begin{aligned}
\bar{\Upsilon}_{12} &= P A_0^T + K_P^T B_2^T + M_P^T,\\
\bar{\Upsilon}_{15} &= \epsilon_b h[P A_0^T + K_P^T B_2^T + M_P^T] - \epsilon_b h P,\\
\bar{\Upsilon}_{17} &= P C_1^T + K_P^T D_{12}^T,\\
\bar{\Upsilon}_{35} &= \epsilon_b h[P A_1^T - M_P^T],\\
\bar{\epsilon}^2 &= 1 + \epsilon_b h^2.
\end{aligned}
$$

We thus arrive at the following theorem:

Theorem 5.3. *Consider the system (5.2a) and (5.3). For a prescribed scalar $\gamma > 0$, and positive tuning scalar $\epsilon_b > 0$, there exists a state-feedback gain that achieves negative J_E for all nonzero $w \in \tilde{l}^2_{\mathcal{F}_k}([0,\infty); \mathcal{R}^q)$, if there exist $n \times n$ matrices $P > 0$, $\bar{R}_1 > 0$, $n \times n$ matrix M_P and a $l \times n$ matrix K_P that satisfy the LMI of (5.30). In the latter case the state-feedback gain is given by:*

$$
K = K_P P^{-1}. \tag{5.31}
$$

In the uncertain case we obtain the following result:

Corollary 5.5.1 *Consider the system (5.2a) and (5.3), where the system matrices lie within the polytope $\bar{\Omega}$ of (5.8). For a prescribed scalar $\gamma > 0$, and positive tuning scalar $\epsilon_b > 0$, there exists a state-feedback gain that achieves negative J_E for all nonzero $w \in \tilde{l}^2_{\mathcal{F}_k}([0,\infty); \mathcal{R}^q)$, if there exist $n \times n$ matrices*

$P > 0$, $\bar{R}_1 > 0$, $n \times n$ matrix M_P and a $l \times n$ matrix K_P that satisfy the following set of LMIs:

$$\begin{bmatrix} \bar{\Upsilon}^i_{11} & \bar{\Upsilon}^i_{12} & 0 & 0 & \bar{\Upsilon}^i_{15} & 0 & \bar{\Upsilon}^i_{17} & \bar{\Upsilon}^i_{18} & 0 & \bar{\Upsilon}^i_{1,10} \\ * & -P & A^i_1 P - M_P & M_P & 0 & B^i_1 & 0 & 0 & 0 & 0 \\ * & * & -\bar{R}_1 & 0 & \bar{\Upsilon}^i_{35} & 0 & 0 & 0 & \bar{\epsilon} P F^{i,T} & 0 \\ * & * & * & -\epsilon_b P & -h\epsilon_b M_P^T & 0 & 0 & 0 & 0 & 0 \\ * & * & * & * & -\epsilon_b P & h\epsilon_b B^i_1 & 0 & 0 & 0 & 0 \\ * & * & * & * & * & -\gamma^2 I_q & 0 & 0 & 0 & 0 \\ * & * & * & * & * & * & -I_r & 0 & 0 & 0 \\ * & * & * & * & * & * & * & -P & 0 & 0 \\ * & * & * & * & * & * & * & * & -P & 0 \\ * & * & * & * & * & * & * & * & * & -P \end{bmatrix} < 0, \tag{5.32}$$

$\forall i,\ i = 1, 2, ..., N$, where

$$\begin{aligned} \bar{\Upsilon}^i_{11} &= -P + \bar{R}_1, \\ \bar{\Upsilon}^i_{12} &= P A_0^{i,T} + K_P^T B_2^{i,T} + M_P^T, \\ \bar{\Upsilon}^i_{15} &= \epsilon_b h [P A_0^{i,T} + K_P^T B_2^{i,T} + M_P^T] - \epsilon_b h P, \\ \bar{\Upsilon}^i_{17} &= P C_1^{i,T} + K_P^T D_{12}^{i,T}, \\ \bar{\Upsilon}^i_{18} &= \bar{\epsilon} P D^{i,T}, \\ \bar{\Upsilon}^i_{1,10} &= \bar{\epsilon} K_P^T G^{i,T}, \\ \bar{\Upsilon}^i_{35} &= \epsilon_b h [P A_1^{i,T} - M_P^T], \\ \bar{\epsilon}^2 &= 1 + \epsilon_b h^2 \end{aligned}$$

In the latter case, the state-feedback gain is given by (5.31).

5.5.1 Example – Robust State-Feedback

We consider the system of (5.2a) and (5.3) with the following system matrices:

$$A = \begin{bmatrix} 0.1 & 0.6 \pm a \\ -1 & -0.5 \end{bmatrix}, \quad D = \begin{bmatrix} 0 & 0.63 \\ 0 & 0 \end{bmatrix}, \quad A_1 = \begin{bmatrix} 0 & 0.1 \\ 0 & 0 \end{bmatrix}, \quad F = \begin{bmatrix} 0 & 0.05 \\ 0 & 0 \end{bmatrix},$$

$$B_1 = \begin{bmatrix} -0.225 \\ 0.45 \end{bmatrix}, \quad C_1 = \begin{bmatrix} -0.5 & 04 \\ 0 & 0 \end{bmatrix}, \quad B_2 = \begin{bmatrix} 0.04 \pm b \\ 0.05 \end{bmatrix}, \quad D_{12} = \begin{bmatrix} 0 \\ 0.1 \end{bmatrix},$$

where $G = 0$. Taking $a = 0,\ b = 0$ for the nominal case and applying the result of Theorem 5.3, we obtain for a delay bound of $h = 11$, a near minimum attenuation level of $\gamma = 6.79$ for $\epsilon_b = 0.0001$. The controller gain is $K = [16.099\ \ 5.804]$ and the closed-loop system poles are at 0.5223 0.0119. Taking $a \in [-0.1\ \ 0.1]$ and $b \in [-0.02\ \ 0.02]$, we obtain for the latter delay bound a near minimum attenuation level of $\gamma = 7.68$ for $\epsilon_b = 10^{-6}$. The controller gain is $K = [13.8653\ \ 3.1584]$.

5.6 Delayed Filtering

In this section we address the filtering problem of the delayed state-multiplicative noisy system. We consider the system of (5.2a,b) and (5.3) with $B_2 = 0$, $G = 0$, $D_{12} = 0$ and the general type filter of (5.5). Denoting $\xi_k^T \triangleq [x_k^T \ \hat{x}_k^T]$, $\bar{w}_k^T \triangleq [w_k^T \ n_k^T]$ we obtain the following augmented system:

$$\begin{aligned}
&\xi_{k+1} = \tilde{A}_0\xi_k + \tilde{B}\bar{w}_k + \tilde{A}_1\xi_{k-\tau(k)} + \tilde{D}\xi_k\nu_k + \tilde{F}\xi_{k-\tau(k)}\mu_k, \qquad \xi_l = 0,\ l \le 0,\\
&\tilde{z}_k = \tilde{C}\xi_k,
\end{aligned} \tag{5.33}$$

where

$$\tilde{A}_0 = \begin{bmatrix} A_0 & 0 \\ B_cC_2 & A_c \end{bmatrix}, \ \tilde{B} = \begin{bmatrix} B_1 & 0 \\ 0 & B_cD_{21} \end{bmatrix}, \ \tilde{A}_1 = \begin{bmatrix} A_1 & 0 \\ 0 & 0 \end{bmatrix}, \ \tilde{D} = \begin{bmatrix} D & 0 \\ 0 & 0 \end{bmatrix},$$

$$\tilde{F} = \begin{bmatrix} F & 0 \\ 0 & 0 \end{bmatrix}, \quad \tilde{C} = [C_1 \ \ -C_c]. \tag{5.34}$$

Using the BRL result of Section 5.4 we obtain the following inequality condition:

$$\tilde{\Upsilon} \triangleq \begin{bmatrix}
\tilde{\Upsilon}_{11} & \tilde{\Upsilon}_{12} & 0 & 0 & \tilde{\Upsilon}_{15} & 0 & \tilde{C}^T & \tilde{\Upsilon}_{18} \\
* & -\tilde{Q} & \tilde{Q}\tilde{A}_1 - \tilde{Q}_M & \tilde{Q}_M & 0 & \tilde{Q}\tilde{B} & 0 & 0 \\
* & * & \tilde{\Upsilon}_{33} & 0 & \epsilon_f h[\tilde{A}_1^T\tilde{Q} - \tilde{Q}_M^T] & 0 & 0 & 0 \\
* & * & * & -\epsilon_f\tilde{Q} & -h\epsilon_f\tilde{Q}_M^T & 0 & 0 & 0 \\
* & * & * & * & -\epsilon_f\tilde{Q} & h\epsilon_f\tilde{Q}\tilde{B} & 0 & 0 \\
* & * & * & * & * & -\gamma^2 I_{q+p} & 0 & 0 \\
* & * & * & * & * & * & -I_r & 0 \\
* & * & * & * & * & * & * & -\tilde{Q}
\end{bmatrix} < 0, \tag{5.35}$$

where

$$\begin{aligned}
\tilde{\Upsilon}_{11} &= -\tilde{Q} + \tilde{R}_1,\\
\tilde{\Upsilon}_{12} &= \tilde{A}_0^T\tilde{Q} + \tilde{Q}_M^T,\\
\tilde{\Upsilon}_{15} &= \epsilon_f h[\tilde{A}_0^T\tilde{Q} + \tilde{Q}_M^T] - \epsilon_f h\tilde{Q},\\
\tilde{\Upsilon}_{18} &= \tilde{D}^T\tilde{Q}\sqrt{1 + \epsilon_f h^2},\\
\tilde{\Upsilon}_{33} &= -\tilde{R}_1 + (1 + \epsilon_f h^2)\tilde{F}^T\tilde{Q}\tilde{F}.
\end{aligned}$$

Defining $\tilde{P} = \tilde{Q}^{-1}$, denoting the following partitions

$$\tilde{P} = \begin{bmatrix} X & M^T \\ M & T \end{bmatrix},$$

$$\tilde{Q} = \begin{bmatrix} Y & N^T \\ N & W \end{bmatrix},$$

$$J = \begin{bmatrix} X^{-1} & Y \\ 0 & N \end{bmatrix},$$

we multiply (5.35) by $\hat{J} = diag\{\tilde{P}J, \tilde{P}J, \tilde{P}J, \tilde{P}J, \tilde{P}J, I, I, \tilde{P}J\}$ from the right and by $\hat{J}^T$, from the left. We obtain, denoting $\bar{R}_p = J^T\tilde{P}\tilde{R}_1\tilde{P}J$,

$$\begin{bmatrix} \hat{\Upsilon}_{11} & \hat{\Upsilon}_{12} & 0 & 0 & \hat{\Upsilon}_{15} & 0 & J^T\tilde{P}\tilde{C}^T & \bar{\epsilon}J^T\tilde{P}\tilde{D}^T J & 0 \\ * & -J^T\tilde{P}J & \hat{\Upsilon}_{23} & J^T\tilde{P}\tilde{Q}_M\tilde{P}J & 0 & J^T\tilde{B} & 0 & 0 & 0 \\ * & * & -\bar{R}_p & 0 & \hat{\Upsilon}_{35} & 0 & 0 & 0 & \bar{\epsilon}J^T\tilde{P}\tilde{F}^T J \\ * & * & * & -\epsilon_f J^T\tilde{P}J & \hat{\Upsilon}_{45} & 0 & 0 & 0 & 0 \\ * & * & * & * & -\epsilon_f J^T\tilde{P}J & h\epsilon_b J^T\tilde{B} & 0 & 0 & 0 \\ * & * & * & * & * & -\gamma^2 I_{q+p} & 0 & 0 & 0 \\ * & * & * & * & * & * & -I_r & 0 & 0 \\ * & * & * & * & * & * & * & -J^T\tilde{P}J & 0 \\ * & * & * & * & * & * & * & * & -J^T\tilde{P}J \end{bmatrix} < 0, \qquad (5.36)$$

where

$$\hat{\Upsilon}_{11} = -J^T \tilde{P} J + \bar{R}_p,$$
$$\hat{\Upsilon}_{12} = J^T \tilde{P} \tilde{A}_0^T J + J^T \tilde{P} \tilde{Q}_M^T \tilde{P} J,$$
$$\hat{\Upsilon}_{15} = \epsilon_f h [J^T \tilde{P} \tilde{A}_0^T J + J^T \tilde{P} \tilde{Q}_M^T \tilde{P} J] - \epsilon_f h J^T \tilde{P} J,$$
$$\hat{\Upsilon}_{23} = J^T \tilde{A}_1 \tilde{P} J - J^T \tilde{P} \tilde{Q}_M \tilde{P} J,$$
$$\hat{\Upsilon}_{35} = \epsilon_f h [J^T \tilde{P} \tilde{A}_1^T J - J^T \tilde{P} \tilde{Q}_M^T \tilde{P} J],$$
$$\hat{\Upsilon}_{45} = -h \epsilon_f J^T \tilde{P} \tilde{Q}_M^T \tilde{P} J,$$
$$\bar{\epsilon}^2 = 1 + \epsilon_f h^2.$$

Denoting

$$\bar{X} = X^{-1}, \ \tilde{P}_M = J^T \tilde{P} \tilde{Q}_M^T \tilde{P} J,$$

we obtain:

$$\begin{bmatrix}
-\begin{bmatrix}\bar{X} & \bar{X}\\ \bar{X} & Y\end{bmatrix} + \bar{R}_p & \Psi_{12} & 0 & 0 & \Psi_{15} & 0 & J^T\tilde{P}\tilde{C}^T & \bar{\epsilon} J^T \tilde{P} \tilde{D}^T J & 0 \\
* & -\begin{bmatrix}\bar{X} & \bar{X}\\ \bar{X} & Y\end{bmatrix} & \Psi_{23} & \tilde{P}_M & 0 & J^T\tilde{B} & 0 & 0 & 0 \\
* & * & -\bar{R}_p & 0 & \Psi_{35} & 0 & 0 & 0 & \bar{\epsilon} J^T \tilde{P} \tilde{F}^T J \\
* & * & * & -\epsilon_f \begin{bmatrix}\bar{X} & \bar{X}\\ \bar{X} & Y\end{bmatrix} & \Psi_{45} & 0 & 0 & 0 & 0 \\
* & * & * & * & -\epsilon_f \begin{bmatrix}\bar{X} & \bar{X}\\ \bar{X} & Y\end{bmatrix} & h\epsilon_b J^T \tilde{B} & 0 & 0 & 0 \\
* & * & * & * & * & -\gamma^2 I_{q+p} & 0 & 0 & 0 \\
* & * & * & * & * & * & -I & 0 & 0 \\
* & * & * & * & * & * & * & -\begin{bmatrix}\bar{X} & \bar{X}\\ \bar{X} & Y\end{bmatrix} & 0 \\
* & * & * & * & * & * & * & * & -\begin{bmatrix}\bar{X} & \bar{X}\\ \bar{X} & Y\end{bmatrix}
\end{bmatrix} < 0, \qquad (5.37)$$

where

$$
\begin{aligned}
\Psi_{12} &= J^T \tilde{P} \tilde{A}_0^T J + \tilde{P}_M, \\
\Psi_{15} &= \epsilon_f h [J^T \tilde{P} \tilde{A}_0^T J + \tilde{P}_M] - \epsilon_f h \begin{bmatrix} \bar{X} & \bar{X} \\ \bar{X} & Y \end{bmatrix}, \\
\Psi_{23} &= J^T \tilde{A}_1 \tilde{P} J - \tilde{P}_M, \\
\Psi_{35} &= \epsilon_f h [J^T \tilde{P} \tilde{A}_1^T J - \tilde{P}_M], \\
\Psi_{45} &= -h \epsilon_f \tilde{P}_M, \\
\bar{\epsilon}^2 &= 1 + \epsilon_f h^2.
\end{aligned}
$$

Carrying out the various multiplications and denoting $K_0 = N^T A_c M \bar{X}$, $U = N^T B_c$ and $Z = C_c M \bar{X}$, we obtain the following result:

$$
\left[\begin{array}{ccccccccc}
-\begin{bmatrix} \bar{X} & \bar{X} \\ \bar{X} & Y \end{bmatrix} + \bar{R}_p & \tilde{\Psi}_{12} & 0 & 0 & \tilde{\Psi}_{15} & 0 & \begin{bmatrix} C_1^T - Z^T \\ C_1^T \end{bmatrix} & \begin{bmatrix} D^T \bar{X} & D^T Y \\ D^T \bar{X} & D^T Y \end{bmatrix} & 0 \\
* & -\begin{bmatrix} \bar{X} & \bar{X} \\ \bar{X} & Y \end{bmatrix} & \tilde{\Psi}_{23} & \tilde{P}_M & 0 & \begin{bmatrix} \bar{X} B_1 & 0 \\ Y B_1 & U D_{21} \end{bmatrix} & 0 & 0 & 0 \\
* & * & -\bar{R}_p & 0 & \tilde{\Psi}_{35} & 0 & 0 & 0 & \begin{bmatrix} F^T \bar{X} & F^T Y \\ F^T \bar{X} & F^T Y \end{bmatrix} \\
* & * & * & -\epsilon_f \begin{bmatrix} \bar{X} & \bar{X} \\ \bar{X} & Y \end{bmatrix} & -h \epsilon_f \tilde{P}_M & 0 & 0 & 0 & 0 \\
* & * & * & * & -\epsilon_f \begin{bmatrix} \bar{X} & \bar{X} \\ \bar{X} & Y \end{bmatrix} & h \epsilon_f \begin{bmatrix} \bar{X} B_1 & 0 \\ Y B_1 & U D_{21} \end{bmatrix} & 0 & 0 & 0 \\
* & * & * & * & * & -\gamma^2 I_{q+p} & 0 & 0 & 0 \\
* & * & * & * & * & * & -I & 0 & 0 \\
* & * & * & * & * & * & * & -\begin{bmatrix} \bar{X} & \bar{X} \\ \bar{X} & Y \end{bmatrix} & 0 \\
* & * & * & * & * & * & * & * & -\begin{bmatrix} \bar{X} & \bar{X} \\ \bar{X} & Y \end{bmatrix}
\end{array}\right] < 0, \qquad (5.38)
$$

where

$$\tilde{\Psi}_{12} = \begin{bmatrix} A_0^T\bar{X} & A_0^T Y + C_2^T U^T + K_0^T \\ A_0^T\bar{X} & A_0^T Y + C_2^T U^T \end{bmatrix} + \tilde{P}_M,$$

$$\tilde{\Psi}_{15} = \epsilon_f h \begin{bmatrix} A_0^T\bar{X} & A_0^T Y + C_2^T U^T + K_0^T \\ A_0^T\bar{X} & A_0^T Y + C_2^T U^T \end{bmatrix} + \epsilon_f h\tilde{P}_M - \epsilon_f h \begin{bmatrix} \bar{X} & \bar{X} \\ \bar{X} & Y \end{bmatrix},$$

$$\tilde{\Psi}_{23} = \begin{bmatrix} \bar{X}A_1 & \bar{X}A_1 \\ YA_1 & YA_1 \end{bmatrix} - \tilde{P}_M,$$

$$\tilde{\Psi}_{35} = \epsilon_f h \begin{bmatrix} A_1^T\bar{X} & A_1^T Y \\ A_1^T\bar{X} & A_1^T Y \end{bmatrix} - \epsilon_f h\tilde{P}_M,$$

$$\bar{\epsilon}^2 = 1 + \epsilon_f h^2.$$

We thus arrive at the following theorem:

Theorem 5.4. *Consider the system of (5.2a,b) and (5.3) with* $B_2 = 0$ *and* $G = 0$, $D_{12} = 0$. *For a prescribed scalar* $\gamma > 0$ *and a positive tuning scalar* ϵ_f, *there exists a filter of the structure (5.5) that achieves* $J_F < 0$, *where* J_F *is given in (5.7), for all nonzero* $w \in \tilde{l}^2([0,\infty);\mathcal{R}^q)$, $n \in \tilde{l}^2([0,\infty);\mathcal{R}^p)$, *if there exist* $n \times n$ *matrices* $\bar{X} > 0$, $Y > 0$, $2n \times 2n$ *matrix* $\bar{R}_p > 0$, $n \times n$ *matrices* K_0 *and* U, $2n \times 2n$ *matrix* $\tilde{P}_M$ *and a* $n \times l$ *matrix* Z, *that satisfy (5.38). In the latter case the filter parameters can be extracted using the following equations:*

$$A_c = N^{-T}K_0\bar{X}M^{-1},\ B_c = N^{-T}U,\ C_c = Z\bar{X}M^{-1}. \qquad (5.39)$$

In the uncertain case we obtain the following set of LMIs:

$$
\begin{bmatrix}
-\begin{bmatrix}\bar{X} & \bar{X}\\ \bar{X} & Y\end{bmatrix}+\bar{R}_p & \hat{\Psi}^i_{12} & 0 & 0 & \hat{\Psi}^i_{15} & 0 & \begin{bmatrix}C_1^{i,T}-Z^T\\ C_1^{i,T}\end{bmatrix} & \begin{bmatrix}D^{i,T}\bar{X} & D^{i,T}Y\\ D^{i,T}\bar{X} & D^{i,T}Y\end{bmatrix} & 0\\
* & -\begin{bmatrix}\bar{X} & \bar{X}\\ \bar{X} & Y\end{bmatrix} & \hat{\Psi}^i_{23} & \tilde{P}_M & 0 & \begin{bmatrix}\bar{X}B_1^i & 0\\ YB_1^i & UD_{21}^i\end{bmatrix} & 0 & 0 & 0\\
* & * & -\bar{R}_p & 0 & \hat{\Psi}^i_{35} & 0 & 0 & 0 & \begin{bmatrix}F^{i,T}\bar{X} & F^{i,T}Y\\ F^{i,T}\bar{X} & F^{i,T}Y\end{bmatrix}\\
* & * & * & -\epsilon_f\begin{bmatrix}\bar{X} & \bar{X}\\ \bar{X} & Y\end{bmatrix} & -h\epsilon_f\tilde{P}_M & 0 & 0 & 0 & 0\\
* & * & * & * & -\epsilon_f\begin{bmatrix}\bar{X} & \bar{X}\\ \bar{X} & Y\end{bmatrix} & h\epsilon_b\begin{bmatrix}\bar{X}B_1^i & 0\\ YB_1^i & UD_{21}^i\end{bmatrix} & 0 & 0 & 0\\
* & * & * & * & * & -\gamma^2 I_{q+p} & 0 & 0 & 0\\
* & * & * & * & * & * & -I & 0 & 0\\
* & * & * & * & * & * & * & -\begin{bmatrix}\bar{X} & \bar{X}\\ \bar{X} & Y\end{bmatrix} & 0\\
* & * & * & * & * & * & * & * & -\begin{bmatrix}\bar{X} & \bar{X}\\ \bar{X} & Y\end{bmatrix}
\end{bmatrix} < 0, \tag{5.40}
$$

$\forall i,\ i = 1, 2, ..., N$, where

$$\hat{\Psi}^i_{12} = \begin{bmatrix} A_0^{i,T}\bar{X} & A_0^{i,T}Y + C_2^{i,T}U^T + K_0^T \\ A_0^{i,T}\bar{X} & A_0^{i,T}Y + C_2^{i,T}U^T \end{bmatrix} + \tilde{P}_M,$$

$$\hat{\Psi}^i_{15} = \epsilon_f h \begin{bmatrix} A_0^{i,T}\bar{X} & A_0^{i,T}Y + C_2^{i,T}U^T + K_0^T \\ A_0^{i,T}\bar{X} & A_0^{i,T}Y + C_2^{i,T}U^T \end{bmatrix} + \epsilon_f h\tilde{P}_M - \epsilon_f h \begin{bmatrix} \bar{X} & \bar{X} \\ \bar{X} & Y \end{bmatrix},$$

$$\hat{\Psi}^i_{23} = \begin{bmatrix} \bar{X}A_1^i & \bar{X}A_1^i \\ YA_1^i & YA_1^i \end{bmatrix} - \tilde{P}_M,$$

$$\hat{\Psi}^i_{35} = \epsilon_f h \begin{bmatrix} A_1^{i,T}\bar{X} & A_1^{i,T}Y \\ A_1^{i,T}\bar{X} & A_1^{i,T}Y \end{bmatrix} - \epsilon_f h\tilde{P}_M,$$

$$\bar{\epsilon}^2 = 1 + \epsilon_f h^2.$$

Noting that $XY - M^T N = I$, the filter matrix parameters A_c, B_c, and C_c can be readily found, without any loss of generality, by a singular value decomposition of $I - XY$.

Corollary 5.6.1 *Consider the system of (5.2a,b) and (5.3) with $B_2 = 0$ and $G = 0$, $D_{12} = 0$, where the system matrices lie within the polytope $\bar{\Omega}$ of (5.8). For a prescribed scalar $\gamma > 0$ and a positive tuning scalar ϵ_f, there exists a filter of the structure (5.5) that achieves $J_F < 0$, where J_F is given in (5.7), for all nonzero $w \in \tilde{l}^2([0,\infty); \mathcal{R}^q)$, $n \in \tilde{l}^2([0,\infty); \mathcal{R}^p)$, if there exist matrices $\bar{X} > 0$, $Y > 0$, K_0, U, $\bar{R}_p$, $\tilde{P}_M$ and Z, as in Theorem 5.4, that satisfy (5.40). In the latter case the filter parameters can be extracted using (5.39).*

5.6.1 Example – Filtering

We consider the system of (5.2a,b) and (5.3) with the following system matrices:

$$A = \begin{bmatrix} 0.1 & 0.6 \\ -1 & -0.5 \end{bmatrix}, \quad D = \begin{bmatrix} 0 & 0.063 \\ 0 & 0 \end{bmatrix}, \quad A_1 = \begin{bmatrix} 0 & 0.1 \\ 0 & 0 \end{bmatrix}, \quad F = \begin{bmatrix} 0 & 0.01 \\ 0 & 0 \end{bmatrix},$$

$$B_1 = \begin{bmatrix} -0.225 \\ 0.45 \end{bmatrix}, \quad C_1 = \begin{bmatrix} -0.5 & 04 \\ 0 & 0 \end{bmatrix}, \quad C_2 = \begin{bmatrix} 0 & 1 \end{bmatrix}, \quad D_{21} = \begin{bmatrix} 0.01 \end{bmatrix},$$

where $B_2 = 0$, $D_{12} = 0$, and $G = 0$. Applying the result of Theorem 5.4, we obtain for a delay bound of $h = 12$, a near minimum attenuation level of $\gamma = 4.28$ for $\epsilon_f = 0.001$. The filter matrix parameters are:

$$A_c = \begin{bmatrix} 5.496 & -9.300 \\ 16.406 & -29.483 \end{bmatrix}, \quad B_c = \begin{bmatrix} -9.344 \\ 2.363 \end{bmatrix}, \quad C_c = \begin{bmatrix} 2.1765 & 9.5492 \\ 0 & 0 \end{bmatrix}.$$

5.7 Conclusions

In this chapter the theory of linear H_∞ state-feedback control and filtering of state-multiplicative noisy systems is developed for discrete-time delayed systems, where the stochastic uncertainties are encountered in both the delayed and the non delayed states in the state space model of the system. The delay is assumed to be unknown and time-varying where only the bound on its size is given. Delay dependent analysis and synthesis methods are developed which are based on the input–output approach, in accordance with the approach taken in Chapter 2 for the solution of the continuous-time state-feedback and filtering problems. This approach transforms the delayed system to a nonretarded system with norm-bounded operators.

Sufficient conditions are thus derived for the stability of the system and the existence of a solution to the corresponding BRL. Based on the BRL derivation, the state-feedback control and filtering problems are formulated and solved.

An inherent overdesign is admitted to our solution due to the use of the bounded operators which enable us to transform the retarded system to an norm-bounded one. Some additional overdesign is also admitted in our solution due to the special structure imposed on R_2. The numerical examples at the end of each section demonstrate the efficiency of the proposed results.

6

H_∞-Like Control for Nonlinear Stochastic Systems

6.1 Introduction

We consider the system (1.11). The problem of state-feedback is formulated in the following obvious way. Given an output which is to be regulated (entitled, in this chapter, controlled output):

$$z_t = z_t(x_t, u_t) = \begin{bmatrix} h(x_t, t) \\ u_t \end{bmatrix} \quad (6.1)$$

where $h : R^n \times [0, \infty) \rightarrow R^r$ is a Borel measurable function with $h(0, t) = 0, \quad \forall t \geq 0$. The objective is to synthesize a controller $u_t = u(x_t, t)$ such that, for a given $\gamma > 0$, the following H_∞ criterion is satisfied.

$$\mathcal{E}\{\int_0^T \|z_t\|^2 dt\} \leq \gamma^2 \mathcal{E}\{\|x_0\|^2 + \int_0^T (\|v_t\|^2) dt\} \quad (6.2)$$

for all $T \geq 0$ and for all disturbances v_t in $\mathcal{A}_u$ (provided $\mathcal{A}_u$ is nonempty). Whenever the system (1.11) satisfies the above inequality, it is said to have the L_2-gain property, and we also write L_2-gain$\leq \gamma$.

The above SF control problem may be also treated within the context of stochastic game theory, analogously to the utilization of game theory in the deterministic H_∞ control and estimation (see, e.g. [5]). In fact, there is a significant volume of research work regarding this approach, which is based on what is called risk-sensitive control, see, e.g., [29] and the references therein. The motivation for investigating risk-sensitive problems stems partly from the duality relation that exists between risk-sensitive stochastic control problems and stochastic differential games, which allows to obtain solutions to the stochastic differential games problems, in particular, robust H_∞ control problems (see, e.g. [29], [76]). The work presented in [17] takes a more direct approach to the minimax stochastic games as it utilizes the information state concept and establishes a Hamilton -Jacobi-Bellman equation which is driven by the system's measurement process while this process is taken to be (for

E. Gershon & U. Shaked: *Advanced Topics in Control & Estimation,* LNCIS 439, pp. 115–128.
DOI: 10.1007/978-1-4471-5070-1_6

technical reasons) a cylindrical stochastic process. This work differ from ours (aside from dealing with output feedback), first in its approach, as ours based on the notion of dissipativity and establishes Hamiltonn, Jacobi inequalities rather than equalities, as done in [17], and second in that we deal with infinite time horizon and therefore consider stochastic stability.

We adopt the approach that is based on the notion of stochastic dissipation. In Section 6.3 we state and prove some kind of a bounded real lemma for non linear stochastic systems. In particular, we introduce a certain Hamilton-Jacobi inequality (HJI for short) and we establish necessary and sufficient conditions for the HJI to guarantee a dissipation of the underlying system, which in turn implies the $L_2 - gain$ property of the system.

In Section 6.4 we discuss the stability of the closed-loop system which is subject to the H_∞ controller. In particular, we establish sufficient conditions under which stability in some mean square sense and in probability is guaranteed.

In Section 6.5 we discuss the stationary case with nonlinear bounded uncertainties. Thus we are led to deal with the algebraic HJI. We introduce a certain LMI the solution to which, if exists, solves also the algebraic HJI; this, in turn, implies that the closed-loop system is stable in the mean-square sense and the underlying nonlinear system has the $L_2 - gain$ property.

Applying the definitions of Chapter 1 we obtain the following.

Theorem 6.1.1 *The function V_a of the above definition is a storage function for the system of (1.11) (or equivalently, the system (1.11) is dissipative with respect to the supply rate S) iff $\mathcal{E}\{V_a(x,t)\}$ is finite for all $t \in [0,\infty)$ and for any F_t measurable R^n valued random variable x with $\mathcal{E}\{||x||^2\} < \infty$.*

Proof: Assume $\mathcal{E}\{V_a\}$ is finite for some u and for all $v \in \mathcal{A}_u$ in the sense of the theorem. It is obvious that $V_a(x,t) \geq 0 \quad P - a.e.$ as 0 is a member of the set over which the supremum in (1.18) is taken (take $T = t$). Let now $0 \leq t_0 \leq t_1$ be arbitrarily fixed. Then

$$\mathcal{E}\{V_a(x_{t_0}, t_0)\} = \mathcal{E}\{ \sup_{T \geq t_0,\, v \in \mathcal{A}_u} [-\mathcal{E}\{[\int_{t_0}^{T} S(v_s, z_s)ds]/x_{t_0}\}]\}$$

$$\geq \mathcal{E}[-\mathcal{E}\{[\int_{t_0}^{T} S(v_s, z_s)ds]/x_{t_0}\}] \quad \forall T \geq t_0, v \in \mathcal{A}_u$$

from which we have

$$\mathcal{E}\{V_a(x_{t_0}, t_0)\} \geq \mathcal{E}[-\mathcal{E}\{[\int_{t_0}^{t} S(v_s, z_s)ds]/F_{x_{t_0}}\}] + \mathcal{E}[-\mathcal{E}\{[\int_{t_1}^{T} S(v_s, z_s)ds]/F_{x_{t_0}}\}],$$
$$\forall T \geq t_1, v \in \mathcal{A}_u.$$

Using now elementary properties of conditional expectation this last inequality implies

$$\mathcal{E}\{V_a(x_{t_0},t_0)\} \geq -\mathcal{E}\{\int_{t_0}^{t_1} S(v_s,z_s)ds\} + \mathcal{E}[-\mathcal{E}\{[\int_{t_1}^{T} S(v_s,z_s)ds]/x_{t_1}\}],$$
$$\forall T \geq t_1, v \in \mathcal{A}_u$$

which leads to

$$\begin{aligned} \mathcal{E}\{V_a(x_{t_0},t_0)\} &\geq -\mathcal{E}\{\int_{t_0}^{t_1} S(v_s,z_s)ds\} \\ &+ \mathcal{E}\{ \sup_{T \geq t_1,\ \bar{v} \in \mathcal{A}_u} [-\mathcal{E}\{[\int_{t_1}^{T} S(\bar{v}_s,z_s)ds]/x_{t_1}\}]\}, \\ &= -\mathcal{E}\{\int_{t_0}^{t_1} S(v_s,z_s)ds\} + \mathcal{E}\{V_a(x_{t_1},t_1)\}, \quad \forall v \in \mathcal{A}_u \end{aligned} \tag{6.3}$$

so that the system (1.11) is dissipative, that is V_a is a storage function of this system.

Suppose now that the system (1.11) is dissipative with respect to the supply rate S, that is, it possesses some storage function V that satisfies

$$0 \leq \mathcal{E}\{V(x_T,T)\} \leq \mathcal{E}\{V(x_{t_0},t_0)\} + \mathcal{E}\{\int_{t_0}^{T} S(v_s,z_s)ds\} \tag{6.4}$$

for any $0 \leq t_0 \leq T$. The last inequality implies $-\mathcal{E}\{\int_{t_0}^{T} S(v_s,z_s)ds\} \leq \mathcal{E}\{V(x_{t_0},t_0)\} < \infty$ for all admissible v and for all $T \geq t_0$. Since

$$-\mathcal{E}\{\int_{t_0}^{T} S(v_s,z_s)ds\} = \mathcal{E}[-\mathcal{E}\{[\int_{t_0}^{T} S(v_s,z_s)ds]/x_{t_0}\}] \tag{6.5}$$

it follows that

$$\mathcal{E}\{V_a(x_{t_0},t_0)\} = \mathcal{E}\{ \sup_{T \geq t_0,\ \bar{v} \in \mathcal{A}_u} [-\mathcal{E}\{[\int_{t_0}^{T} S(\bar{v}_s,z_s)ds]/F_{x_{t_0}}\}]\} < \infty. \tag{6.6}$$

This completes the proof. □

6.2 Stochastic H_∞ SF Control

In this section we consider the stochastic system (1.11). The H_∞ SF control objective is to synthesize a state-feedback $u_t = u(x_t,t)$ such that for a prescribed $\gamma > 0$ the controlled output (6.1) satisfies (6.2) for all $T \geq 0$ and for all $v \in \mathcal{A}_u$ where $h : R^n \times [0,T) \to R^r$ is a continuous function in both arguments. In this case, we say that the closed-loop system has an L_2-gain$\leq \gamma$.

As has been mentioned before, we take an approach that is based on the dissipation of stochastic systems, and establish a connection between the dissipation of a stochastic system and a certain HJI. In fact, it will be shown that a necessary and sufficient condition for a stochastic system to be dissipative with respect to a prescribed supply rate is that there exists a storage function which satisfies this HJI. Also, it will be shown that dissipation implies the desired objective, i.e. L_2-gain$\leq \gamma$. We then consider the case of infinite time horizon and establish conditions under which the H_∞ controller is a stabilizing one. We also consider the time-invariant case, which yields an algebraic HJI which, in turn, yields a stabilizing time-invariant controller. Stability will be established in both probability and mean square sense.

We first state the following trivial fact.

Lemma 6.2.1 *Suppose there is a controller $u_t = u(x_t, t)$ such that the system (1.11) is dissipative with respect to the supply rate $S(v, z) = \gamma^2 \|v\|^2 - \|z\|^2$ and assume that the associated storage function satisfies $\mathcal{E}\{V(x_0, t)\} \leq \gamma^2 \mathcal{E} \|x_0\|^2$ for all $t \geq 0$. Then, the closed-loop system (1.11) has an L_2-gain$\leq \gamma$.*

Utilizing now the of stochastic dissipation concept, we prove the following:

Theorem 6.2.1 *Consider the system described by (1.11) with the controlled output of (6.1), and the supply rate $S(v, z) = \gamma^2 \|v\|^2 - \|z\|^2$. Then the following hold:*

A. *Let $\gamma > 0$ be fixed, and let $V(x, t) \in C^{2,1}$ be a positive function satisfying*

$$\bar{U} = \gamma^2 I - \frac{1}{2} U(x, t) \geq \alpha I$$

for some $\alpha > 0$, and for all x, t, where $U(x, t)$ is defined by

$$U(x, t) = [g_2(x, t)]^T V_{xx}(x, t) g_2(x, t). \tag{6.7}$$

Let also
$D(x, t) = I + \frac{1}{2} \bar{g}^T(x, t) V_{xx}(x, t) \bar{g}(x, t)$.
Assume the following HJI is satisfied

$$\begin{aligned} &V_t(x, t) + V_x(x, t) f(x, t) - \tfrac{1}{4} V_x(x, t) g(x, t) D^{-1}(x, t) g^T(x, t) V_x^T(x, t) \\ &+ \tfrac{1}{4} V_x(x, t) g_1(x, t) \bar{U}^{-1}(x, t) g_1^T(x, t) V_x^T(x, t) \\ &+ \tfrac{1}{2} Tr\{G^T(x, t) V_{xx}(x, t) G(x, t)\} + h^T(x, t) h(x, t) \leq 0 \; \forall x \in R^n, \qquad \forall t \geq 0 \end{aligned} \tag{6.8}$$

Then, for $u_t = -\frac{1}{2} D^{-1}(x, t) g^T(x, t) V_x^T(x, t)$ the system (1.11) is dissipative with respect to the supply rate $S(v, z)$ (provided $\mathcal{A}_u$ is nonempty).

B. *Assume that for some control $u_t = l(x_t, t)$ the system (1.11) is dissipative with respect to the supply rate $S(v, z)$ for some storage function $V \in C^{2,1}$ which is assumed to satisfy $\bar{U} \geq \alpha I$ for all x, t. Assume also that $v_t(x) = \frac{1}{2} \bar{U}^{-1}(x, t) g_1^T(x, t) V_x^T(x, t) \in \mathcal{A}_u$. Then $V(x, t)$ satisfies the HJI for all $x \in R^n$ and for all $t \geq 0$.*

Proof:
A. Assume there is a positive function $V(x,t)$ that satisfies the HJI, and let

$$u_t(x) = -\frac{1}{2}D^{-1}(x,t)g^T(x,t)V_x^T(x,t).$$

We first show that the following inequality implies dissipation, and then complete the proof of part **A** by showing that this inequality is implied by the HJI with the storage function V. Assume

$$LV(x,t) + \|z(x,t)\|^2 - \gamma^2\|v\|^2 \leq 0 \tag{6.9}$$

for all $x \in R^n$, for all $v \in R^{m_1}$, and for all $t \geq 0$, where LV is the infinitesimal generator of the system (1.11) with respect to V, that is

$$\begin{aligned} LV(x,t) &= V_t(x,t) + V_x(x,t)[f(x,t)) + g(x,t)u + g_1(x,t)v] \\ &+\tfrac{1}{2}Tr\{G^T(x,t)V_{xx}(x,t)G(x,t)\}+\tfrac{1}{2}v^T\{[g_2(x,t)]^T V_{xx}(x,t)g_2(x,t)\}v \\ &+\tfrac{1}{2}u^T\{[\bar{g}(x,t)]^T V_{xx}(x,t)\bar{g}(x,t)\}u. \end{aligned} \tag{6.10}$$

Using now the Ito formula in the derivation of the stochastic differential of $V(x_t,t)$, where x_t is the solution of the differential equation of (1.11), one finds that

$$\begin{aligned} dV(x_t,t) &= LV(x_t,t)dt + V_x(x_t,t)G(x_t,t)dW_t^1 \\ &+V_x(x_t,t)g_2(x_t,t)v_t dW_t^2 V_x(x_t,t)\bar{g}(x_t,t)u_t dW_t, \ P\text{-a.e.}, \end{aligned}$$

which may be written in the following integral form:

$$\begin{aligned} V(x_t,t) &= V(x_0,0) + \textstyle\int_0^t LV(x_s,s)ds + \int_0^t V_x(x_s,s)G(x_s,s)dW_s^1 \\ &+\textstyle\int_0^t V_x(x_s,s)g_2(x_s,s)v_s dW_s^2 + \int_0^t V_x(x_t,t)\bar{g}(x_t,t)u_t dW_t \quad P-a.e. \end{aligned} \tag{6.11}$$

where $v \in \mathcal{A}_u$ ($\mathcal{A}_u$ is assumed to be nonempty) and x_t is the solution to (1.11). Using the inequality of (6.9) in (6.11) it is found that:

$$\begin{aligned} V(x_t,t) &\leq V(x_0,0) + \textstyle\int_0^t(\gamma^2\|v_s\|^2 - \|z_s\|^2)ds + \int_0^t V_x(x_s,s)G(x_s,s)dW_s^1 \\ &+ \textstyle\int_0^t V_x(x_s,s)g_2(x_s,s)v_s dW_s^2 + \int_0^t V_x(x_t,t)\bar{g}(x_t,t)u_t dW_t \quad P-a.e. \end{aligned}$$

Taking expectation of both sides of the last inequality and recall that

$$\begin{aligned} &\mathcal{E}\{\textstyle\int_0^t V_x(x_s,s)G(x_s,s)dW_s^1 + \int_0^t V_x(x_s,s)g_2(x_s,s)v_s dW_s^2 \\ &+ \textstyle\int_0^t V_x(x_t,t)\bar{g}(x_t,t)u_t dW_t\} \quad = 0, \end{aligned}$$

we then arrive at

$$\mathcal{E}\{V(x_t,t)\} \leq \mathcal{E}\{V(x_0,0)\} + \mathcal{E}\{\int_0^t (\gamma^2\|v_\sigma\|^2 - \|z_\sigma\|^2)d\sigma\} \tag{6.12}$$

for all $v \in \mathcal{A}_u$, and for all $t \geq 0$.

Thus, the system is dissipative for $u_t(x) = -\frac{1}{2}D^{-1}(x,t)g^T(x,t)V_x^T(x,t)$ (which yields the inequality (6.9)). Completing the left hand side of the inequality (6.9) to squares, yields

$$\begin{aligned}
&LV_t(x,t)+V_x[f(x,t)) + g_1(x,t)v+g(x,t)u]+\|z\|^2-\gamma^2\|v\|^2\\
&+\tfrac{1}{2}Tr\{G^T(x,t)V_{xx}(x,t)G(x,t)\} + \tfrac{1}{2}v^T\{[g_2(x,t)]^T V_{xx}(x,t)g_2(x,t)\}v^T\\
&+\tfrac{1}{2}u^T\{[\bar{g}(x,t)]^T V_{xx}(x,t)\bar{g}(x,t)\}u^T\\
&=V_t(x,t)+V_x(x,t)f(x,t)-u^{*T}D(x,t)u^*+v^{*T}\bar{U}(x,t)v^* + h(x,t)^T h(x,t)\\
&+\tfrac{1}{2}Tr\{G^T(x,t)V_{xx}(x,t)G(x,t)\}+ (u-u^*)^T D(x,t)(u-u^*)\\
&+(v-v^*)^T\bar{U}(x,t)(v-v^*)
\end{aligned}$$

where:

$$u^* = -\frac{1}{2}D^{-1}(x,t)g^T(x,t)V_x^T(x,t), \text{ and } v^* = -\frac{1}{2}\bar{U}(x,t)g_1^T(x,t)V_x^T(x,t)$$

Using $u = -u^*$ together with the hypothesis on the HJI, yields the inequality (6.9), which in turn, implies the dissipation of the system for this u.

B. Assume $u = l(x,t)$ renders the system of (1.11) dissipative with respect to the supply rate $S(v,z) = \gamma^2\|v\|^2 - \|z\|^2$ for some continuous function l, and for a storage function $V \in C^{2,1}$, i.e.

$$\mathcal{E}\{V(x_t,t)\} \leq \mathcal{E}\{V(x_s,s)\} + \mathcal{E}\{\int_s^t (\gamma^2\|v_\sigma\|^2 - \|z_\sigma\|^2)d\sigma\} \tag{6.13}$$

for all x_s which is F_s- measurable, square-integrable, and for all $t \geq s$ and $v \in \mathcal{A}_u$. Fix now $x \in R^n$, take $x_s = x \in R^n$, and choose $v \in \mathcal{A}_u$ to be an arbitrary constant function in R^{m_1}. In view of the last inequality the following holds for all $t > s$:

$$\frac{1}{t-s}\mathcal{E}\{V(x_t,t) - V(x_s,s)\} \leq \frac{1}{t-s}\mathcal{E}\{\int_s^t (\gamma^2\|v_\sigma\|^2 - \|z_\sigma\|^2)d\sigma\}. \tag{6.14}$$

Application of the Ito formula to $V(x_t,t)$ yields

$$\begin{aligned}
&\frac{1}{t-s}\mathcal{E}\{V(x_s,s) + \int_s^t (LV)(x_\sigma,\sigma)d\sigma + \int_s^t V_x(x_\sigma,\sigma)G(x_\sigma,\sigma)dW_\sigma^1\\
&+\textstyle\int_s^t V_x(x_\sigma,\sigma)g_2(x_\sigma,\sigma)v_\sigma dW_\sigma^2 + \displaystyle\int_s^t V_x(x_t,t)\bar{g}(x_t,t)u_t dW_t - V(x_s,s)\}\\
&\leq \frac{1}{t-s}\mathcal{E}\{\int_s^t (\gamma^2\|v_\sigma\|^2 - \|z_\sigma\|^2)d\sigma.
\end{aligned} \tag{6.15}$$

By the assumed nature of v (being a constant), by the continuity of u (in x and t) and the $P-$a.s. continuity of x_t (recall: x_t is a solution to (1.11)), and by the Fubini theorem ([37]), together with the following fact:

$$\begin{aligned}&\mathcal{E}\{\textstyle\int_s^t V_x(x_\sigma,\sigma)G(x_\sigma,\sigma)dW_\sigma^1 + \int_s^t V_x(x_\sigma,\sigma)g_2(x_\sigma,\sigma)v_\sigma dW_\sigma^2\\ &+ \textstyle\int_s^t V_x(x_t,t)\bar{g}(x_t,t)u_t dW_t\} = 0,\end{aligned} \tag{6.16}$$

the last inequality yields

$$\frac{1}{t-s}\int_s^t \mathcal{E}\{(LV)(x_\sigma,\sigma)\}d\sigma \leq \frac{1}{t-s}\int_s^t \mathcal{E}\{(\gamma^2\left\|v_\sigma\right\|^2 - \left\|z_\sigma\right\|^2)d\sigma \quad \forall t>s. \tag{6.17}$$

This inequality implies, as $t \downarrow s$:

$$\mathcal{E}\{LV(x_s,s)\} \leq \mathcal{E}\{\gamma^2||v_s||^2 - ||z_s||^2\} = \mathcal{E}\{\gamma^2||v_s||^2 - ||z_s(x_s,u_s(x_s))||^2\}.$$

As x_s and v_s are taken to be arbitrary deterministic vectors in R^n and R^{m_1}, respectively, the latter inequality yields $(LV)(x,s) \leq \gamma^2\left\|v\right\|^2 - \left\|z_s\right\|^2$, for all $x \in R^n$ and for all $v \in R^{m_1}$. Completion to squares then yields

$$\begin{aligned}&V_t(x,t)+V_x(x,t)f(x,t)-\tfrac{1}{4}V_x(x,t)g(x,t)g^T(x,t)V_x^T(x,t) + v^{*T}\bar{U}(x,t)v^*\\ &\tfrac{1}{2}Tr\{G^T(x,t)V_{xx}(x,t)G(x,t)\} + h^T(x,t)h(x,t) + (u-u^*)^T D(x,t)(u-u^*)\\ &+(v-v^*)^T\bar{U}(x,t)(v-v^*) \leq 0.\end{aligned} \tag{6.18}$$

Introducing $v_t(x) = v^*$ into the last inequality yields the HJI. □

6.3 The Infinite-Time Horizon Case: A Stabilizing Controller

In this section we consider the infinite-time horizon case. The problem of synthesizing an H_∞ controller that renders a stable closed-loop system is investigated. Conditions under which the closed-loop system is asymptotically stable, in both the probability sense and the mean-square sense, will be discussed.

We first recall few facts from the theory of stochastic stability (see e.g. [67]). We remark that in what follows we consider only global stability. Obviously, local stability results may also be achieved, in a similar way.

Definition 6.3.1 *Consider the stochastic system*

$$dx_t = f(x_t,t)dt + G(x_t,t)dW_t \tag{6.19}$$

with $f(0,t) = G(0,t) = 0$ *for all* $t \geq 0$, *and assume that* f, G *satisfy the usual Lipschitz conditions that guarantee a unique strong solution relative to the filtered probability space* $(\Omega, F, \{F_t\}_{t\geq 0}, P)$ *where* $\{F_t\}$ *is generated by the Wiener Process. The solution* x_t *is said to be stable in probability if for any* $\varepsilon > 0$ $\lim_{x\to 0} P\{\sup_{t\geq 0}\|x_t\| > \varepsilon\} = 0$.

Definition 6.3.2 *The solution x_t of (6.19) is said to be globally asymptotically stable in probability if it is stable in probability, and if $P\{\lim_{t\to\infty} x_t = 0\} = 1$ for any initial state $x \in R^n$.*

A sufficient condition for a global stability in probability is given by the theorem below.

Theorem 6.3.1 *([67])Assume there exists a positive function $V(x,t) \in C^{2,1}$, with $V(0,t) = 0$, so that $(LV)(x,t) < 0$ for all $x \in R^n$ and for all $t \geq 0$. Assume also that $\inf_{t>0} V(x,t) \to \infty$ as $\|x\| \to \infty$. Then, the system of (6.19) is globally asymptotically stable in probability.*

Definition 6.3.3 *[stability in the mean square sense] The system of (6.19) is said to be globally exponentially stable in the mean square sense if $\mathcal{E}\{||x_t||^2\} \leq k\mathcal{E}\{\|x_s\|^2\}\exp\{-\alpha(t-s)\}$ for all $0 \leq s \leq t$, and for some positive numbers k and α.*

Theorem 6.3.2 *([67])Assume there exists a positive function $V(x,t) \in C^{2,1}$, with $V(0,t) = 0$. Then the system of (6.19) is globally exponentially stable if there are positive numbers k_1, k_2, k_3 such that the following hold.*

$$k_1||x||^2 \leq V(x,t) \leq k_2||x||^2, \quad (LV)(x,t) \leq -k_3||x||^2 \text{ for all } t \geq 0. \tag{6.20}$$

We now state and prove two lemmas regarding the stability of the closed-loop system.

Lemma 6.3.1 *[Stability in probability] Assume there exists a positive function $V(x,t) \in C^{2,1}$ such that $\inf_{t>0} V(x,t) \to \infty$ as $\|x\| \to \infty$, and assume $V(x,t)$ satisfies the HJI (6.8), and $h^T(x,t)h^T(x,t) > 0$ for all x and for all $t \geq 0$. Then, the homogeneous closed-loop system ($v = 0$) with $u = -\frac{1}{2}g^T(x,t)V_x^T(x,t)$ is globally asymptotically stable in probability.*

Proof: Note that for $v = 0$ the HJI (6.8) reduces to

$$\begin{aligned} &V_t(x,t) + V_x(x,t)f(x,t) - \tfrac{1}{4}V_x(x,t)g(x,t)D^{-1}(x,t)g^T(x,t)V_x^T(x,t) \\ &+\tfrac{1}{2}Tr\{G_1^T(x,t)V_{xx}(x,t)G_1(x,t)\} + h^T(x,t)h(x,t) \leq 0 \;\forall x \in R^n, \;\forall t \geq 0. \end{aligned} \tag{6.21}$$

Thus, we have $(LV)(x,t) \leq -h^T(x,t)h(x,t) < 0$. □

Remark 6.1. Assume there exists a solution V to (6.21). A sufficient condition for V to be positive is the zero state observability of the system with respect to z, which is defined as follows. The system (6.19) is zero state observable at a time s if for $v = 0$, $z(t) = 0$ for all $s \leq t$ implies $x(t) = 0$ for all $s \leq t$. That this implies the desired positiveness of V is readily seen from the dissipation inequality (1.17). In fact, the dissipation inequality implies

$$\mathcal{E}\{\int_s^t ||z(\sigma)||^2 d\sigma\} \leq V(x,s) \text{ for all } t \geq s \text{ and for all } x \in R^n. \tag{6.22}$$

Assume now that for some $s > 0$ and x, $V(x,s) = 0$. Then $\mathcal{E}\{\int_s^t ||z(\sigma)||^2 d\sigma\} = 0$ for all $t \geq s$, which implies $z(t) = 0$ $P - a.e.$ for all $t \geq s$. The zero state observability implies now that $x(t) = 0$ for all $t \geq s$, and in particular $x = 0$ $(x = x(s))$. Therefore, we conclude that $V(x,s) = 0$ implies $x = 0$, so that V is positive.

The next lemma establishes the stability of the closed-loop system in the mean square sense.

Lemma 6.3.2 *Assume there exists a positive function $V(x,t) \in C^{2,1}$, with $V(0,t) = 0$ for all $t \geq 0$, which satisfies the HJI of (6.8) for some $\gamma > 0$.In addition, let V satisfy*

$$k_1||x||^2 \leq V(x,t) \leq k_2||x||^2 \tag{6.23}$$

for all $t \geq 0$ and for some positive numbers k_1, k_2. Furthermore, assume that for some k_3, $h^T(x,t)h(x,t) \geq k_3||x||^2$ for all $t \geq 0$. Then the closed-loop system of (6.19) with $v = 0$ and $u = -\frac{1}{2}D^{-1}(x,t)g^T(x,t)V_x^T(x,t)$ is exponentially stable in the mean square sense, and has the property of $L_2 -$ gain $\leq \gamma$, that is

$$\mathcal{E}\{\int_0^\infty \|z_t\|^2 dt\} \leq \gamma^2 \mathcal{E}\{\|x_0\|^2 + \int_0^\infty (\|v_t\|^2) dt\} \tag{6.24}$$

for all nonanticipative stochastic processes v that satisfy $\mathcal{E}\{\int_0^\infty (\|v_t\|^2)dt\} < \infty$, and whenever x_0 satisfies $\mathcal{E}\{V(x_0,0)\} \leq \gamma^2 \mathcal{E}\{||x_0||^2\}$.

Proof: By Theorem 6.2.1 the existence of the HJI (6.8) implies the dissipation inequality

$$\mathcal{E}\{V(x_t,t)\} \leq \mathcal{E}\{V(x_0,0)\} + \mathcal{E}\{\int_0^t (\gamma^2 \|v_\sigma\|^2 - \|z_\sigma\|^2) d\sigma\} \tag{6.25}$$

for all $t \geq 0$, which implies$\mathcal{E}\{\int_0^\infty \|z_t\|^2 dt\} \leq \gamma^2 \mathcal{E}\{\|x_0\|^2 + \int_0^\infty (\|v_t\|^2)dt\}$, provided $\mathcal{E}\{\int_0^\infty (\|v_t\|^2)dt\} < \infty$ and $\mathcal{E}\{V(x_0,0)\} \leq \gamma^2 \mathcal{E}\{||x_0||^2\}$. Now, for $v = 0$, the HJI reduces to the one used in the proof of Lemma 6.3.1 which implies that $(LV)(x,t) \leq -h^T(x,t)h(x,t) \leq -k_3||x||^2$, where LV is the infinitesimal generator of the homogeneous closed-loop system with $u = -\frac{1}{2}D^{-1}(x,t)g^T(x,t)V_x^T(x,t)$, that is

$$\begin{aligned}(LV)(x,t) =& V_t(x,t) + V_x(x,t)f(x,t)\\ &-\tfrac{1}{2}V_x(x,t)g(x,t)D^{-1}(x,t)g^T(x,t)V_x^T(x,t) + \tfrac{1}{2}Tr\{G_1^T(x,t)V_{xx}(x,t)G_1(x,t)\}.\end{aligned} \tag{6.26}$$

By the last inequality and by (6.23) it follows that the system of (6.19) is exponentially stable in the mean-square sense. □

We consider now the time invariant case:

$$dx_t = f(x_t)dt + g(x_t)u_t dt + \bar{g}(x_t)u_t dW_t + g_1(x_t)v_t dt + g_2(x_t)v_t dW_t^2 + G(x_t)dW_t^1. \tag{6.27}$$

It is easy now to prove a time invariant analog of Theorem 6.2.1 as a simple corollary. However, in order to facilitate the development of what follows in the next section, we choose to adopt a slightly different point of view. In fact, the following lemma which provides a sufficient condition for the existence of the L_2-gain property, may be proved.

Lemma 6.3.3 *Consider the system (6.27) with the controlled output* $z(x,u(x)) = \begin{bmatrix} h(x) \\ u(x) \end{bmatrix}$, *where* $u(x) = l(x)$, $l : R^n \rightarrow R^m$ *is a continuous function. Suppose there is a positive function* $V(x) \in C^2$ *which satisfies* $\bar{U}(x) \geq \alpha I$ *for some* $\alpha > 0$, *for some* $\gamma > 0$, *and for all* x. *Assume the following HJI is satisfied*

$$\begin{aligned} &V_x(x)[f(x) + g(x)l(x)] + \tfrac{1}{4}V_x(x)g_1(x)\bar{U}^{-1}g_1^T(x)V_x(x) \\ &+\tfrac{1}{2}Tr\{G^T(x)V_{xx}(x)G(x)\} + h^T(x)h(x) \leq 0 \; \forall x \in R^n. \end{aligned} \tag{6.28}$$

Assume also $\frac{1}{2}\bar{U}^{-1}(x)g_1^T(x)V_x^T(x) \in \mathcal{A}_l$. *Then, the system of (6.27) is* L_2-*gain* $\leq \gamma$, *that is*

$$\mathcal{E}\{\int_0^t \|z_s\|^2 ds\} \leq \gamma^2 \mathcal{E}\{\|x_0\|^2 + \int_0^t (\|v_s\|^2)ds\} \tag{6.29}$$

for all $t \in [0,\infty)$, *for all* $v \in \mathcal{A}_l$, *and provided* $\mathcal{E}\{\|x_0\|^2\} \geq \mathcal{E}\{V(x_0)\}$. *Moreover, assume also that* $V(x) > 0$ *for all* $x \neq 0$ *and* $V(0) = 0$. *Then, the system (6.27) is asymptotically stable in probability. In addition, if* V *satisfies:* $\beta_1||x||^2 \leq V(x) \leq \beta_2||x||^2$ *for some positive* β_1 *and* β_2, *the system (6.27) is also exponentially stable in the mean square sense.*

An example is now in order. We introduce the following one which, although is one dimensional, it is fairly important as it includes, as a special case, the family of bilinear systems.

6.3.1 Example

Consider the following scalar system:

$$dx_t = [-ax + g(x)u + bv]dt + \bar{g}(x)udW_t + GxdW_t^1$$

where $z = col\{cx,\ u\}$ and where a is a positive scalar and W_t, W_t^1 are independent scalar Wiener processes. We assume that: $G^2 < 2a$. Choosing $V(x) = x^2 p$ where p is a positive scalar, the resulting HJI is:

$$-2apx^2 - \frac{p^2 g(x)^2 x^2}{1 + px^2 \bar{g}(x)^2} + \gamma^{-2} p^2 x^2 b^2 + x^2 G^2 p + c^2 x^2 \leq 0.$$

The requirement that there exists $p > 0$ that satisfies $\gamma^{-2} b^2 p^2 + (G - 2a)p + c^2 \leq 0$ will then be a sufficient condition for the existence of a state feedback control input $u = -\frac{pg(x)}{1 + p\bar{g}(x)^2} x$ that achieves L_2-gain $< \gamma$ whenever $v \in \mathcal{A}_u$ which contains at least all the functions v which satisfy $||v|| \leq M||x||$ for some $M > 0$. Internal exponential stability in the mean square sense is also achieved by this controller (see the remark below).

The latter inequality has a real solution if $(2a - G^2)^2 > 4b^2 c^2 \gamma^{-2}$. The minimum value of γ is thus $\frac{2bc}{2a - G^2}$. For this value of γ the resulting p is $p = \frac{2c^2}{2a - G^2}$.

Remark 6.2. There is a question as to whether the above differential equation has a solution which satisfies $E|x_t|^2 < \infty$. The answer for this is the following one.

The HJI above implies: $LV(x) - \gamma^2 |v|^2 + |u(x)|^2 + c^2 |x|^2 \leq 0$ for all $v : |v| \leq M$ for some $M > 0$, $\forall x \in R$, and for the u introduced above. If, for example $\frac{pg^2(x)}{1 + p\bar{g}(x)^2}$ is bounded, one has, for some $\bar{M} > 0$: $LV(x) \leq \bar{M} px^2 = \bar{M} V(x) \quad \forall x \in R$. It is also easy to show that the above differential equation coefficients satisfies sufficient conditions for the regularity, provided, for example, that $g(x), \bar{g}(x)$ are locally Lipschitz (see, e.g. [67]). These imply that the above system possesses a unique strong solution which satisfies $E|x_t|^2 < \infty$ for all $t \geq 0$, provided $E|x_0|^2 < \infty$. The exponential stability in the mean square is a consequence of Theorem 6.3.2.

6.4 Norm-Bounded Uncertainty in the Stationary Case

As an application of the above theory, we consider now the special case of time-invariant stochastic system with norm bounded uncertainties:

$$dx_t = f(x_t) x_t dt + G(x_t) x_t dW_t + g_1(x_t) v_t dt + g(x_t) u_t dt \tag{6.30}$$

where $f(x) = (A + HF(x)E_1)$, $G(x) = (A_1 + HF(x)E_3)$, $g_1(x) = B_1$, $g(x) = B_2 + HF(x)E_2$. A, A_1, B_1 and B_2 are matrices of appropriate dimensions.

We view the nonlinear part $F(x)$ as an uncertainty which is to be present in the stochastic model of (6.30). The nonlinear part $F(x)$ is assumed to be bounded, namely

$$F(x)^T F(x) \leq I \tag{6.31}$$

The sufficient condition, in terms of the appropriate HJI, for (6.30) to satisfy L_2-gain $\leq \gamma$ is

$$\begin{aligned} &V_x(x)f(x) - \tfrac{1}{4}V_x(x)g(x)g^T(x)V_x^T(x) + \tfrac{1}{4\gamma^2}V_x(x)g_1(x)g_1^T(x)V_x(x) \\ &+\tfrac{1}{2}Tr\{G_1^T(x)V_{xx}(x)G_1(x)\} + h^T(x)h(x) \le 0,\ \forall x \in R^n \end{aligned} \tag{6.32}$$

In what follows we seek a state-feedback control $u = Kx$, where K is a constant matrix of the appropriate dimensions and we consider the case where $z(x) = C_1x+DKx$. We construct a certain linear matrix inequality, a solution to which will be shown to satisfy (6.32). Applying $V(x) = x^TPx$, where P is a positive definite matrix in $\mathcal{R}^{n\times n}$, it readily follows from (6.32) that

$$\begin{aligned} &x^T\left[P(A+B_2K+HF(x)(E_1+E_2K)) + (A^T+K^TB_2^T+(E_1^T+K^TE_2^T)\right. \\ &F(x)^TH^T)P+\gamma^{-2}PB_1B_1^TP + (A_1^T+E_3^TF^T(x)H^T)P(A_1+HF(x)E_3) \\ &\left.+ (C_1+K^TD^T)(C_1+DK)\right]x \le 0 \end{aligned} \tag{6.33}$$

In a matrix inequality form we obtain, using Schur complement formula ([15]), the following:

$$\begin{bmatrix} \Gamma_{11} & QA_1^T+QE_3^TF^TH^T & B_1 & QC_1^T+Y^TD^T \\ * & -Q & 0 & 0 \\ * & * & -\gamma^2I & 0 \\ * & * & * & -I \end{bmatrix} \le 0 \tag{6.34}$$

where:

$\Gamma_{11} \triangleq QA^T+AQ+B_2Y+Y^TB_2^T+HF(E_1Q+E_2Y)+(QE_1^T+Y^TE_2^T)F^TH^T$, $Q \triangleq P^{-1}$ and $Y \triangleq KQ$.

The latter inequality can also be written as:

$$\Gamma + \Phi_1F\Phi_2^T + \Phi_2F^T\Phi_1^T + \Phi_3F\Phi_4^T + \Phi_4F^T\Phi_3^T \le 0 \tag{6.35}$$

where

$$\begin{aligned} &\Phi_1 = \left[H^T\ 0\ 0\ 0\right]^T, \Phi_2 = \left[E_1Q+E_2Y\ 0\ 0\ 0\right]^T, \Phi_3 = \left[0\ H^T\ 0\ 0\right]^T, \\ &\Phi_4 = \left[E_3Q\ 0\ 0\ 0\right]^T, \end{aligned} \tag{6.36}$$

and where

$$\Gamma = \begin{bmatrix} QA^T+AQ+B_2Y+Y^TB_2^T & QA_1^T & B_1 & QC_1^T+Y^TD^T \\ * & -Q & 0 & 0 \\ * & * & -\gamma^2I & 0 \\ * & * & * & -I \end{bmatrix}. \tag{6.37}$$

Using the fact that for any two matrices α and β of compatible dimensions and for any positive scalar ε the following holds: $\alpha\beta^T + \beta\alpha^T \le \varepsilon\alpha\alpha^T + \varepsilon^{-1}\beta\beta^T$, together with the bound of (6.31), we obtain the following:

$$\Gamma_1 + \varepsilon_1\Phi_1\Phi_1^T + \varepsilon_1^{-1}\Phi_2\Phi_2^T + \varepsilon_2\Phi_3\Phi_3^T + \varepsilon_2^{-1}\Phi_4\Phi_4^T \le 0 \tag{6.38}$$

for some positive scalars ε_1 and ε_2. We thus obtain the following.

Theorem 6.4.1 *Consider the system of (6.30). Given the scalar* $0 < \gamma$, *there exists a state-feedback controller* $u = Kx$ *that globally stabilizes the closed-loop exponentially, in the mean-square sense, and achieves* $L_2 - gain \leq \gamma$ *if there exist* $0 < Q$ *in* $R^{n \times n}$, Y *in* $R^{p \times n}$ *and positive scalars* ε_1 *and* ε_2 *so that the following LMI is satisfied:*

$$\begin{bmatrix} \tilde{\Gamma}_{11} & QE_3^T & QA_1^T & B_1 & QC_1^T+Y^TD^T & \varepsilon_1 H & QE_1^T+Y^TE_2^T & 0 \\ * & -\varepsilon_2 I & 0 & 0 & 0 & 0 & 0 & 0 \\ * & * & -Q & 0 & 0 & 0 & 0 & \varepsilon_2 H \\ * & * & * & -\gamma^2 I & 0 & 0 & 0 & 0 \\ * & * & * & * & -I & 0 & 0 & 0 \\ * & * & * & * & * & -\varepsilon_1 I & 0 & 0 \\ * & * & * & * & * & * & -\varepsilon_1 I & 0 \\ * & * & * & * & * & * & * & -\varepsilon_2 I \end{bmatrix} \leq 0. \tag{6.39}$$

where $\tilde{\Gamma}_{11} \triangleq QA^T+AQ+B_2Y+Y^TB_2^T$. *If a solution to this LMI exists, the state-feedback gain is given by*

$$K = YQ^{-1}. \tag{6.40}$$

6.4.1 Example

We consider the following model for a single degree of freedom inverted pendulum with the the multiplicative white noise ω:

$$ml^2\ddot{\theta} - mgl\sin(\theta) + (\varsigma + \omega)\dot{\theta} + k\theta = u + 2v \tag{6.41}$$

where k is the spring coefficient and ς is damping coefficient. The signal v is a deterministic disturbance acting on the control input u and ω is the stochastic uncertainty in the damping. In this model, θ is the inclination angle of the pendulum, l and m are its length and mass, respectively, and g is the gravitation coefficient. The state space representation of this model is given by:

$$\begin{bmatrix} \dot{x}_1 \\ \dot{x}_2 \end{bmatrix} = \begin{bmatrix} 0 & \frac{1}{ml^2} \\ mgl\frac{\sin(x_1)}{x_1} - k & -\frac{\varsigma}{ml^2} \end{bmatrix} \begin{bmatrix} x_1 \\ x_2 \end{bmatrix} + \begin{bmatrix} 0 \\ 1 \end{bmatrix} u + \begin{bmatrix} 0 \\ 2 \end{bmatrix} v + \begin{bmatrix} 0 \\ -\frac{x_2}{ml^2} \end{bmatrix} \omega$$

where $\theta = \begin{bmatrix} 1 & 0 \end{bmatrix} \begin{bmatrix} x_1 \\ x_2 \end{bmatrix}$. The Ito formalism of this is:

$$dx = [A(x) + HF(x)E_1]xdt + \begin{bmatrix} 0 \\ 2 \end{bmatrix} vdt + \begin{bmatrix} 0 \\ -\frac{x_2}{ml^2} \end{bmatrix} dW_t + \begin{bmatrix} 0 \\ 1 \end{bmatrix} udt \tag{6.42}$$

where $H = \begin{bmatrix} 0 & mgl \end{bmatrix}^T$, $E_1 = \begin{bmatrix} 1 & 0 \end{bmatrix}$ and $F(x) = x_1^{-1} sin(x_1)$. In the notations of (6.30) we have:

$$A_1 = \begin{bmatrix} 0 & 0 \\ 0 & -\frac{1}{ml^2} \end{bmatrix}, \; B_1 = \begin{bmatrix} 0 \\ 2 \end{bmatrix}, \; B_2 = \begin{bmatrix} 0 \\ 1 \end{bmatrix}, \; E_2 = 0. \tag{6.43}$$

For $m = 0.5kg$, $l = 0.7meter$, $k = 0.5Newton/meter$ and $\varsigma = 0.25$ we solve (6.39). Requiring $\gamma = 0.35$, and choosing: $C_1 = \begin{bmatrix} 1 & 0 \\ 0 & 0 \end{bmatrix}$, $D = \begin{bmatrix} 0 \\ 0.1 \end{bmatrix}$, we find that the LMI of Theorem 6.4.1 is satisfied by

$$Q = \begin{bmatrix} 0.830 & -0.381 \\ -0.381 & 1.943 \end{bmatrix}, \quad Y = -\begin{bmatrix} 0.313 & 112.98 \end{bmatrix} \quad \text{and} \quad \rho = 2.5. \tag{6.44}$$

The latter leads to $K = -\begin{bmatrix} 29.74 & 63.97 \end{bmatrix}$. The corresponding function $V = x^T Q^{-1} x$ will then satisfy the HJI of (6.32).

6.5 Conclusions

H_∞ control theory has been extended, via the concept of dissipative stochastic systems, to accommodate nonlinear stochastic systems. In particular, the state-feedback problem is solved for this type of systems. In addition, a particular class of systems is considered which consists of norm-bounded nonlinearities. For this class, a certain LMI is introduced so that its solution satisfies the HJI of (6.8), and therefore provides a state-feedback controller that stabilizes the stochastic closed-loop system, exponentially in the mean square sense, and also renders a $L_2 - gain$ system.

This theory of H_∞ control for nonlinear stochastic systems is closely related to the control and estimation problems of uncertain systems when the underlying system is linear with uncertainties modelled as multiplicative noise (almost invariably assumed to be a Wiener process for technical reasons). However, there are instances when uncertainties of more general nature would be more appropriate. One common approach to treat such uncertainties is to consider the noise as the output of a linear system driven by a Wiener process. Augmenting the system state to accommodate both systems, a nonlinear system is obtained which can be treated by the theory developed in this chapter.

7

Non Linear Systems – H_∞-Type Estimation

7.1 Introduction

In this chapter we develop an H_∞ estimation theory for nonlinear time-invariant stochastic systems of the following form.

$$dx_t = f(x_t)dt + g_1(x_t)v_t dt + g_2(x_t)v_t dW_t^2 + G(x_t)dW_t^1 \qquad (7.1)$$

$$dy_t = h_2(x_t)dt + g_3(x_t)v_t dt + G_2(x_t)dW^3 \qquad (7.2)$$

where $\{x_t\}_{t\geq 0}$ is a solution to (7.1) with the initial condition x_0, an exogenous disturbance $\{v_t\}_{t\geq 0}$, and Wiener processes $\{W_t^1\}_{t\geq 0}$, $\{W_t^2\}_{t\geq 0}$, and $\{W_t^3\}_{t\geq 0}$. Also, y_t is an observations vector in R^p which is corrupted with noise (a Wiener process $\{W_t^3\}_{t\geq 0}$), and contains an uncertain component (a stochastic process $\{v_t\}_{t\geq 0}$). The above system is the time-invariant version of the system (1.11) and (1.11) for a zero control input u_t.

We assume that the assumptions of Section 1.3.1 hold:

The estimation problem is formulated in the following way. Given an output which is to be regulated (entitled, in this chapter, controlled output):

$$z(x, \hat{x}) = h(x) - h(\hat{x}) \qquad (7.3)$$

where $h : R^n \to R^r$ is a continuous function, synthesize an estimator $\{\hat{x}_t\}$ such that for a given $\gamma > 0$, the following H_∞ criterion is satisfied.

$$\mathcal{E}\{\int_0^T \|z_t\|^2 dt\} \leq \gamma^2 \mathcal{E}\{\|x_0\|^2 + \int_0^T (\|v_t\|^2)dt\} \qquad (7.4)$$

for all $T \geq 0$ and for all disturbances v_t in $\mathcal{A}$ (provided $\mathcal{A}$ is nonempty). Whenever the system (7.1) satisfies the above inequality, it is said to have the L_2-gain property, and we also write L_2-gain$\leq \gamma$.

This problem may be treated within the context of stochastic game theory analogously to the utilization of the game theory in the deterministic H_∞ control and estimation (see, e.g., [5]). In this connection, there is also a recent

E. Gershon & U. Shaked: *Advanced Topics in Control & Estimation*, LNCIS 439, pp. 129–138.
DOI: 10.1007/978-1-4471-5070-1_7

work ([17]) that deals with the problem of stochastic minimax dynamic games using the information state concept, where the objective is a control synthesis (rather than estimation). The latter work characterizes the information state by means of a solution to a certain Hamilton-Jacobi equation.

We adopt here the approach that is based on the notion of stochastic dissipation that has been introduced in Section 1.3.1. Applying inequality (6.18) of the latter section to the time-invariant system (7.1), realizing that here:

$$V_t = 0, \quad g = 0, \quad D = I, \quad \bar{U}(x) = \gamma^2 I - \frac{1}{2}U(x) \text{ and } U(x) = g_2(x)^T V_{xx} g_2 x, \tag{7.5}$$

we obtain the following:

Lemma 7.1.1 *Consider the system (7.1) with the controlled output $z = h(x)$. Then, the satisfaction of the of the HJI (6.8) implies*

$$\begin{aligned} &V_x[f(x) + g_1(x)v] + \|z\|^2 - \gamma^2 \|v\|^2 + \tfrac{1}{2}Tr\{G^T(x)V_{xx}(x)G(x)\} \\ &+\tfrac{1}{2}v^T\{[g_2(x)]^T V_{xx}(x)g_2(x)\}v \\ &\leq -\left\| [\gamma^2 I - \tfrac{1}{2}U(x)]^{\frac{1}{2}} v - [\gamma^2 I - \tfrac{1}{2}U(x)]^{-\frac{1}{2}} g_1^T(x) V_x^T(x) \right\|^2 . \end{aligned}$$

7.2 Stochastic H_∞ Estimation

We consider now the system (7.1) , together with the observations (7.2) and the controlled output (7.3). Here we take the approach of prescribing a filter structure (see, e.g., [73] for the deterministic casc) as follows:

$$d\hat{x}_t = f(\hat{x}_t)dt + K(\hat{x}_t)(dy_t - h_2(\hat{x}_t)dt), \tag{7.6}$$

where $K(\hat{x}_t)$ is the estimator gain, an $n \times r$ matrix,

Using now y_t of the observations equation (7.2) in (7.6), we arrive at the following augmented system.

$$dx_t^e = f^e(x_t^e, K)dt + g_1^e(x_t^e, K)v_t dt + g_2^e(x_t)v_t dW_t^2 + G^e(x_t^e)d\bar{W}_t^1, \tag{7.7}$$

where

$$x_t^e = \begin{bmatrix} x_t \\ \hat{x}_t \end{bmatrix}, \bar{W}_t^1 = \begin{bmatrix} W_t^1 \\ W_t^3 \end{bmatrix}, \tag{7.8}$$

$$f^e(x_t^e) = \begin{bmatrix} f(x_t) \\ f(\hat{x}_t) + K(\hat{x}_t)(h_2(x_t) - h_2(\hat{x}_t)) \end{bmatrix} \tag{7.9}$$

$$\begin{aligned} g_1^e(x_t^e) &= \begin{bmatrix} g_1(x_t) \\ K(\hat{x}_t)g_3(x_t) \end{bmatrix} \\ g_2^e(x_t) &= \begin{bmatrix} g_2(x_t) \\ 0 \end{bmatrix} \end{aligned} \tag{7.10}$$

$$G^e(x_t^e) = \begin{bmatrix} G(x_t) & 0 \\ 0 & K(\hat{x}_t)G_2(x_t) \end{bmatrix}. \tag{7.11}$$

We now have the following theorem.

Theorem 7.1. *Consider the stochastic system (7.1) together with the augmented system (7.7) and the controlled output (7.2). Assume there are: a positive function* $\tilde{W} : R^{2n} \to R^+$ *and a matrix* $K(\hat{x})$, *which satisfy the following HJI for some* $\gamma > 0$, *and that* $\gamma^2 I - \frac{1}{2}[g_2^e(x^e)]^T \tilde{W}_{x^e x^e}(x^e) g_2^e(x^e) \geq \alpha I$ *for some positive number* α, *and for all* $x^e \in R^{2n}$.

$$\begin{aligned} &\tilde{W}_{x^e}(x^e) f^e(x^e) + \tfrac{1}{4}\tilde{W}_{x^e}(x^e) g_1^e(x^e)[\gamma^2 I - \tfrac{1}{2}\bar{U}(x^e)]^{-1} g_1^e(x^e)^T \tilde{W}_{x^e}^T(x^e) \\ &+\tfrac{1}{2}Tr\left\{(G^e)^T \tilde{W}_{x^e x^e}(x^e) G^e(x^e)\right\} + h^e(x^e)^T h^e(x^e) \leq 0, \end{aligned} \tag{7.12}$$

where

$$\bar{U}(x^e) = [g_2^e(x^e)]^T \tilde{W}_{x^e x^e}(x^e) g_2^e(x^e), \; h^e(x^e) = x - \hat{x}. \tag{7.13}$$

Then, the augmented system (7.7) is dissipative with respect to the supply rate $\gamma^2||v||^2 - ||z||^2$, *with* $\tilde{W}(x^e)$ *as a storage function. Therefore, the augmented system is* L_2*-gain* $\leq \gamma$. *(i.e. the inequality (7.4) is satisfied).*

Proof: Application of Lemma 7.1.1 and (7.12) yields

$$L\left\{\tilde{W}(x^e)\right\} + ||h^e(x^e)||^2 - \gamma^2||v||^2 \leq 0,$$

where L is the infinitesimal generator of the augmented system with respect to $\tilde{W}(x^e)$, that is

$$L\left\{\tilde{W}(x^e)\right\} = \tilde{W}_{x^e}[f^e(x^e) + g_1^e(x^e)v] + \tfrac{1}{2}Tr\left\{(G(x^e))^T \tilde{W}_{x^e x^e} G^e(x^e)\right\}.$$

This implies that $\tilde{W}(x^e)$ is a storage function for the augmented system with the supply rate $||z||^2 - \gamma^2||v||^2$, which implies that the augmented system is L_2-gain$\leq \gamma$.

Remark 7.1. As in the deterministic case, it is difficult to establish, in general, conditions under which there exists a matrix $K(\hat{x}_t)$ so that the HJI (7.12) is satisfied. The part of the latter inequality that contains K is given by:

$$\begin{aligned} &\Gamma(K) = \tilde{W}_{\hat{x}}(x^e)K(\hat{x}_t)(h_2(x_t) - h_2(\hat{x}_t)) + \tfrac{1}{4}(\tilde{W}_{\hat{x}}(x^e)K(\hat{x}_t)g_3(x_t) + \tilde{W}_x(x^e)g_1(x_t)) \\ &(\gamma^2 I - 1\tfrac{1}{2}\bar{U}(x^e))^{-1}(\tilde{W}_{\hat{x}}(x^e)K(\hat{x}_t)g_3(x_t) + \tilde{W}_x(x^e)g_1(x_t))^T \\ &= \tfrac{1}{4\gamma^2}[\tilde{W}_{\hat{x}}(x^e)K(\hat{x}_t) - \Psi(x^e)]R_a(x_t)[K^T(\hat{x}_t)\tilde{W}_{\hat{x}}^T(x^e) - \Psi^T(x^e)] \\ &- \tfrac{1}{4\gamma^2}[\Psi(x^e)R_a(x_t)\Psi^T(x^e) - \tilde{W}_x(x^e)g_1(x_t)(I - \tfrac{1}{2\gamma^2}\bar{U}(x^e))^{-1}g_1^T(x_t)\tilde{W}_x^T(x^e)], \end{aligned} \tag{7.14}$$

where

$$\Psi(x^e) = -[2\gamma^2(h_2(x_t) - h_2(\hat{x}_t))^T + \tilde{W}_x(x^e)g_1(x_t)$$
$$(I - \frac{1}{2\gamma^2}g_2^T(x_t)\tilde{W}_{xx}(x^e)g_2(x_t))^{-1}g_3^T(x_t)]R_a^{-1}(x_t),$$

and

$$R_a(x_t) = g_3(x_t)(I - \frac{1}{2\gamma^2}g_2^T(x_t)\tilde{W}_{xx}(x^e)g_2(x_t))^{-1}g_3^T(x_t).$$

The gain matrix $K(\hat{x}_t)$ that minimizes $\Gamma(K)$, and thus leads to a minimum left hand side in (7.12), is clearly the one that satisfies $\tilde{W}_{\hat{x}}(x^e)K(\hat{x}_t) = \Psi(x^e)$. Unfortunately, this last equation may not possess a solution for K which depends only on $\hat{x}_t$. One way to circumvent this difficulty is to choose $K(\hat{x}_t)$ s.t.

$$\tilde{W}_{\hat{x}}(x^e)K(\hat{x}_t) = \Psi(x^e) + \Phi(x^e), \tag{7.15}$$

where $\Phi(x^e)$ is a function that allows a solution K^* for (7.15) that is independent of x_t. For this choice of $K^*(\hat{x}_t)$ the above $\Gamma(K)$ becomes the following.

$$\Gamma(K^*) = \frac{1}{4\gamma^2}[\Phi(x^e)R_a(x_t)\Phi^T(x^e) - \Psi(x^e)R_a(x_t)\Psi^T(x^e)$$
$$+\tilde{W}_x(x^e)g_1(x_t)(I - \frac{1}{2\gamma^2}\tilde{U}(x^e))^{-1}g_1^T(x_t)\tilde{W}_x^T(x^e)]. \tag{7.16}$$

Assuming an existence of a solution $K(\hat{x}_t)$ to (7.15) for some function $\Phi(x^e)$, we have established the following theorem.

Theorem 7.2. *Consider the augmented system (7.7) and the objective vector (7.2). Assume that there are: a positive function $\tilde{W} : R^{2n} \to R^+$ in C^2 and a matrix $K(\hat{x})$, which satisfy (7.15). In addition, let $\tilde{W}$ satisfy the following HJI.*

$$\tilde{W}_t(x^e) + \tilde{W}_x(x^e)[f(x_t) + g_1(x_t)v_t^*(x_t)] + \tilde{W}_{\hat{x}}(x^e)[f(\hat{x}_t) + g_1(\hat{x}_t)v_t^*(\hat{x}_t)]$$
$$+\frac{1}{2}Tr\left\{G(x)^T\tilde{W}_{xx}(x^e)G(x)\right\} + \frac{1}{2}v^{*T}(x_t)g_2^T(x_t)\tilde{W}_{xx}(x^e)g_2(x_t)v^*(x_t)$$
$$+h^e(x^e)^Th^e(x^e) + \frac{1}{4\gamma^2}[\Phi(x^e)R_a(x_t)\Phi^T(x^e) - \Psi(x^e)R_a(x_t)\Psi^T(x^e)$$
$$+\tilde{W}_x(x^e)g_1(x_t)(I - \frac{1}{2\gamma^2}g_2^T(x_t)\tilde{W}_{xx}(x^e)g_2(x_t))^{-1}g_1^T(x_t)\tilde{W}_x^T(x^e)] \le 0 \; \forall x^e \in R^{2n}. \tag{7.17}$$

Then, the augmented system is dissipative with respect to the supply rate $\gamma^2||v||^2 - ||z||^2$, with the storage function $\tilde{W}(x^e)$, and therefore it has an L_2-gain $\le \gamma$.

7.2.1 Stability

Various types of asymptotic stability may be considered now. In this subsection we consider global asymptotic stability in probability and mean square sense. A comprehensive account of stochastic systems' stability may be found in [67]. We recall first some sufficient conditions for global asymptotic stability of the stochastic system given by

$$dx_t = f(x_t)dt + G(x_t)dW_t \tag{7.18}$$

with $f(0) = G(0) = 0$, and assume that f, G satisfy conditions that guarantee a unique strong solution relative to the filtered probability space $(\Omega, F, \{F_t\}_{t\geq 0}, P)$.

Sufficient conditions for a global stability in probability and in the mean square sense are summarized in the following two theorems.

Theorem 7.3. *([67])Assume there exists a positive function $V(x) \in C^2$, with $V(0) = 0$, so that $(LV)(x) < 0$ for all $x \in R^n$. Assume also that $V(x) \to \infty$ as $\|x\| \to \infty$. Then, the system of (7.18) is globally asymptotically stable in probability.*

Theorem 7.4. *([67])Assume there exists a positive function $V(x) \in C^2$, with $V(0) = 0$. Then the system of (7.18) is globally exponentially stable if there are positive numbers k_1, k_2, k_3 such that the following hold.*

$$k_1||x||^2 \leq V(x) \leq k_2||x||^2$$
$$(LV)(x) \leq -k_3||x||^2.$$

As a consequence of the last two theorems we have the following results.

Lemma 7.2.1 *Assume that there is a positive function $\tilde{W} \in C^2$, $\tilde{W} : R^{2n} \to R^+$, satisfying the HJI (7.12) with a strict inequality, so that $\tilde{W}(x^e) \to \infty$ as $||x^e|| \to \infty$. Then, the closed loop system is internally globally asymptotically stable in probability.*

Lemma 7.2.2 *Assume that for some k_3, $h_2^T(x)h_2(x) \geq k_3||x||^2$ for all $x \in R^n$. Assume also that there is a positive function $\tilde{W} \in C^2$. and $\tilde{W} : R^{2n} \to R^+$ with*

$$k_4||\hat{x}||^2 \leq \tilde{W}(x^e) \leq k_5||x^e||^2 \quad x^e \in R^{2n} \tag{7.19}$$

which satisfies the following algebraic HJI for some $\gamma > 0$:

$$\begin{aligned}&\tilde{W}_{x^e}(x^e)f^e(x^e) + \tfrac{1}{2}Tr\left\{(G^e)^T\tilde{W}_{x^ex^e}(x^e)G^e(x^e)\right\}\\ &+(h^e(x^e)^T h^e(x^e) \leq -Q(x^e) \quad \forall x^e \in R^{2n}\end{aligned} \tag{7.20}$$

for some positive function $Q(x^e)$ with the property that $(h^e(x^e,t)^T h^e(x^e,t) + Q(x^e) \geq k_6||\hat{x}||^2$ for all $x^e \in R^{2n}$, and for some $k_6 > 0$. Then the augmented system (7.7) with $v = 0$ is exponentially stable in the mean square sense, and has the property of $L_2 - gain \leq \gamma$, that is

$$\mathcal{E}\{\int_0^\infty \|z_t\|^2 dt\} \leq \gamma^2 \mathcal{E}\{\|x_0\|^2 + \int_0^\infty (\|v_t\|^2)dt\}$$

for all non-anticipative stochastic processes v that satisfy $\mathcal{E}\{\int_0^\infty (\|v_t\|^2)dt\} < \infty$, and whenever x_0 satisfies $\mathcal{E}\{V(x_0)\} \leq \gamma^2\mathcal{E}\{||x_0||^2\}$.

7.3 Norm-Bounded Uncertainty

The above results were based on the assumption that the system's parameters are completely known. In the present section we consider the case where these parameters are uncertain. We consider the following system:

$$\begin{aligned} dx_t &= (f(x_t) + \Delta f)dt + (g_1(x_t) + \Delta g)v_t dt + g_2(x_t)v_t dW_t^2 + G(x_t)dW_t^1 \\ z(t) &= h(x_t), \end{aligned} \tag{7.21}$$

where $\{x_t\}_{t\geq 0} \in R^n$ is a solution to (7.21) with the initial condition x_0, $\{v_t\}_{t\geq 0} \in R^{m_1}$ is an exogenous disturbance and $\{W_t^1\}_{t\geq 0} \in R^\ell$, $\{W_t^2\}_{t\geq 0} \in R^1$ are Wiener processes. Also, $z_t \in R^r$ is the output vector to be regulated. The matrices Δf and Δg are assumed to be continuous functions of x_t and t that possess the following structure:

$$\left[\Delta f(x)\ \Delta g(x)\right] = H_1(x)F(x)\left[E_1(x)\ E_2(x)\right] \tag{7.22}$$

where $H_1 : R^n \times \to R^{n\times n_1}$, $E_1 : R^n \to R^{n_2}$, $E_2 : R^n \to R^{n_2\times m_1}$ and $F : R^n \to R^{n_1\times n_2}$ are continuous functions and where

$$F(x)F^T(x) \leq I_{n_1}, \ \forall x \in R^n \ . \tag{7.23}$$

Applying the BRL condition of (7.1.1 by replacing f and g_1 by $f + \Delta f$ and $g_1 + \Delta g$, respectively, the corresponding HJI becomes:

$$\begin{aligned} &V_x(x)(f(x) + H_1(x)F(x)E_1(x) + \tfrac{1}{4}V_x(x)(g_1(x) + H_1(x)F(x)E_2(x))[\gamma^2 I \\ &-\tfrac{1}{2}U(x)]^{-1}(g_1^T(x) + E_2^T(x)F^T(x)H_1^T(x))V_x(x) \\ &+\tfrac{1}{2}Tr\{G^T(x)V_{xx}(x)G(x,t)\} + (h^T(x) + E_1^T(x)F^T(x)H_2^t(x))h(x) \leq 0 \\ &\forall x \in R^n, \qquad \forall t \geq 0 \end{aligned}$$

where $U(x)$ is defined in (7.5). Applying Schur complement to the latter, and denoting:

$$\bar{\Gamma}(x) = \begin{bmatrix} V_t(x) + V_x(x)f(x) + \frac{1}{2}Tr\{G^T(x)V_{xx}(x)G(x)\} & V_x(x)g_1(x) & h^T(x) \\ * & -[4\gamma^2 I - 2U(x)] & 0 \\ * & * & -I_r \end{bmatrix} \tag{7.24}$$

the following result is obtained.

Theorem 7.5. *Consider the system described by (7.21)-(7.23)) and the supply rate $S(v,z) = \gamma^2\|v\|^2 - \|z\|^2$. Assume that the following HJI is satisfied, for all $x \in R^n$, for some positive scalar ε:*

$$\begin{bmatrix} \bar{\Gamma}(x) & \begin{bmatrix} V_x(x)H_1(x) \\ 0 \\ 0 \end{bmatrix} & \varepsilon\begin{bmatrix} \frac{1}{2}E_1^T(x) \\ E_2^T(x) \\ 0 \end{bmatrix} & \varepsilon\begin{bmatrix} E_1^T(x) \\ 0 \\ 0 \end{bmatrix} \\ * & -\varepsilon I_{n_1} & 0 & 0 \\ * & * & -\varepsilon I_{n_2} & 0 \\ * & * & * & -\varepsilon I_{n_2} \end{bmatrix} \leq 0, \tag{7.25}$$

where $\bar{\Gamma}(x)$ is defined in (7.24). Then, the system (7.21) is dissipative with respect to the supply rate $S(v,z)$ (provided $\mathcal{A}$ is nonempty).

Considering next the case where the measurement is given by:

$$dy_t = h_2(x_t,t)dt + g_3(x_t,t)v_t dt + G_2(x_t,t)dW^3. \tag{7.26}$$

Once the BRL is obtained for systems with norm-bounded uncertainties the estimation problem can be solved as follows. Consider the following system

$$\begin{aligned} dx_t &= (f(x_t,t) + \Delta f)dt + g_1(x_t,t)v_t dt + g_2(x_t,t)v_t dW_t^2 + G(x_t,t)dW_t^1 \\ dy_t &= (h_2(x_t,t) + \Delta h_2)dt + g_3(x_t,t)v_t dt + G_2(x_t,t)dW^3 \end{aligned} \tag{7.27}$$

$$z(t) = h(x_t,t) - h(\hat{x}_t,t) \tag{7.28}$$

where $\{x_t\}_{t\geq 0} \in R^n$ is a solution to (7.21) with the initial condition x_0, $\{v_t\}_{t\geq 0} \in R^{m_1}$, $\{y_t\}_{t\geq 0} \in R^p$ is the measured output and $\{W_t^1\}_{t\geq 0} \in R^{\ell}$, $\{W_t^2\}_{t\geq 0} \in R^1$ and $\{W_t^3\}_{t\geq 0} \in R_3^m$ are Wiener processes. Also, $z_t \in R^r$ is the objective vector. The matrices Δf and Δh_2 are assumed to be continuous functions of x_t and t that possess the following structure:

$$\begin{bmatrix} \Delta f(x,t) \\ \Delta h_2(x,t) \end{bmatrix} = \begin{bmatrix} H_1(x,t) \\ H_3(x,t) \end{bmatrix} F(x,t)E_1(x,t), \tag{7.29}$$

where $H_1(x,t) : R^n \times [0,\infty) \to R^{n\times n_1}$, $H_3(t) : R^n \times [0,\infty) \to R^{p\times n_y}$, $E_1(x,t) : R^n \times [0,\infty) \to R^{n_2}$, and $F(x,t) : R^n \times [0,\infty) \to R^{n_1\times n_2}$ are continuous functions and where $F(x,t)$ satisfies (7.23).

Denoting:

$$\Delta f^e = \left\{ \begin{bmatrix} H_1(x_t,t) \\ 0 \end{bmatrix} + \begin{bmatrix} 0 \\ K(\hat{x}_t,t)H_3(t) \end{bmatrix} \right\} F(x_t,t)E_1(x_t,t)$$

the BRL of Theorem 7.5 is applied to the latter definition and to (7.7). Denoting:

$$\Gamma^e = \begin{bmatrix} \tilde{W}_t(x^e,t) + \tilde{W}_{x^e}(x^e,t)f^e(x^e,t) \\ +\frac{1}{2}Tr\{(G^e(x^e,t))^T \tilde{W}_{x^e x^e}(x^e,t)G^e(x^e,t)\} & * & * \\ g_1^{eT}(x^e,t)\tilde{W}_{x^e}^T(x^e,t) & -[4\gamma^2 I - 2\bar{U}(x^e,t)] & * \\ h^e(x^e,t) & 0 & -I \end{bmatrix}, \tag{7.30}$$

where $U(x^e,t)$ and $h^e(x^e,t)$ are defined in (7.13a,b) the following inequality is obtained, $\forall x^e \in R^{2n}$ and $\forall t \geq 0$,

$$\begin{bmatrix} \Gamma^e & * & * \\ [\tilde{H}_1^T(x,t)W_x^T(x^e,t) + H_3^T(\hat{x},t)K^T(\hat{x}_t,t)\tilde{W}_{\hat{x}}(x^e,t)^T \; 0 \; 0] & -\bar{\varepsilon}I_{n_1} & * \\ \bar{\varepsilon}\left[\frac{1}{2}E_1(x,t) \; 0 \; 0\right] & 0 & -\bar{\varepsilon}I_{n_1} \end{bmatrix} \leq 0 \tag{7.31}$$

where $\bar{\varepsilon}$ is a tuning scalar. The solution to the estimation problem with the norm-bounded uncertainty described in (7.27)–(7.29), is described as follows.

Theorem 7.6. *Consider the stochastic system (7.27)–(7.29) together with the observer system (7.6). Assume that there exist: a positive function $\tilde{W}: R^{2n} \times [0,\infty) \to R^+$ a matrix $K(\hat{x},t)$ and a scalar $\bar{\varepsilon}$ that satisfy (7.31) for some $\gamma \ge 0$, and that $\gamma^2 I - \frac{1}{2}[g_2^e(x^e,t)]^T \tilde{W}_{x^e x^e}(x^e,t) g_2^e(x^e,t) \ge \alpha I$ for some positive number α, and for all $x^e \in R^{2n}$. Then, the system that relates the input v_t to the estimation error z_t is dissipative with respect to the supply rate $\gamma^2||v||^2 - ||z||^2$, it possesses a storage function defined as $S(x^e,t) = \tilde{W}(x^e,t)$, and has an L_2-gain $\le \gamma$.*

Remark 7.2. Similar to the arguments of Remark 7.1 and Theorem 7.2, in the case where $G_2 \equiv 0$, the last theorem can be used to obtain a minimizing K that depends only on $\hat{x}$ and t by converting the matrix inequality (7.31) into a scalar HJI using Schur complement. A completion to squares with respect to $\tilde{W}_{\hat{x}}(x^e,t)K(\hat{x}_t,t)$ can be obtained. Using then (7.15) a result similar to the one obtained in Theorem 7.2 can be achieved. The case where G_2 is not zero can be solved applying the method of Section 7.2

7.3.1 Example

Consider the following 2nd order system:

$$\begin{aligned} &dx{=}Axdt{+}H_1F(x)Exdt{+}Bwdt{+}GxdW^1, \quad x(0){=}0 \\ &dy{=}(C_2{+}H_2F(x)E)xdt{+}\rho sin2(x_2{-}x_1)dt{+}kwdt{+}G_2xdW^2 \end{aligned} \tag{7.32}$$

where $x{=}col\{x_1,\ x_2\}$, x_1, x_2 and y are scalar functions of time, $F(x)$ is a matrix that satisfies (7.23) and W^1 and W^2 are scalar Wiener processes. We consider the following estimator:

$$d\hat{x} = A\hat{x}dt + K(\hat{x})[dy - C_2\hat{x}dt], \quad \hat{x}(0) = 0 \tag{7.33}$$

and the objective vector

$$z = C_1(x - \hat{x}).$$

We define

$$x^e = col\{x,\ \hat{x}\}, \quad \bar{W} = col\{W^1,\ W^2\}.$$

and obtain:

$$\begin{aligned} dx^e = &\begin{bmatrix} (A{+}H_1FE)xdt \\ A\hat{x}dt{+}K(\hat{x})(C_2e{+}H_2FEx{+}\rho\phi)dt \end{bmatrix} + \begin{bmatrix} B \\ K(\hat{x})k \end{bmatrix} wdt \\ &{+}diag\{Gx, K(\hat{x})G_2x\}d\bar{W}, \end{aligned}$$

where

$$\phi = sin2(x_2 - x_1) - sin(2(\hat{x}_2 - \hat{x}_1)$$
$$2cos(x_2 + \hat{x}_2 - x_1 - \hat{x}_1)sinc(x_2 - \hat{x}_2 - x_1 + \hat{x}_1)(x_2 - \hat{x}_2 - x_1 + \hat{x}_1)$$
$$= 2U(x)(x_2 - \hat{x}_2 - x_1 + \hat{x}_1) = 2U(x)\begin{bmatrix} -1 & 1 \end{bmatrix} e$$

and where $|U(x)| \leq 1 \; \forall x \in \mathcal{R}^2$.

We choose

$$\tilde{W}(x^e) = e^T Pe + x^T Qx = (x^e)^T \begin{bmatrix} P+Q & -P \\ -P & P \end{bmatrix} x^e. \tag{7.34}$$

And consider an estimator gain K that is independent of the estimate $\hat{x}$. The corresponding HJI becomes:

$$\begin{aligned} &e^T[P(A-KC_2-2K\rho U[-1\;1])+(A^T-C_2^TK^T-2\rho[-1\;1]^TUK^T)P \\ &+\gamma^{-2}P(B-Kk)(B^T-kK^T)P+C_1^TC_1]e \\ &+x^T[QA+A^TQ+QH_1FE+E^TF^TH_1^TQ+\gamma^{-2}QBB^TQ \\ &+(G^T-G_2^TK^T)P(G-KG_2)+G^TQG]x \\ &+2e^T[P(H_1-KH_2)FE+\gamma^{-2}P(B-Kk)B^TQ]x \leq 0. \end{aligned} \tag{7.35}$$

We consider, next the products

$$2\begin{bmatrix} e^T & x^T \end{bmatrix} \begin{bmatrix} P(H_1-KH_2) \\ QH_1 \end{bmatrix} F \begin{bmatrix} 0 & E \end{bmatrix} \begin{bmatrix} e \\ x \end{bmatrix}$$

and $-4e^TPK\rho U\begin{bmatrix} -1 & 1 \end{bmatrix} e$. Applying the fact that $2\alpha^T\beta \leq \epsilon\alpha^T\alpha + \epsilon^{-1}\beta^T\beta$ for any matrices α and β of the appropriate dimensions and for any positive scalar ϵ we obtain from (7.35) the following requirement that will guarantee the estimation performance.

$$\begin{aligned} &e^T[P(A-KC_2)+(A^T-C_2^TK^T)P+\epsilon_1^{-1}PKK^TP \\ &+4\epsilon_1\rho^2\begin{bmatrix} 1 & -1 \\ -1 & 1 \end{bmatrix}+\gamma^{-2}P(B-Kk)(B^T-kK^T)P \\ &+\epsilon_2^{-1}P(H_1-KH_2)(H_1^T-H_2^TK^T)+C_1^TC_1]e \\ &+x^T[QA+A^TQ+\epsilon_2^{-1}QH_1H_1^TQ+\epsilon_2E^TE+\gamma^{-2}QBB^TQ \\ &+(G^T-G_2^TK^T)P(G-KG_2)+G^TQG]x \\ &+2e^T[\epsilon_2^{-1}P(H_1-KH_2)H_1^TQ+\gamma^{-2}P(B-Kk)B^TQ]x \leq 0. \end{aligned} \tag{7.36}$$

Denoting

$$Y \triangleq PK$$

the latter inequality can be written as:

$$
\begin{aligned}
&\begin{bmatrix} e^T x^T \end{bmatrix} \Bigg[\begin{bmatrix} \bar{a}+\epsilon_1^{-1} Y Y^T & 0 \\ 0 & \bar{b}+(G^T P - G_2^T Y^T) P^{-1} (P G - Y G_2) \end{bmatrix} \\
&+\gamma^{-2} \begin{bmatrix} PB - Yk \\ QB \end{bmatrix} \begin{bmatrix} B^T P - k Y^T & B^T Q \end{bmatrix} + \epsilon_2^{-1} \begin{bmatrix} PH_1 - Y H_2 \\ QH_1 \end{bmatrix} \\
&\begin{bmatrix} H_1^T P - H_2^T Y^T & H_1^T Q \end{bmatrix} \Bigg] \begin{bmatrix} e \\ x \end{bmatrix} \le 0,
\end{aligned}
$$

where:

$$
\begin{aligned}
\bar{a} &= PA + A^T P - Y C_2 - C_2^T Y^T + 4\epsilon_1 \rho^2 \begin{bmatrix} 1 & -1 \\ -1 & 1 \end{bmatrix} + C_1^T C_1 \\
\bar{b} &= QA + A^T Q + G^T Q G + \epsilon_2 E^T E.
\end{aligned}
\tag{7.37}
$$

For prescribed positive tuning parameters ϵ_1 and ϵ_2, the last inequality can be written as the following LMI, in the decision variables P, Q and Y, by applying the Schur complement formula.

$$
\begin{bmatrix}
\begin{bmatrix} \bar{a} & 0 \\ 0 & \bar{b} \end{bmatrix} & \begin{bmatrix} PB - Yk \\ QB \end{bmatrix} & \begin{bmatrix} PH_1 - YH_2 \\ QH_1 \end{bmatrix} & \begin{bmatrix} Y \\ 0 \end{bmatrix} & \begin{bmatrix} 0 \\ G^T P - G_2^T Y^T \end{bmatrix} \\
* & -\gamma^2 & 0 & 0 & 0 \\
* & * & -\epsilon_2 I & 0 & 0 \\
* & * & * & -\epsilon_1 & 0 \\
* & * & * & * & -P
\end{bmatrix} \le 0.
\tag{7.38}
$$

If for a prescribed estimation performance parameter γ there exist $0 < P$ and $0 < Q$ that for some value of the tuning parameters ϵ_1 and ϵ_2 satisfy (7.38) then the estimator of (7.33) will achieve the required estimation performance where $K(\hat{x}) = P^{-1} Y$.

Remark 7.3. In (7.32) the vector $F(x)Ex$ could represent an uncertain, norm-bounded, nonlinear part of the system's dynamics or a known nonlinear part that resides in a cone. An example for the latter is $F(x) = sin(x_1)$.

7.4 Conclusions

An H_∞ estimation procedure has been developed for achieving a prescribed attenuation level of disturbance rejection, that acts on a nonlinear system with multiplicative noise. A sufficient condition is obtained for the above that is expressed by a nonlinear matrix inequality. In the simplified nonlinear example that is treated in the example, a linear matrix inequality is obtained, the solution of which guarantees the required attenuation level.

8

Non Linear Systems – Measurement Output-Feedback Control

8.1 Introduction and Problem Formulation

We consider the system that is described by (1.11) and (1.12). We also consider the following controlled output:

$$z_t = \begin{bmatrix} h(x_t, t) \\ u_t \end{bmatrix}, \quad t \in [0, \infty), \tag{8.1}$$

where $h : R^n \times [0, \infty) \to R^r$ is a Borel measurable function, and let $\mathcal{Y}_t = \{y_s : s \leq t\}$. In Chapters 6 and 7 we have considered the corresponding problems of H_∞state-feedback and H_∞state estimation and obtained the sufficient conditions that guarantee the dissipation of the corresponding treated systems and assure a given performance level. In the present chapter we apply the results that have been obtained in the last two sections to the problem of H_∞output-feedback control of the above mentioned stochastic nonlinear system.

Let $\alpha(\cdot, \cdot)$ be positive Borel function on $R^n \times [0, \infty)$ (where $R^n \times [0, \infty)$ is endowed with the Borel σ-algebra). In what follows it will be assumed that $\mathcal{E}\{\alpha(x, t)\} < \infty$ for all $t \in [0, \infty)$ and for all F-measurable, R^n-valued random variables which satisfy $\mathcal{E}\{||x||^2\} < \infty$. The control objective is to find an output-feedback controller $u_t = u(\mathcal{Y}_t, t)$ such that, for a given $\gamma > 0$, the following H_∞ criterion is satisfied.

$$\mathcal{E}\{\int_{t_1}^{t_2} ||z_t||^2 dt\} \leq \gamma^2 \mathcal{E}\{\alpha(x_{t_1}, t_1) + \int_{t_1}^{t_2} \gamma^2 ||v_t||^2 dt\} \tag{8.2}$$

for all $0 \leq t_1 < t_2$, for all F_0-measurable x_0 with $\mathcal{E}\{||x_0||^2\} < \infty$, and for all disturbances v_t in $\mathcal{A}_u$ (provided $\mathcal{A}_u$ is nonempty). Whenever the system (1.11) satisfies the above inequality, it is said to possess an L_2-gain that is less than or equal to γ. Note that for the infinite time-horizon $t_2 = \infty$, and it is required that v satisfies: $\int_0^\infty ||v_t||^2 dt < \infty$.

E. Gershon & U. Shaked: *Advanced Topics in Control & Estimation*, LNCIS 439, pp. 139–154.
DOI: 10.1007/978-1-4471-5070-1_8

8.2 Stochastic H_∞ OF Control

We consider now the system (1.11) together with the observations (1.12) and the controlled output (8.1). As the state x_t of the plant is not available, we follow the common practice (the certainty equivalence approach) of replacing the state that is to be processed by the controller with an estimator output $\hat{x}_t$. A natural choice of an estimator (see, e.g. [73] for the deterministic case) is:

$$d\hat{x}_t = f(\hat{x}_t,t)dt + g(\hat{x}_t,t)u_t^*(\hat{x}_t)dt + g_1(\hat{x}_t,t)v_t^*(\hat{x}_t) + K(\hat{x}_t,t)(dy_t - h_2(\hat{x}_t,t)dt \\ -g_3(\hat{x}_t,t)v_t^*(\hat{x}_t)dt), \tag{8.3}$$

where $K(\hat{x}_t,t)$ is the estimator gain, an $n \times r$ matrix,

$$u_t^*(\hat{x}) = -\frac{1}{2}[I + \frac{1}{2}\bar{g}^T(\hat{x},t)V_{xx}(\hat{x},t)\bar{g}(\hat{x},t)]^{-1}g^T(\hat{x},t)V_x^T(\hat{x},t)$$

and

$$v_t^*(\hat{x}_t) = \frac{1}{2}[\gamma^2 I - \frac{1}{2}U(\hat{x}_t,t)]^{-1}g_1^T(\hat{x}_t,t)V_x^T(\hat{x}_t,t).$$

Using y_t of (1.12) in (8.3), we arrive at the following augmented system.

$$dx_t^e = f^e(x_t^e,K,t)dt + g_1^e(x_t^e,K,t)[v_t - v_t^*(x_t)]dt + g_2^e(x_t,t)[v_t - v_t^*(x_t)]dW_t^2 \\ +G^e(x_t^e,t)d\bar{W}_t^1, \tag{8.4}$$

where $K = K(\hat{x},t)$, while

$$x_t^e = col\{x_t,\ \hat{x}_t\},\ \bar{W}_t^1 = col\{W_t,\ W_t^1,\ W_t^2,\ W_t^3\}, \tag{8.5}$$

$$f^e(x_t^e,K,t) \\ = \begin{bmatrix} f(x_t,t) + g_1(x_t,t)v_t^*(x_t) + g(x_t,t)u_t^*(\hat{x}_t) \\ f(\hat{x}_t,t) + g(\hat{x}_t,t)u_t^*(\hat{x}_t) + g_1(\hat{x}_t,t)v_t^*(\hat{x}_t) + K(\hat{x}_t,t)(\tilde{h}_2(x_t,t) - \tilde{h}_2(\hat{x}_t,t)) \end{bmatrix}, \tag{8.6}$$

$$g_1^e(x_t^e,K,t) = col\{g_1(x_t,t),\ K(\hat{x}_t,t)g_3(x_t,t)\},\ g_2^e(x_t,t) = col\{g_2(x_t,t),\ 0\}, \tag{8.7}$$

$$G^e(x_t^e,v_t) = \begin{bmatrix} \bar{g}(x,t)u^*(\hat{x}) & G(x_t,t) & g_2(x_t,t)v_t^*(x_t) & 0 \\ 0 & 0 & 0 & K(\hat{x}_t,t)G_2(x_t,t) \end{bmatrix}, \tag{8.8}$$

and

$$\tilde{h}_2(x,t) = h_2(x,t) - g_3(x,t)v_t^*(x)\ \tilde{h}_2(\hat{x},t) = h_2(\hat{x},t) - g_3(\hat{x},t)v_t^*(\hat{x}). \tag{8.9}$$

Note that occasionally (whenever found convenient) we will use the abbreviation $f^e(x_t^e,t)$ for $f^e(x_t^e,K,t)$. We now have the following theorem.

Theorem 8.1. *Consider the stochastic system (1.11) together with the augmented system (1.12) and the controlled output (8.1). Assume there is a positive function* $V : R^n \times [0,\infty) \to R^+$*, with* $V \in C^{2,1}$ *so that it satisfies the HJI (6.8) of Theorem 6.2.1. Assume also that there are: a positive function* $\tilde{W} : R^{2n} \times [0,\infty) \to R^+$ *and a matrix* $K(\hat{x},t)$ *which satisfy the following HJI for some* $\gamma > 0$.

$$\begin{aligned}&\tilde{W}_t(x^e,t) + \tilde{W}_{x^e}(x^e,t)f^e(x^e,t) + \tfrac{1}{4}\tilde{W}_{x^e}(x^e,t)g_1^e(x^e,t)[\gamma^2 I - \tfrac{1}{2}\bar{U}(x^e,t)]^{-1}\\&g_1^e(x^e,t)^T\tilde{W}_{x^e}^T(x^e,t) + \tfrac{1}{2}Tr\left\{(G^e)^T\tilde{W}_{x^ex^e}(x^e,t)G^e(x^e,t)\right\}\\&+h^e(x^e,t)^T h^e(x^e,t) \le 0 \quad \forall x^e \in R^{2n}\end{aligned} \tag{8.10}$$

where

$$\bar{U}(x^e,t) = [g_2^e(x^e,t)]^T\tilde{W}_{x^ex^e}(x^e,t)g_2^e(x^e,t),\ h^e(x^e,t) = u^*(x) - u^*(\hat{x}) \tag{8.11}$$

such that

$$\gamma^2 I - \frac{1}{2}[g_2^e(x^e,t)]^T\tilde{W}_{x^ex^e}(x^e,t)g_2^e(x^e,t) \ge \alpha I$$

for some positive number α*, and for all* $x^e \in R^{2n}$*. Then, the closed-loop system with the control*

$$u_t^*(\hat{x}) = -\frac{1}{2}[I + \frac{1}{2}\bar{g}^T(\hat{x},t)V_{xx}(\hat{x},t)\bar{g}(\hat{x},t)]^{-1}g^T(\hat{x},t)V_{xx}(\hat{x},t)$$

is dissipative with respect to the supply rate $\gamma^2||v||^2 - ||z||^2$*, it possesses a storage function*

$$S(x^e,t) = V(x,t) + \tilde{W}(x^e,t),$$

and it has an L_2*-gain* $\le \gamma$.

Proof: Application of Theorem 6.2.1 yields

$$\begin{aligned}&V_t(x,t)+V_x[f(x,t)) + g_1(x,t)v+g(x,t)u^*(\hat{x})]+\|z\|^2-\gamma^2\|v\|^2\\&+\tfrac{1}{2}Tr\{G^T(x,t)V_{xx}(x,t)G(x,t)\}\\&+\tfrac{1}{2}u^{*T}(\hat{x})\bar{g}^T(x,t)V_{xx}(x,t)\bar{g}(x,t)u^*(\hat{x})\}\\&\le \left\|\tfrac{1}{2}[I + \tfrac{1}{2}\bar{g}^T(x,t)V_{xx}(x,t)\bar{g}(x,t)]^{-1}g^T(x,t)V_x^T(x,t)+ u^*(\hat{x})\right\|^2\\&-\left\|v - \tfrac{1}{2\gamma^2}g_1^T(x,t)V_x^T(x,t)\right\|^2\\&= ||u^*(x) - u^*(\hat{x})||^2 - \gamma^2||v - v^*(x)||^2 = ||h^e(x^e)||^2 - \gamma^2||r||^2\end{aligned} \tag{8.12}$$

where $r = r(x) = v - v^*(x)$.

Define

$$S(x^e,t) = V(x,t) + \tilde{W}(x^e,t).$$

Thus, S is positive definite and satisfies:

$$S(0,t)=0 \quad \forall t \geq 0.$$

Obviously, the infinitesimal generator of the process x^e satisfies:

$$L\{S(x^e,t)\} = L\{V(x,t)\} + L\{\tilde{W}(x^e,t)\}$$

where:

$$L\{V(x,t)\} = V_t(x,t) + V_x(x,t)\{f(x,t) + g_1(x,t)v + g(x,t)u^*(\hat{x})\}$$
$$+\tfrac{1}{2}Tr\{G^T(x,t)V_{xx}(x,t)G(x,t)\} + \tfrac{1}{2}u^{*T}(\hat{x})\bar{g}^T(x,t)V_{xx}(x,t)\bar{g}(x,t)u^*(\hat{x})$$
$$L\left\{\tilde{W}(x^e,t)\right\} = \tilde{W}_t(x^e,t) + \tilde{W}_{x^e}[f^e(x^e,t) + g_1^e(x^e,t)r]$$
$$+\tfrac{1}{2}Tr\left\{(G(x^e,t))^T\tilde{W}_{x^e x^e}G^e(x^e,t)\right\}.$$

In view of (8.12) we have:

$$L\{V(x,t)\} + ||z||^2 - \gamma^2||v||^2 \leq ||h^e(x^e)||^2 - \gamma^2||r||^2.$$

By the HJI (8.10) it follows that

$$L\left\{\tilde{W}(x^e,t)\right\} + ||h^e(x^e)||^2 - \gamma^2||r||^2 \leq 0.$$

Therefore,

$$L\{S(x^e,t)\} + ||z||^2 - \gamma^2||v||^2 = L\{V(x^e,t)\} + ||z||^2 - \gamma^2||v||^2 + L\left\{\tilde{W}(x^e,t)\right\} \leq 0.$$

This implies that $S(x^e,t)$ is a storage function for the closed-loop system with the supply rate $||z||^2 - \gamma^2||v||^2$, which implies that the closed-loop system has an L_2-gain property (i.e., the inequality (8.2) is satisfied)

Remark 8.1. As in the deterministic case it is not easy to find a matrix $K(\hat{x},t)$ depending on $\hat{x}$ and t alone, such that HJI (8.10) is satisfied (see [73] for the deterministic case where a local solution is offered). In what follows we offer two ways for approaching this problem which yield, in certain cases, a satisfactory solution for $K(\hat{x},t)$.

Consider first the case for which $G_2 = 0$. Then, the part of the latter inequality that contains K is given by:

$$\Gamma(K) = \tilde{W}_{\hat{x}}(x^e,t)K(\hat{x},t)(\tilde{h}_2(x,t) - \tilde{h}_2(\hat{x},t)) + \tfrac{1}{4}(\tilde{W}_{\hat{x}}(x^e,t)K(\hat{x},t)g_3(x,t)$$
$$+\tilde{W}_x(x^e,t)g_1(x,t))(\gamma^2 I - \tfrac{1}{2}\tilde{U}(x^e,t))^{-1}(\tilde{W}_{\hat{x}}(x^e,t)K(\hat{x},t)g_3(x,t)$$
$$+\tilde{W}_x(x^e,t)g_1(x,t))^T$$
$$= \tfrac{1}{4\gamma^2}[\tilde{W}_{\hat{x}}(x^e,t)K(\hat{x},t) - \Psi(x^e,t)]R_a(x,t)[K^T(\hat{x},t)\tilde{W}_{\hat{x}}^T(x^e,t) - \Psi^T(x^e,t)]$$
$$-\tfrac{1}{4\gamma^2}[\Psi(x^e,t)R_a(x,t)\Psi^T(x^e,t) - \tilde{W}_x(x^e,t)g_1(x,t)(I - \tfrac{1}{2\gamma^2}\tilde{U}(x^e,t))^{-1}$$
$$g_1^T(x,t)\tilde{W}_x^T(x^e,t)], \tag{8.13}$$

where

$$\Psi(x^e,t) = -[2\gamma^2(\tilde{h}_2(x,t) - \tilde{h}_2(\hat{x},t))^T + \tilde{W}_x(x^e,t)g_1(x,t)\hat{R}(x,t)]R_a^{-1}(x,t)$$
$$\hat{R}(x,t) = (I - \tfrac{1}{2\gamma^2} g_2^T(x,t)\tilde{W}_{xx}(x^e;t)g_2(x,t))^{-1} g_3^T(x,t)$$

and $R_a(x,t) = g_3(x,t)\hat{R}(x,t)$.

The gain matrix $K(\hat{x},t)$ that minimizes $\Gamma(K)$, and thus leads to a minimum left hand side in (8.10), is clearly one that satisfies $\tilde{W}_{\hat{x}}(x^e;t)K(\hat{x},t) = \Psi(x^e;t)$. Unfortunately, the latter equation may not possess a solution for K which depends only on $\hat{x}$ and t. One way to circumvent this difficulty is to choose $K(\hat{x},t)$ s.t.

$$\tilde{W}_{\hat{x}}(x^e;t)K(\hat{x},t) = \Psi(x^e;t) + \Phi(x^e;t), \tag{8.14}$$

where $\Phi(x^e,t)$ is a function that allows a solution K^* for (8.14) which is independent of x. For this choice of $K^*(\hat{x},t)$ the above $\Gamma(K)$ becomes the following.

$$\Gamma(K^*) = \tfrac{1}{4\gamma^2}[\Phi(x^e;t)R_a(x,t)\Phi^T(x^e;t) - \Psi(x^e;t)R_a(x,t)\Psi^T(x^e;t)$$
$$+\tilde{W}_x(x^e;t)g_1(x,t)(I - \tfrac{1}{2\gamma^2}\tilde{U}(x^e,t))^{-1}g_1^T(x,t)\tilde{W}_x^T(x^e;t)].$$

Assuming an existence of a solution $K(\hat{x},t)$ to (8.14) for some function $\Phi(x^e,t)$, we have established the following theorem.

Theorem 8.2. *Consider the stochastic system (1.11) together with the augmented system (8.4) and the controlled output (8.1). Assume there is a positive function $V : R^n \times [0,T] \to R^+$, with $V \in C^{2,1}$ so that it satisfies the HJI (6.8) of Theorem 6.2.1. Assume also that there are: a positive function $\tilde{W} : R^{2n} \times [0,T] \to R^+$ in $C^{2,1}$ and a matrix $K(\hat{x},t)$, which satisfy (8.14). In addition, let $\tilde{W}$ satisfy the following HJI.*

$$\tilde{W}_t(x^e,t) + \tilde{W}_x(x^e,t)[f(x,t) + g_1(x,t)v_t^*(x) + g(x,t)u_t^*(\hat{x})]$$
$$+\tilde{W}_{\hat{x}}(x^e,t)[f(\hat{x},t) + g_1(\hat{x},t)v_t^*(\hat{x}) + g(\hat{x},t)u^*(\hat{x})]$$
$$+\tfrac{1}{2}Tr\left\{G(x,t)^T\tilde{W}_{xx}(x^e,t)G(x,t)\right\} + \tfrac{1}{2}u^{*T}(\hat{x})\bar{g}^T(x,t)\tilde{W}_{xx}(x^e,t)\bar{g}(x,t)u^*(\hat{x})$$
$$+\tfrac{1}{2}v^{*T}(x)g_2^T(x,t)\tilde{W}_{xx}(x^e,t)g_2(x,t)v^*(x) + h^e(x^e,t)^T h^e(x^e,t)$$
$$+\tfrac{1}{4\gamma^2}[\Phi(x^e;t)R_a(x,t)\Phi^T(x^e;t) - \Psi(x^e;t)R_a(x,t)\Psi^T(x^e;t)$$
$$+\tilde{W}_x(x^e;t)g_1(x,t)(I - \tfrac{1}{2\gamma^2}g_2^T(x,t)\tilde{W}_{xx}(x^e;t)g_2(x,t))^{-1}g_1^T(x,t)\tilde{W}_x^T(x^e;t)] \leq 0$$
$$\forall x^e \in R^{2n}$$

$$\tilde{W}(x^e,0) = \gamma^2||x - \hat{x}||^2. \tag{8.15}$$

Then, the closed-loop system is dissipative with respect to the supply rate $\gamma^2||v||^2 - ||z||^2$, with the storage function defined as

$$S(x^e,t) = V(x,t) + \tilde{W}(x^e,t),$$

and therefore has an L_2-gain $\leq \gamma$.

Remark 8.2. In Theorem 8.1 above, it was assumed that $K(\hat{x},t)$ exists so that (8.10) is satisfied. Finding such K is a difficult task, in general. The method that is offered above allows a solution (that may be conservative) for K which is independent of the system's state x. It is noted that the inequality (8.15) is only sufficient, but by no means necessary.

8.2.1 Example

In order to demonstrate the applicability of the above approach we consider the following special case of (1.11) and (1.12):

$$\begin{aligned} dx_t &= Ax_t dt + B_1 v_t dt + \bar{G} x_t dW_t \\ dy_t &= C_2 x_t dt + \rho \cdot sin(2(x_2 - x_1))dt + \kappa v_t dt, \end{aligned} \tag{8.16}$$

where $\{x_t\}_{t\geq 0} \in \mathcal{R}^2$ and $y_t \in \mathcal{R}^1$. It is assumed that (6.8) is satisfied with $V(x) = x^T P x$. Choosing $\tilde{W}(x^e,t) = x^{eT} Q x^e$, eq.(8.13) now reads:

$$\begin{aligned} \Psi = &-2\gamma^2\kappa^{-1}\left[C_2(x-\hat{x}) + 2\rho sin(x_2 - \hat{x}_2 - x_1 + \hat{x}_1)cos(x_2 + \hat{x}_2 - x_1 - \hat{x}_1)\right] \\ &+2\kappa^{-1}(x-\hat{x})^{-1}PB_1 - \kappa^{-1}x^{eT}Q\tilde{I}B_1 \end{aligned} \tag{8.17}$$

where $\tilde{I} = [I_2 \; 0]^T$. Denoting also $\hat{I} = [I_2 \; -I_2]^T$ and $\Upsilon = col\{-1,1,1,-1\}$, we find that: $\Psi = x^{eT}\bar{\Psi}$ where

$$\begin{aligned} \bar{\Psi} = \Big[&-2\gamma^2\kappa^{-2}\hat{I}C_2^T - 4\gamma^2\kappa^{-2}\rho\Upsilon sinc(x_2 - \hat{x}_2 - x_1 + \hat{x}_1)cos(x_2 + \hat{x}_2 - x_1 - \hat{x}_1) \\ &+2\kappa^{-1}(\hat{I}PB_1 - Q\tilde{I}B_1)\Big]. \end{aligned} \tag{8.18}$$

For the above choice of $\tilde{W}$ we find that $\tilde{W}_x = x^{eT}Q\tilde{I}$ and $\tilde{W}_{\hat{x}} = x^{eT}Q\tilde{\tilde{I}}$ where $\tilde{\tilde{I}} = [0 \; I_2]^T$.

Denoting $\Phi = x^{eT}\bar{\Phi}$, (8.15) reduces now to

$$\begin{aligned} &x^{eT}\Big[2Q\left[\tilde{I}(A + \gamma^{-2}B_1B_1^TP) - \tilde{I}B_2B_2^TP + \tilde{\tilde{I}}(A + \gamma_2 B_1B_1^TP - B_2B_2^TP)\right] \\ &+\bar{G}^T\tilde{I}^TQ\tilde{I}\bar{G} + \hat{I}PB_2B_2^TP\hat{I} + 4Q\tilde{I}B_1B_1^T\tilde{I}^TQ\Big]x^e + \tfrac{\kappa^2}{4\gamma^2}x^{eT}\left[\bar{\Phi}\bar{\Phi}^T - \bar{\Psi}\bar{\Psi}\right]x^e \leq 0 \end{aligned} \tag{8.19}$$

Eq. (14) becomes: $\tilde{\tilde{I}}K(\hat{x}) = Q^{-1}\left[\bar{\Psi} + \bar{\Phi}\right]$ and then

$$K(\hat{x}) = 2Q_2^{-1}\kappa^{-1}(\gamma^2\kappa^{-1}C_2^T - PB_1).$$

Replacing $\bar{\Phi}\bar{\Phi}^T$ in (8.19) with its upper bound $16\gamma^4\kappa^{-4}\rho^2\begin{bmatrix} 1 & -1 \\ -1 & 1 \end{bmatrix}$ and eliminating x^e from both sides of (8.19), the resulting inequality in Q_1 and Q_2

is independent of x and $\hat{x}$. The requirement in regard to Q_2 then becomes a simple Lyaponov inequality, while the result for Q_1 can be found using linear matrix inequality solver.

8.2.2 The Case of Nonzero G_2

In the case where $G_2(x,t)$ is not identically zero, $G^e(x^e, v_t)$ of (8.4e) should be replaces by:
$G^e(x^e, v_t) = diag\{G(x,t),\ K(\hat{x},t)G_2(x,t)\}$
and the term $\frac{1}{2}Tr\{G_2(x,t)^T K(\hat{x},t)^T \tilde{W}_{\hat{x}\hat{x}}(x^e,t)K(\hat{x},t)G_2(x,t)\}$
should thus be added to $\Gamma(K)$ of (8.13). The latter term does not allow completion of the terms in $\Gamma(K)$ to squares and thus solutions for K that achieve the prescribed L_2-gain bound can be obtained by one of the following methods.

Solution no. 1: One may solve for K in (8.14). In this case, due to the additional term in $\Gamma(K)$, the positive term
$\frac{1}{2}Tr\Big\{G_2(x_t,t)^T[\Psi^T(x^e,t)+\Phi^T(x^e,t)]\tilde{W}^{-1}_{\hat{x}\hat{x}}(x^e,t)[\Psi(x^e,t)+\Phi(x^e,t)]G_2(x_t,t)\Big\}$
is added to the left side of (8.15).

Solution no. 2: Since the above solution entails an over design one may consider, similarly to the method used in linear gain scheduling with uncertainty in the input or the output matrices (see, e.g., [41], p. 7-4), the following modified system which contains a simple linear lowpass component of large bandwidth between the measured output and the point where the noise signal v_t is applied. The effect of this component, which will be a part of the controller, on the solution is negligible whenever its bandwidth is very large in comparison with the system 'bandwidth'. We may now formulate the above in the following way. Define a new state ζ as

$$d\zeta_t = -\rho\zeta_t dt + \rho h_2(x_t,t)dt + \rho G_2(x_t,t)dW_t^3$$

and we measure now the filter output corrupted with the deterministic noise $g_2(x_t,t)v_t dt$ that is,

$$d\bar{y}_t = \zeta_t dt + g_3(x_t,t)v_t dt.$$

Replacing x_t in x_t^e of (8.4b) with $col\{x_t, \zeta_t\}$ and defining $\bar{x}_t$ to be the estimator of the latter augmented vector we obtain that $d\bar{x}_t$ satisfies the following:

$$d\bar{x}_t \triangleq \begin{bmatrix} d\hat{x}_t \\ d\hat{\zeta}_t \end{bmatrix} = \begin{bmatrix} f(\hat{x}_t,t) \\ \rho h_2(\hat{x}_t,t) - \rho\hat{\zeta}_t \end{bmatrix} dt + \begin{bmatrix} g(\hat{x}_t,t) \\ 0 \end{bmatrix} u_t^*(\hat{x}_t)dt$$

$$+ \begin{bmatrix} g_1(x_t,t) \\ 0 \end{bmatrix} v_t^*(\hat{x}_t)dt + \begin{bmatrix} K(\hat{x}_t,\hat{\zeta}_t,t) \\ \rho K_1(\hat{x}_t,\hat{\zeta}_t,t) \end{bmatrix} (d\bar{y}_t - \hat{\zeta}_t dt - g_3(\hat{x}_t,t)v_t^*(\hat{x}_t)).$$

Denoting then:

$$\bar{f}^e(x_t^e,t) = col\{f(x_t,t)+g_1(x_t,t)v_t^*(x_t)+g(x_t,t)u_t^*(\hat{x}_t),\ -\rho(\zeta_t-h_2(x_t,t)),$$
$$f(\hat{x}_t,t)+g(\hat{x}_t,t)u_t^*(\hat{x}_t)+g_1(\hat{x}_t,t)v_t^*(\hat{x}_t)+K(\hat{x}_t,t)[\zeta_t-\hat{\zeta}_t+g_3(x_t,t)v_t^*(x_t)$$
$$-g_3(\hat{x}_t,t)v_t^*(\hat{x}_t)],$$
$$\rho(h_2(\hat{x}_t,t)-\hat{\zeta}_t)+\rho K_1(\hat{x}_t,\hat{\zeta}_t,t)[\zeta_t-\hat{\zeta}_t+g_3(x_t,t)v_t^*(x_t)-g_3(\hat{x}_t,t)v_t^*(\hat{x}_t)]\}$$

$$\bar{g}_1(\bar{x}_t^e,t) = col\{g_1(x_t,t),\ 0,\ K(\hat{x}_t,\hat{\zeta}_t,t)g_3(x_t,t),\ \rho K_1(\hat{x}_t,\hat{\zeta}_t,t)g_3(x_t,t)\},$$

$$g_2^e(\bar{x}_t^e,t) = col\{g_2(x_t,t),\ 0,\ 0,\ 0\},$$

$$\bar{G}^e(\bar{x}_t^e,t) = \begin{bmatrix} \bar{g}(x,t)u^*(\hat{x}) & G(x_t,t) & g_2(x_t,t)v_t^*(x_t) & 0 \\ 0 & 0 & 0 & \rho G_2(x_t,t) \\ 0 & 0 & 0 & 0 \\ 0 & 0 & 0 & 0 \end{bmatrix}$$

and defining $W_t^e = col\{W_t, W_t^1, W_t^2, W_t^3\}$ we obtain:

$$d\bar{x}_t^e = \bar{f}^e(\bar{x}_t^e,t)dt + \bar{g}_1(\bar{x}_t^e,t)[v_t(x_t) - v_t^*(x_t)] + \bar{G}^e(x_t^e,t)dW_t^e$$

and (8.20)

$$\bar{x}_t^e = col\{x_t, \zeta_t, \bar{x}_t\}.$$

Applying the result of Theorem 8.2 to the augmented system (8.20) we obtain that the minimizing gain matrix $col\{K, \rho K_1\}$ satisfies

$$\bar{W}_{\bar{x}}(\bar{x}_t^e t)\begin{bmatrix} K(\hat{x}_t,t) \\ \rho K_1(\hat{x}_t,t) \end{bmatrix} = \bar{\Psi}(\bar{x}_t^e t) + \bar{\Phi}(\bar{x}_t^e t)$$

and

$$\bar{\Psi}(\bar{x}_t^e t) = -[2\gamma^2(\zeta_t-\hat{\zeta}_t+g_3(x_t,t)v_t^*(x_t)-g_3(\hat{x}_t,t)v_t^*(\hat{x}_t))^T$$
$$+\bar{W}_x(\bar{x}^e,t)g_1(x_t,t)(I - \tfrac{1}{2\gamma^2}g_2^T(x_t,t)\tilde{W}_{xx}(x_t^e t)g_2(x_t,t))^{-1}g_3^T(x_t,t)]R_a^{-1},$$
(8.21)

where $\Phi(x^e,t)$ is a function that allows a solution $col\{K, K_1\}$ for (8.21) that is independent of x_t, and where $\bar{W}(\bar{x}_t^e t)$ is a positive function $\bar{W}: R^{2(n+p)} \times [0,\infty) \to R^+$ that satisfies the following inequality.

$$\bar{W}_t(\bar{x}^e,t)+\bar{W}_x(\bar{x}^e,t)[f(x_t,t)+g_1(x_t,t)v_t^*(x_t)+g(x_t,t)u_t^*(\hat{x}_t)]$$
$$-\rho\bar{W}_\zeta[\zeta_t-h_2(x_t,t)]+\rho\bar{W}_{\hat{\zeta}_t}[h_2(\hat{x}_t,t)-\hat{\zeta}_t]+\tfrac{1}{2}Tr\left\{G(x,t)^T\tilde{W}_{xx}(\bar{x}^e,t)G(x,t)\right\}$$
$$+\bar{W}_{\hat{x}}(\bar{x}^e,t)[f(\hat{x}_t,t)+g_1(\hat{x}_t,t)v_t^*(\hat{x}_t)+g(\hat{x}_t,t)u^*(\hat{x}_t)]$$
$$+\tfrac{\rho^2}{2}Tr\left\{G_2(x,t)^T\tilde{W}_{\zeta\zeta}(\bar{x}^e,t)G_2(x,t)\right\} + h^e(x^e,t)^Th^e(x^e,t)$$
$$+\bar{W}_x(\bar{x}_t^e t)g_1(x_t,t)(I - \tfrac{1}{2\gamma^2}g_2^T(x_t,t)\tilde{W}_{xx}(x_t^e t)g_2(x_t,t))^{-1}g_1^T(x_t,t)\bar{W}_x^T(\bar{x}_t^e t)]$$
$$+\tfrac{1}{4\gamma^2}[\bar{\Phi}(\bar{x}_t^e t)R_a(x_t,t)\bar{\Phi}^T(\bar{x}_t^e t)-\bar{\Psi}(\bar{x}_t^e t)R_a(x_t,t)\bar{\Psi}^T(\bar{x}_t^e t) \le 0 \forall x^e \in R^{2n}.$$

8.3 Norm-Bounded Uncertainty

The above results were based on the assumption that the system's parameters are completely known. In the present section we consider the case where these parameters are uncertain. We consider the following system:

$$dx_t = (f(x_t,t) + \Delta f)dt + (g_1(x_t,t) + \Delta g)v_t dt + g_2(x_t,t)v_t dW_t^2 + G(x_t,t)dW_t^1 \\ + g(x_t,t)u_t dt + \bar{g}(x_t)u_t dW_t \tag{8.22}$$

$$z(t) = h(x_t,t) + \Delta h \tag{8.23}$$

where $u_t \in R^m$ is the control input which is taken to be identically zero, where $\{x_t\}_{t\geq 0} \in R^n$ is a solution to (8.22) with the initial condition x_0, $\{v_t\}_{t\geq 0} \in R^{m_1}$ is an exogenous disturbance and $\{W_t^1\}_{t\geq 0} \in R^\ell$, $\{W_t^2\}_{t\geq 0} \in R^1$ are Wiener processes. Also, $z_t \in R^r$ is the output vector to be regulated. The matrices Δf, Δg and Δh are assumed to be continuous functions of x_t and t that possess the following structure:

$$\begin{bmatrix} \Delta f(x,t) & \Delta g(x,t) \end{bmatrix} = H_1(x,t)F(x,t)\begin{bmatrix} E_1(x,t) & E_2(x,t) \end{bmatrix},$$

$$\Delta h(x,t) = H_2(x,t)F(x,t)E_1(x,t),$$

where $H_1(x,t) : R^n \times [0,\infty) \to R^{n\times n_1}$, $H_2(x,t) : R^n \times [0,\infty) \to R^{r\times n_1}$, $E_1(x,t) : R^n \times [0,\infty) \to R^{n_2}$, $E_2(x,t) : R^n \times [0,\infty) \to R^{n_2\times m_1}$ and $F(x,t) : R^n \times [0,\infty) \to R^{n_1\times n_2}$ are continuous functions and where

$$F(x,t)F^T(x,t) \leq I_{n_1}, \quad \forall x \in R^n \text{ and } t \in [0,\ \infty). \tag{8.24}$$

Applying the BRL obtained in Theorem 6.2.1 by replacing f, g_1 and h by $f + \Delta f$, $g_1 + \Delta g$ and $h + \Delta h$, respectively, the corresponding HJI becomes:

$$\begin{aligned} & V_t(x,t) + V_x(x,t)[f(x,t) + H_1(x,t)F(x,t)E_1(x,t)] + \tfrac{1}{4}V_x(x,t)(g_1(x,t) \\ & + H_1(x,t)F(x,t)E_2(x,t))[\gamma^2 I - \tfrac{1}{2}U(x,t)]^{-1}(g_1^T(x,t) + E_2^T(x,t)F^T(x,t)H_1^T(x,t)) \\ & V_x(x,t) + \tfrac{1}{2}Tr\{G^T(x,t)V_{xx}(x,t)G(x,t)\} \\ & + (h^T(x,t) + E_1^T(x,t)F^T(x,t)H_2^t(x,t))(h(x,t) + H_2(x,t)F(x,t)E_2(x,t)) \\ & \leq\ 0\ \forall x \in R^n, \qquad \forall t \geq 0, \end{aligned}$$

where $U(x,t)$ is defined in (6.7). Applying Schur complement to the latter, and denoting:

$$\bar{\Gamma}(x,t) \\ = \begin{bmatrix} \begin{matrix} V_t(x,t) + V_x(x,t)f(x,t) \\ + \tfrac{1}{2}Tr\{G^T(x,t)V_{xx}(x,t)G(x,t)\} \end{matrix} & V_x(x,t)g_1(x,t) & h^T(x,t) \\ * & -[4\gamma^2 I - 2U(x,t)] & 0 \\ * & * & -I_r \end{bmatrix} \tag{8.25}$$

the following inequality is obtained;

$$\bar{\Gamma}(x,t)+\begin{bmatrix}0\\0\\H_2(x,t)\end{bmatrix}F(x,t)\begin{bmatrix}E_1(x,t) & 0 & 0\end{bmatrix}+\begin{bmatrix}E_1^T(x,t)\\0\\0\end{bmatrix}F^T(x,t)\begin{bmatrix}0 & 0 & H_2^T(x,t)\end{bmatrix}$$

$$+\begin{bmatrix}V_x(x,t)H_1(x,t)\\0\\0\end{bmatrix}F(x,t)\begin{bmatrix}\frac{1}{2}E_1(x,t) & E_2(x,t) & 0\end{bmatrix}$$

$$+\begin{bmatrix}\frac{1}{2}E_1^T(x,t)\\E_2^T(x,t)\\0\end{bmatrix}F^T(x,t)\begin{bmatrix}H_1^T(x,t)V_x^T(x,t) & 0 & 0\end{bmatrix}\leq 0\ \forall x\in R^n,\qquad \forall t\geq 0.$$

Using the fact that for any two matrices α and β of compatible dimensions and for any positive scalar ε the following holds:

$$\alpha\beta^T+\beta\alpha^T\leq\varepsilon\alpha\alpha^T+\varepsilon^{-1}\beta\beta^T,$$

together with the bound of (8.24), we obtain the following:

Theorem 8.3. *Consider the system described by (8.22)–(8.24), where $u_t\equiv 0$, with the controlled output of (8.23), and the supply rate $S(v,z)=\gamma^2\|v\|^2-\|z\|^2$. Suppose there is a positive function $V(x,t)\in C^{2,1}$. Let $V(x,t)$ satisfy $\gamma^2 I-\frac{1}{2}U(x,t)\geq\alpha I$ for some $\alpha>0$, and for all x, t, where $U(x,t)$ is defined in (6.7) and assume that the following HJI is satisfied for some positive scalars ε_1 and ε_2:*

$$\begin{bmatrix}\bar{\Gamma}(x,t) & \begin{bmatrix}V_x(x,t)H_1(x,t)\\0\\0\end{bmatrix} & \varepsilon_1\begin{bmatrix}\frac{1}{2}E_1^T(x,t)\\E_2^T(x,t)\\0\end{bmatrix} & \begin{bmatrix}0\\0\\H_2(x,t)\end{bmatrix} & \varepsilon_2\begin{bmatrix}E_1^T(x,t)\\0\\0\end{bmatrix}\\ * & -\varepsilon_1 I_{n_1} & 0 & 0 & 0\\ * & * & -\varepsilon_1 I_{n_2} & 0 & 0\\ * & * & * & -\varepsilon_2 I_{n_1} & 0\\ * & * & * & * & -\varepsilon_2 I_{n_2}\end{bmatrix}\leq 0\quad (8.26)$$

$\forall x\in R^n$ and $\forall t\geq 0$ where $\bar{\Gamma}(x,t)$ is defined in (8.25). Then, the system (8.22) is dissipative with respect to the supply rate $S(v,z)$.

The above result assumed $u_t\equiv 0$ in (8.22). If this is not the case and if

$$z(t)=col\{h(x_t,t)+\Delta h,\ u_t\}\tag{8.27}$$

the following result replaces the one of Theorem 6.2.1.

Lemma 8.3.1 *Consider the system described by (8.22)–(8.24) with the controlled output of (8.27), and the supply rate $S(v,z)=\gamma^2\|v\|^2-\|z\|^2$. Suppose there is a positive function $V(x,t)\in C^{2,1}$. Let $V(x,t)$ satisfy $\gamma^2 I-\frac{1}{2}U(x,t)\geq$*

αI for some $\alpha > 0$, and for all x, t, where $U(x,t)$ is defined in (6.7) and assume that (8.26) is satisfied for some positive scalars ε_1 and ε_2 where to the first block on the diagonal of $\bar{\Gamma}(x,t)$ the term

$$-\frac{1}{4}V_x(x,t)g(x,t)[I+\frac{1}{2}\bar{g}^T(x,t)V_{xx}(x,t)\bar{g}(x,t)]^{-1}g^T(x,t)V_x^T(x,t)$$

is added. Then, for

$$u_t = -\frac{1}{2}[I+\frac{1}{2}\bar{g}^T(x,t)V_{xx}\bar{g}(x,t)]^{-1}g^T(x,t)V_x^T(x,t)$$

the system (8.22) is dissipative with respect to the supply rate $S(v,z)$.

Considering next the case where the measurement is given by:

$$dy_t = h_2(x_t,t)dt + g_3(x_t,t)v_t dt + G_2(x_t,t)dW^3.$$

Once the BRL is obtained for systems with norm-bounded uncertainties the output-feedback problem can be solved as follows. Consider the following system.

$$\begin{aligned} dx_t &= (f(x_t,t)+\Delta f)dt + g_1(x_t,t)v_t dt + g_2(x_t,t)v_t dW_t^2 + G(x_t,t)dW_t^1 \\ &\quad + g(x_t,t)u_t dt + \bar{g}(x_t)u_t dW_t \\ dy_t &= (h_2(x_t,t)+\Delta h_2)dt + g_3(x_t,t)v_t dt + G_2(x_t,t)dW^3 \\ z(t) &= col\{h(x_t,t)+\Delta h,\ u_t\}, \end{aligned} \tag{8.28}$$

where $u_t \in R^m$ is the control input, $\{x_t\}_{t\geq 0} \in R^n$ is a solution to (8.22) with the initial condition x_0, $\{v_t\}_{t\geq 0} \in R^{m_1}$, $\{y_t\}_{t\geq 0} \in R^p$ is the measured output and $\{W_t^1\}_{t\geq 0} \in R^\ell$, $\{W_t^2\}_{t\geq 0} \in R^1$ and $\{W_t^3\}_{t\geq 0} \in R_3^m$ are Wiener processes. Also, $z_t \in R^r$ is the output vector to be regulated. The matrices Δf, Δh_2 and Δh are assumed to be continuous functions of x_t and t that possess the following structure:

$$\begin{aligned} &col\{\Delta f(x,t),\ \Delta h(x,t),\ \Delta h_2(x,t)\} = col\{H_1(x,t),\ H_2(x,t), \\ &H_3(x,t)\}F(x,t)E_1(x,t) \end{aligned} \tag{8.29}$$

where $H_1(x,t) : R^n \times [0,\infty) \to R^{n\times n_1}$, $H_2(x,t) : R^n \times [0,\infty) \to R^{r\times n_1}$, $H_3(t) : R^n \times [0,\infty) \to R^{p\times n_y}$, $E_1(x,t) : R^n \times [0,\infty) \to R^{n_2}$, and $F(x,t) : R^n \times [0,\infty) \to R^{n_1\times n_2}$ are continuous functions and where $F(x,t)$ satisfies (8.24).

Denoting:

$$\hat{\Gamma}(x,t) = \begin{bmatrix} \hat{\Gamma}_{11}(x,t) & V_x(x,t)g_1(x,t) & h^T(x,t) \\ * & -[4\gamma^2 I - 2U(x,t)] & 0 \\ * & * & -I_r \end{bmatrix}$$

with

$$\hat{\Gamma}_{11}(x,t) = V_t(x,t)+V_x(x,t)f(x,t)+\frac{1}{2}Tr\{G^T(x,t)V_{xx}(x,t)G(x,t)\}$$
$$-\frac{1}{4}V_x(x,t)g(x,t)[I+\frac{1}{2}\bar{g}^T(x,t)V_{xx}(x,t)\bar{g}(x,t)]^{-1}g^T(x,t)V_x^T(x,t)$$

we obtain the following inequality that corresponds to the HJI (8.26).

$$\begin{bmatrix} \hat{\Gamma}(x,t) & \begin{bmatrix} V_x(x,t)H_1(x,t) \\ 0 \\ 0 \end{bmatrix} & \varepsilon_1 \begin{bmatrix} \frac{1}{2}E_1^T(x,t) \\ 0 \\ 0 \end{bmatrix} & \begin{bmatrix} 0 \\ 0 \\ H_2(x,t) \end{bmatrix} & \varepsilon_2 \begin{bmatrix} E_1^T(x,t) \\ 0 \\ 0 \end{bmatrix} \\ * & -\varepsilon_1 I_{n_1} & 0 & 0 & 0 \\ * & * & -\varepsilon_1 I_{n_2} & 0 & 0 \\ * & * & * & -\varepsilon_2 I_{n_1} & 0 \\ * & * & * & * & -\varepsilon_2 I_{n_2} \end{bmatrix} \leq 0$$

$\forall x \in R^n$, $\forall t \geq 0$.

Denoting then:

$$\Delta f^e = \begin{bmatrix} H_1(x_t,t) \\ 0 \end{bmatrix} F(x_t,t)E_1(x_t,t) + \begin{bmatrix} 0 \\ K(\hat{x}_t,t)H_3(t) \end{bmatrix} F(x_t,t)E_1(x_t,t)$$

the BRL of Theorem 8.3 is applied to the latter definition and to (8.4). Denoting

$$\Gamma^e = \begin{bmatrix} \Upsilon(x^e,t) & \tilde{W}_{x^e}(x^e,t)g_1^e(x^e,t) & h^{eT}(x^e,t) \\ * & -[4\gamma^2 I - 2\bar{U}(x^e,t)] & 0 \\ * & * & -I \end{bmatrix},$$

where
$\Upsilon(x^e,t) = \tilde{W}_t(x^e,t)+\tilde{W}_{x^e}(x^e,t)f^e(x^e,t) + \frac{1}{2}Tr\{(G^e(x^e,t))^T\tilde{W}_{x^ex^e}(x^e,t)G^e(x^e,t)\}$

and where $U(x^e,t)$ and $h^e(x^e,t)$ are defined in (8.11a,b) the following inequality is obtained $\forall x^e \in R^{2n}$ and $\forall t \geq 0$;

$$\begin{bmatrix} \Gamma^e & \begin{bmatrix} \tilde{W}_x(x^e,t)H_1(x,t)+\tilde{W}_{\hat{x}}(x^e,t)K(\hat{x}_t,t)H_3(\hat{x},t) \\ 0 \\ 0 \end{bmatrix} & \bar{\varepsilon}\begin{bmatrix} \frac{1}{2}E_1^T(x,t) \\ 0 \\ 0 \end{bmatrix} \\ * & -\bar{\varepsilon}I_{n_1} & 0 \\ * & * & -\bar{\varepsilon}I_{n_1} \end{bmatrix} \leq 0, \tag{8.30}$$

where $\bar{\varepsilon}$ is a decision variable in $\mathcal{R}^1$. The solution to the output-feedback problem with the norm-bounded uncertainty described in (8.28)–(8.29) is described as follows.

Theorem 8.4. *Consider the stochastic system (8.28)–(8.29) together with the augmented system (8.4). Assume there exist a positive function* $V : R^n \times [0,\infty) \rightarrow R^+$, *with* $V \in C^{2,1}$ *and scalars* ε_1 *and* ε_2 *that satisfy the HJI*

(8.26) of Theorem 8.3. Assume also that there exist: a positive function $\tilde{W} : R^{2n} \times [0, \infty) \to R^+$ *a matrix* $K(\hat{x}, t)$ *and a scalar* $\bar{\varepsilon}$ *that satisfy (8.30) for some* $\gamma \geq 0$*, and that*

$$\gamma^2 I - \frac{1}{2}[g_2^e(x^e, t)]^T \tilde{W}_{x^e x^e}(x^e, t) g_2^e(x^e, t) \geq \alpha I$$

for some positive number α*, and for all* $x^e \in R^{2n}$ *. Then, the closed-loop system with the control*

$$u_t^*(\hat{x}) = -\frac{1}{2}[I + \frac{1}{2}\bar{g}^T(\hat{x}, t) V_{xx}(\hat{x}, t) \bar{g}(\hat{x}, t)]^{-1} g^T(\hat{x}, t) V_{xx}(\hat{x}, t)$$

and with the observer (8.3) is dissipative with respect to the supply rate $\gamma^2||v||^2 - ||z||^2$*, it possesses a storage function defined as* $S(x^e, t) = V(x, t) + \tilde{W}(x^e, t)$*, and has an* L_2*-gain* $\leq \gamma$*.*

Remark 8.3. Similar to the arguments of Remark 8.1 and Theorem 8.2, in the case where $G_2 \equiv 0$, the last theorem can be used to obtain a minimizing K that depends only on $\hat{x}$ and t by converting the matrix inequality (8.30) into a scalar HJI using Schur complement. A completion to squares with respect to $\tilde{W}_{\hat{x}}(x^e, t) K(\hat{x}_t, t)$ can be obtained. Using then (8.14) a result similar to the one obtained in Theorem (8.3) can be achieved. The case where G_2 is not zero can be solved applying the method of Section 8.2.1.

8.4 Infinite-Time Horizon Case: A Stabilizing H_∞ Controller

In this section we consider the infinite-time horizon case. The problem of synthesizing an H_∞ controller that renders a stable closed-loop system is investigated. Conditions under which the closed-loop system is asymptotically stable, in both the probability sense and the mean-square sense are discussed.

We now state and prove two lemmas regarding the internal stability of the closed-loop system.

Lemma 8.4.1 *[Stability in probability]*
Assume there exists a positive function $V(x, t) \in C^{2,1}$ *such that* $\inf_{t>0} V(x, t) \to \infty$ *as* $\|x\| \to \infty$*, satisfying the HJI (6.8) with* $h^T(x, t) h(x, t) > 0$ *for all* x *and for all* $t \geq 0$*. Assume also that there is a positive function* $\tilde{W} \in C^{2,1}$*,* $\tilde{W} : R^{2n} \to R^+$*, satisfying the HJI (8.10) with a strict inequality, so that*

$$\inf_{t > 0,\ x \in R^n} \tilde{W}(x^e, t) \to \infty \text{ as } ||x^e|| \to \infty.$$

Then, the closed loop system is internally globally asymptotically stable in probability.

Proof: Note that for $v=0$ the HJI (6.8) reduces to

$$\begin{aligned}&V_t(x,t)+V_x(x,t)f(x,t)-\tfrac{1}{4}V_x(x,t)g(x,t)g^T(x,t)V_x^T(x,t)\\&+\tfrac{1}{2}Tr\{G_1^T(x,t)V_{xx}(x,t)G_1(x,t)\}+h^T(x,t)h(x,t)\le 0\;\forall x\in R^n,\;\forall t\ge 0\end{aligned}$$

which implies $(LV)(x,t)\le -h^T(x,t)h(x,t)<0$. On the other hand for $v=0$ the HJI (8.10) becomes

$$\begin{aligned}&\tilde{W}_t(x^e,t)+\tilde{W}_{x^e}(x^e,t)f^e(x^e,t)\\&+\tfrac{1}{2}Tr\left\{(G^e)^T\tilde{W}_{x^ex^e}(x^e,t)G^e(x^e,t)\right\}+(h^e(x^e,t)^Th^e(x^e,t)<0\quad\forall x^e\in R^{2n}.\end{aligned}$$

This implies $L_0\{\tilde{W}(x^e,t)\}<0$.

Summarizing, one has now

$$L_0\{S(x^e,t)\}=L_0\{V(x,t)\}+L_0\{\tilde{W}(x^e,t)\}<0$$

which implies, by Theorem 1.3.1, that the closed loop system is internally asymptotically stable in the probability sense.

The next lemma establishes the stability of the closed-loop in the mean square sense.

Lemma 8.4.2 *Assume there exists a positive function $V(x,t)\in C^{2,1}$, with $V(0,t)=0$ for all $t\ge 0$, which satisfies the HJI of (6.8) for some $\gamma>0$. In addition, let V satisfy*

$$k_1||x||^2\le V(x,t)\le k_2||x||^2 \tag{8.31}$$

for all $t\ge 0$ and for some positive numbers k_1,k_2. Furthermore, assume that for some k_3, $h_2^T(x,t)h_2(x,t)\ge k_3||x||^2$ for all $t\ge 0$ and for all $x\in R^n$. Assume also that there is a positive function $\tilde{W}\in C^{2,1}$. and $\tilde{W}:R^{2n}\to R^+$ with

$$k_4||\hat{x}||^2\le \tilde{W}(x^e,t)\le k_5||x^e||^2\quad x^e\in R^{2n},\quad t\ge 0 \tag{8.32}$$

which satisfies the following algebraic HJI:

$$\begin{aligned}&\tilde{W}_{x^e}(x^e,t)f^e(x^e,t)+\tfrac{1}{2}Tr\left\{(G^e)^T\tilde{W}_{x^ex^e}(x^e,t)G^e(x^e,t)\right\}+(h^e(x^e,t)^Th^e(x^e,t)\\&\le -Q(x^e)\;\forall x^e\in R^{2n}\end{aligned}$$

for some positive function $Q(x^e)$ with the property that

$$h^e(x^e,t)^Th^e(x^e,t)+Q(x^e)\ge k_6||\hat{x}||^2\;\text{for all}\,t\ge 0,\quad\text{for all}\;x^e\in R^{2n},$$

and for some $k_6>0$. Then, the closed-loop system (8.4) with $v=0$ and $u=-\frac{1}{2}g^T(x,t)V_x^T(\hat{x},t)$ is exponentially stable in the mean square sense, and has the property of L_2- gain $\le\gamma$, that is

$$\mathcal{E}\{\int_0^\infty \|z_t\|^2 dt\} \leq \gamma^2 \mathcal{E}\{\|x_0^e\|^2 + \int_0^\infty (\|v_t\|^2) dt\} \tag{8.33}$$

for all non-anticipative stochastic processes v *that satisfy* $\mathcal{E}\{\int_0^\infty (\|v_t\|^2)dt\} < \infty$, *and whenever* x_0 *satisfies* $\mathcal{E}\{\tilde{W}(x_0^e, 0)\} \leq \gamma^2 \mathcal{E}\{||x_0^e||^2\}$.

Proof: The proof is a simple application of Theorem 1.3.2

We consider now the time-invariant case with infinite-time horizon. Developing an analog theory is straightforward, so we omit proofs and detailed discussions.

The time-invariant version of (1.11) and (1.12) is given by

$$dx_t = f(x_t)dt + g(x_t)u_t dt + g_1(x_t)v_t dt + G(x_t)dW_t^1$$

$$dy_t = h_2(x_t)dt + g_3(x_t)v_t dt + G_2(x_t)dW^3$$

with the controlled output

$$z = col\{h(x),\ u(x)\}.$$

The time-invariant closed-loop system is now defined by:

$$dx_t^e = f^e(x_t^e)dt + g_1^e(x_t^e)[v_t - v_t^*(x_t)]dt + G^e(x_t^e)dW_t^e. \tag{8.34}$$

We state now, without a proof the time-invariant analog of Theorems regarding stability.

Lemma 8.4.3 *[Stability in probability: the invariant case]*
Assume there exists a positive function $V(x) \in C^2$ *such that* $V(x) \to \infty$ *as* $\|x\| \to \infty$, *and assume* $V(x)$ *satisfies the algebraic HJI:*

$$\begin{aligned} &V_x(x)f(x) - \tfrac{1}{4}V_x(x)g(x)g^T(x)V_x^T(x) + \tfrac{1}{4\gamma^2}V_x(x)g_1(x)g_1^T(x)V_x(x) \\ &+\tfrac{1}{2}Tr\{G^T(x)V_{xx}(x)G(x)\} + h_2^T(x)h_2(x) \leq 0\ \forall x \in R^n, \qquad \forall t \geq 0 \end{aligned} \tag{8.35}$$

and $h_2^T(x,t)h_2^T(x,t) > 0$ *for all* x *and for all* $t \geq 0$. *Also assume that there is a positive function* $\tilde{W} \in C^2$. *and* $\tilde{W}: R^{2n} \to R^+$ *with* $\inf_{x \in R^n} \tilde{W}(x^e, t) \to \infty$ *as* $||\hat{x}|| \to \infty$ *which satisfies the HJI*

$$\begin{aligned} &\tilde{W}_{x^e}(x^e)f^e(x^e) + \tfrac{1}{4\gamma^2}\tilde{W}_{x^e}(x^e)g_1^e(x^e)(g_1^e)^T(x^e)\tilde{W}_{x^e}^T(x^e) \\ &+\tfrac{1}{2}Tr\left\{(G^e(x^e))^T\tilde{W}_{x^e x^e}(x^e)G^e(x^e)\right\} + h^e(x^e)^T h^e(x^e) < 0 \quad \forall x^e \in R^{2n}. \end{aligned}$$

Then, the closed loop system is internally globally asymptotically stable in probability, and has the property of L_2*-gain* $\leq \gamma$.

Lemma 8.4.4 *Assume there exists a positive function* $V(x) \in C^2$, *with* $V(0) = 0$, *which satisfies the HJI of (8.35) for some* $\gamma > 0$. *In addition, let* V *satisfy*

$$k_1||x||^2 \leq V(x) \leq k_2||x||^2$$

for all $t \geq 0$ *and for some positive numbers* k_1, k_2. *Furthermore, assume that for some* k_3, $h_2^T(x)h_2(x) \geq k_3||x||^2$ *for all* $x \in R^n$. *Assume also that there is a positive function* $\tilde{W} \in C^2$ *and* $\tilde{W} : R^{2n} \to R^+$ *with*

$$k_4||\hat{x}||^2 \leq \tilde{W}(x^e) \leq k_5||x^e||^2 \quad x^e \in R^{2n},$$

which satisfies the following algebraic HJI:

$$\tilde{W}_{x^e}(x^e)f^e(x^e) + \frac{1}{2}Tr\left\{(G^e)^T\tilde{W}_{x^ex^e}(x^e)G^e(x^e)\right\} + h^e(x^e)^Th^e(x^e)$$
$$\leq -Q(x^e)\ \forall x^e \in R^{2n}$$

for some positive function $Q(x^e)$ *with the property that* $h^e(x^e)^Th^e(x^e) + Q(x^e) \geq k_6||\hat{x}||^2$ *for all* $x^e \in R^{2n}$, *and for some* $k_6 > 0$. *Then, the closed-loop system (8.34), with* $v = 0$ *and* $u = -\frac{1}{2}g^T(x)V_x^T(\hat{x})$, *is exponentially stable in the mean square sense, and has the property of* $L_2 -$ *gain* $\leq \gamma$, *that is*

$$\mathcal{E}\{\int_0^\infty \|z_t\|^2\, dt\} \leq \gamma^2\mathcal{E}\{||x_0^e||^2 + \int_0^\infty \|v_t\|^2\, dt\} \tag{8.36}$$

for all nonanticipative stochastic processes v *that satisfy* $\mathcal{E}\{\int_0^\infty(\|v_t\|^2)dt\} < \infty$, *and whenever* x_0 *satisfies* $\mathcal{E}\{\tilde{W}(x_0^e, 0)\} \leq \gamma^2\mathcal{E}\{||x_0^e||^2\}$.

8.5 Conclusions

In this chapter, we have formulated and solved an H_∞ output-feedback control problem for nonlinear stochastic systems. The theory which has been introduced in this work was facilitated by the concept of stochastic dissipation. As we utilized a certainty equivalence principle, we have facilitated the establishment of sufficient conditions, in terms of a pair of HJI, the solution of which guarantees a controller that renders the underlying closed-loop system an internal asymptotic stability in the mean square sense and L_2-gain that is less than or equal to a prescribed attenuation level. We note that we bring in Chapter 13, Section 13.4, an example, which concerns the output-feedback stabilization of an inverted pendulum.

9

l_2-Gain and Robust SF Control of Discrete-Time NL Stochastic Systems

9.1 Introduction

In recent years there has been a growing interest, as is reflected in the various published research works, in the extension of H_∞ control and estimation theory to accommodate stochastic systems (see e.g., [22], [26], [27], [44], [47], [69]). Most of the research that has been done this area deals with linear systems. The works of [6] and [17] investigate the problem of H_∞ control for continuous-time nonlinear stochastic systems. While [17] has taken the differential games approach to achieve an H_∞ output control, [6] and [8] present an H_∞ control theory for continuous-time systems based on the concept of stochastic dissipation. The work of Ugrinovskii [114] is also related to our work as it deals with filtering of uncertain linear stochastic systems where the uncertainty is characterized in terms of probability distribution.

The main thrust for these efforts stems from the attempt to model system uncertainties as a stochastic process, in particular, as a white noise, or formally as a Wiener process. This has led to the development of an H_∞ control and estimation theory for stochastic linear systems with multiplicative noise.

In analogy to the continuous-time case (see [6], [8]), we first establish conditions under which the system (1.19) possesses a storage function (that is, the system is dissipative). We introduce below a particular positive function, which is shown to be a storage function, provided its expected value is finite.

Definition 9.1.1 *Consider the system (1.19). Given an integer $j \geq 0$ and let x be an F_j-measurable, R^n valued random variable defined on (Ω, F, P) with $\mathcal{E}\{||x||^2\} < \infty$. Let $S(z_i, v_i)$ be as in Definition 1. Define*

$$V_j^a(x) = \sup_{\substack{l \geq j \\ v \in \mathcal{A}}} [-\mathcal{E}\{[\sum_{i=j}^{l-1} S(z_i, v_i)]|x\}] \tag{9.1}$$

where $\{x_i\}_{i \geq j}$ is the solution to (1.19) with the initial condition $x_j = x$.

E. Gershon & U. Shaked: *Advanced Topics in Control & Estimation*, LNCIS 439, pp. 155–163.
DOI: 10.1007/978-1-4471-5070-1_9

Remark 9.1. We note that in the deterministic setup V_j^a is called an available storage.

Theorem 9.1 below characterizes the smallest storage function of a system. For the continuous time counterpart, see Chapter 6.

Theorem 9.1. *Given a Borel function $S(z,v)$, the function $V^a = \{V_i\}_{0 \le i \le N}$, as defined by (9.1), is a storage function for the system (1.19) (or equivalently, the system (1.19) is dissipative with respect to the supply rate S) iff $\mathcal{E}\{V_i^a(x)\}$ is finite for all $i \ge 0$ and for all F_i-measurable, R^n valued random variable x with $\mathcal{E}\{||x||^2\} < \infty$. Moreover, if V_j is any other storage function for the system (1.19) which is associated with the same supply rate, then $V_j^a(x) \le V_j(x)$ for all $j \ge 0$ and for all $x \in R^n$.*

Proof: Assume first that $E\{V_j^a(x)\}$ is finite for all $j \ge 0$, and for all x which satisfies the hypothesis of the theorem. It is obvious that $V_j^a(x) \ge 0$ $P-a.e.$ as $l = j$ yields the smallest candidate: $V_j^a(x) = 0$ P-a.e. Let now k be arbitrarily fixed as long as it satisfies $0 \le j \le k$. Then,

$$E\{V_j^a(x_j)\} = E\{\sup{}_{N \ge l \ge j, v \in \mathcal{A}} [-\mathcal{E}\{[\textstyle\sum_j^{l-1} S(z_i,v_i)|x_j\}]\}$$
$$\ge E\{-\mathcal{E}\{[\textstyle\sum_{i=j}^{l} S(z_i,v_i)]|x_j\}\}$$

for all $l : l \ge j$ and for all $v \in \mathcal{A}$, which implies:

$$E\{V_j^a(x_j)\} \ge E\{-\mathcal{E}\{[\textstyle\sum_j^{k-1} S(z_i,v_i)]|x_j\}\} + E\{-\mathcal{E}\{[\textstyle\sum_k^{l-1} S(z_i,v_i)]|x_j\}\}$$
$$\forall l > k, \quad \forall v \in \mathcal{A}.$$

Now using elementary properties of the conditional expectation, and the fact that the last inequality holds for all $v \in \mathcal{A}$, we obtain the following.

$$\mathcal{E}\{V_j^a(x_j)\} \ge -\mathcal{E}\{\textstyle\sum_{i=j}^{k-1} S(z_i,v_i)\} + E[\sup_{\substack{l \ge k \\ v \in \mathcal{A}}} -\mathcal{E}\{[\textstyle\sum_{i=k}^{l-1} S(z_i,v_i)]|x_k\}]$$
$$= -\mathcal{E}\{\textstyle\sum_{i=j}^{k-1} S(z_i,v_i)\} + \mathcal{E}\{V_k^a(x_k)\} \quad \forall k \ge j, v \in \mathcal{A}$$

This establishes the system's dissipativity.

Suppose now that the system is dissipative with respect to the supply rate S, that is, it possesses a storage function V which satisfies $0 \le \mathcal{E}\{V_k(x_k)\} \le \mathcal{E}\{V_j(x_j)\} + \mathcal{E}\{\sum_{i=j}^{k-1} S(z_i,v_i)\}$ for all $k > j \ge 0$. This implies:

$$-\mathcal{E}\{\sum_{i=j}^{k-1} S(z_i,v_i)\} \le \mathcal{E}\{V_j(x_j)\} < \infty \quad \forall v \in \mathcal{A}. \tag{9.2}$$

Since $-\mathcal{E}\{\sum_{i=j}^{k-1} S(z_i,v_i)\} = \mathcal{E}\{-\mathcal{E}\{[\sum_{i=j}^{k-1} S(z_i,v_i)]|x_j\}\}$ it follows that

$$\mathcal{E}\{V_j^a(x_j)\} = \mathcal{E}\{ \sup_{k \geq j, v \in \mathcal{A}} [-\mathcal{E}\{[\sum_{i=j}^{k-1} S(z_i, v_i)]|x_j\}]\} \tag{9.3}$$

is finite for all $j : j \geq 0$. In view of (9.1) and by (9.2), it follows that for an arbitrary $x \in R^n$

$$V_j^a(x) \leq V_j(x) \quad \forall j \geq 0$$

This completes the proof. □

Having proved Theorem 9.1 we now state and prove what we call the BRL for stochastic nonlinear systems.

Theorem 9.2. *(BRL) Consider the system (1.19) with the controlled output (1.20) and with $u_k = 0$, $\forall k \geq 0$.*

*(**a**) Let $\gamma > 0$, and let $\{V_k\}_{k \geq 0}$ be a family of positive real valued functions: V_k: $R^n \to R^+$ satisfying the following HJI:*

$$V_k(x) \geq \sup_{v \in R^{m_1}} \{||z_k||^2 - \gamma^2||v||^2 + E_{\omega_k}\{V_{k+1}[f_k(x, v, \omega_k)]\}\}, \tag{9.4}$$

for all $x \in R^n$ and for all $v \in R^{m_1}$.

Assume also that $\mathcal{E}\{V_i(x)\} < \infty \quad \forall i \geq 0$ and for all F-measurable x with $\mathcal{E}\{||x||^2\} < \infty$. Then the system (1.19) has l_2-gain$\leq \gamma$.

*(**b**) Assume the system (1.19) has l_2-gain$\leq \gamma$ for some $\gamma > 0$. Then, there is a family $\{V_k\}_{k \geq 0}$ of positive functions which satisfy the HJI of (9.4).*

Proof:

(**a**) Let $\{v_k\}_{k \geq 0}$ be any sequence of random variables in $\mathcal{A}$, and let x_0 satisfy the above hypothesis. Let $\{x_k\}_{k \geq 0}$ be the solution to (1.19). Note that ω_k is independent of $\{x_k, v_k,\}$ as ω_k is a white sequence, and $\{x_k, v_k,\}$ are F_k-measurable. This implies

$$E_{\omega_k}\{V_{k+1}[f_k(x_k, v_k, \omega_k)]\} = \mathcal{E}\{V_{k+1}[f_k(x_k, v_k, \omega_k)]|x_k, v_k\}$$

Therefore, the HJI yields

$$V_k(x_k) \geq ||z_k(x_k)||^2 - \gamma^2||v_k||^2 + \mathcal{E}\{V_{k+1}[f_k(x_k, v_k, \omega_k)]|x_k, v_k\}$$

P-a.e. Applying now the expectation operation to both sides of the last inequality, we obtain

$$\mathcal{E}\{V_k(x_k)\} \geq \mathcal{E}\{||z_k(x_k)||^2\} - \gamma^2\mathcal{E}\{||v_k||^2\} + \mathcal{E}\{V_{k+1}(x_{k+1})\}$$

which clearly implies dissipativity. Taking now $\beta_k = V_k$, this yields l_2-gain$\leq \gamma$.

(**b**) Note that since the system has l_2-gain$\leq \gamma$, it follows that

$$\sum_{i=j}^{k-1} \mathcal{E}\{||z_i||^2\} - \gamma^2 \sum_{i=j}^{k-1} \mathcal{E}\{||v_i||^2\} < \infty$$

for all $0 \leq j \leq k$ and for all $\{v_i\}_{i\geq 0} \in \mathcal{A}$. Therefore, by Theorem 9.1, there is a family of positive functions $\{V_k\}_{k\geq 0}$ so that the system (1.19) is dissipative with respect to the supply rate $S(z,v) = \gamma^2||v||^2 - ||z||^2$. Fix $0 \leq k$, let $x \in R^n, v \in R^{m_1}$. and take $x_k = x$, $v_k = v$, then the dissipation implies

$$E_\omega\{V_{k+1}(f_k(x,v,\omega_k))\} \leq \mathcal{E}\{V_k(x)\} + \gamma^2\mathcal{E}\{||v||^2\} - \mathcal{E}\{||z_k||^2\} \tag{9.5}$$

for all $x \in R^n$, and for all $v \in R^{m_1}$. This obviously implies the inequality (9.4).

□

Remark 9.2. No boundary conditions have been imposed on $\{V_k\}_k$. In order to obtain a particular solution to (9.4) such boundary conditions should be specified. In the sequel, we focus on the stationary case where such conditions are irrelevant.

The question of stability is now in order. The particular definition of the l_2-gain property in this work is closely related to the system stability in various probability senses.

We first note that whenever the system (1.19) has l_2-gain$\leq \gamma$, and in the absence of exogenous disturbances, the following holds.

$$\mathcal{E}\{\sum_{i=0}^{\infty} ||z_k||^2\} \leq \gamma^2[\mathcal{E}\{||x_0||^2\}] \tag{9.6}$$

which implies $\lim_{k\to\infty} \mathcal{E}\{||z_k||^2\} \to 0$, from which it also follows that $\lim_{k\to\infty} \mathcal{E}\{||h_k(x_k)||^2\} \to 0$. So $h_i(x_i)$ converges to zero in the mean square sense and hence, also in probability. Assume now that $h_i(x) = 0$ implies $x = 0$ for all $i \in N^+$ and for all $x \in R^n$. Since As $h_i(x_i)$ converges to zero in probability it follows that the sequence $\{x_i\}_{i\geq 0}$ converges to zero in probability, that is, the system is stable in the probability sense.

Obviously, in the case for which $||h_i(x)|| \geq \alpha_i||x||$, the system is stable in the mean square sense also.

As an application of the BRL (Th. 9.2), we consider now the time-invariant case, that is:

$$x_{k+1} = f(x_k, v_k, u_k, \omega_k) \tag{9.7}$$

with the controlled output (1.20), and with a time invariant h. As the underlying system is now time invariant, the l_2-gain property as defined previously is now equivalent to

$$\mathcal{E}\{\sum_{i=0}^{k} ||z_i||^2\} \leq \gamma^2\mathcal{E}\{||x_0||^2\} + \mathcal{E}\{\sum_{i=0}^{k} [\gamma^2||v_i||^2 - ||z_i|^2|]\} \quad \forall k \geq 0.$$

We will assume in the sequel that $\sum_{i=1}^{\infty} \mathcal{E}\{||v_i||^2\} < \infty$. For the time invariant case, Theorem 9.1 is valid when the available storage function is taken to be time invariant as follows:

$$V^a(x) = \sup_{k \geq 0, v \in \mathcal{A}} [-\mathcal{E}\{[\sum_{i=0}^{k-1} S(z_i, v_i)] | x\}].$$

In this case one easily establishes, analogously to the time-varying case, a BRL which is a trivial corollary of Theorem 9.2. It yields a time-invariant algebraic HJI, the counterpart of the well known linear stochastic case. This corollary is stated below.

Corollary 9.1.1 *Consider the system (9.7) with the controlled output (1.20), and let* $u_k = 0 \quad \forall k \geq 0$.

*(**a**) Let* $\gamma > 0$*, and let* V *be a positive real valued function:* $V : R^n \to R^+$ *satisfying the following Hamilton-Jacobi Algebraic Inequality (entitled HJAI)*

$$V(x) \geq \sup_{v \in R^n} \{||z(x)||^2 - \gamma^2 ||v||^2 + E_{\omega_k}\{V[f(x, v, \omega_k)]\}\} \tag{9.8}$$

for all $x \in R^n$*. Assume also:* $\mathcal{E}\{V(x_0)\} \leq \gamma^2 \mathcal{E}\{||x_0||^2\}$ *for all* F_0*-measurable random variables* x_0 *with* $\mathcal{E}\{||x_0||^2\} < \infty$*. Then, the system of (9.7) has* l_2*-gain*$\leq \gamma$.

*(**b**) Assume that there is a control signal* $\{u_k\}_{k \geq 0}$ *which makes the* l_2*-gain of the system (9.7) less than or equal to* γ *for some* $\gamma > 0$ *and for all* F_0*-measurable* x_0 *with* $\mathcal{E}\{||x_0||^2\} < \infty$*. Then, there is a positive function* V *which satisfies the HJI (9.8).*

9.2 Sufficient Conditions for l_2- Gain$\leq \gamma$: A Special Case

In this section we consider a special case of (1.19) which is characteristic of a large class of mechanical systems. For this class of systems we establish a more tractable sufficient conditions in terms of a HJI. Thus, we consider the following system:

$$x_{k+1} = f_k(x_k, v_k, \omega_k), \quad z_k = h_k(x_k)$$

In what follows we assume a special case of the above nonlinear system:

$$x_{k+1} = f_k(x_k) + g_k^1(x_k) v_k + g_k^2(x_k) v_k \omega_k^2 + G_k(x_k) \omega_k^1 \tag{9.9}$$

where $\{\omega_k\} \triangleq \left\{ \left[(\omega_k^1)^T \ \omega_k^2 \right]^T \right\}$ is an R^{l+1}-valued white noise sequence with uncorrelated components defined on the probability space (Ω, F, P), with the covariance $\mathcal{E}\{\omega_k \omega_k^T\} = \bar{R} = diag\{r_1, r_2, ..., r_l, r\} = diag\{R, r\}$. Obviously, $R = \mathcal{E}\{\omega_k^1 \omega_k^{1T}\}$ and $r = \mathcal{E}\{(\omega_k^2)^2\}$.

The exogenous disturbance $\{v_k\}_{k \geq 0}$ is assumed to be $\{F_k\}_{k \geq 0}$-adapted, that is v_k is F_{k-1}-measurable for all k, where F_k is the minimal σ-algebra generated by $\{\omega_i : i \leq k-1\}$. Moreover, it is assumed that $\sum_{k=1}^{\infty} ||v_k||^2 < \infty$.

The functions $f_k : R^n \rightarrow R^n$, $g_k^1 : R^n \rightarrow R^{n\times m_1}$, $g_k^2 : R^n \rightarrow R^{n\times m_1}$, and $G_k : R^n \rightarrow R^{n\times l}$ are all assumed to be continuous. It is further assumed that the above functions are of the following structure. $G_k(x)$ is taken to be an $n \times l$ matrix, so that

$$G_k(x) = [\ G_k^1(x)x \quad G_k^2(x)x \cdots G_k^l(x)x\ ] \tag{9.10}$$

where $G_k^i(x)$ are $n \times n$ matrices. It follows that
$G_k(x)\omega_k = \sum_{i=1}^l G_k^i(x)x\omega_k^i$.

It is also assumed that: $f_k(x) = F_k(x)x$ and $h_k(x) = H_k(x)x$. Application of the BRL yields the following.

Lemma 9.2.1 *Consider the system (9.9), with $f_k, g_k^1, g_k^2, G_k, h_k$ as defined above*

*(**a**) Let $\{Q_k(x)\}_{k\geq 0}$ be a family of symmetric positive matrices for all $x \in R^n$ which satisfies*

$$(i)\phi_k(x) \triangleq [\gamma^2 I - g_k^{1T}(x)Q_{k+1}(x)g_k^1(x) - rg_k^{2T}(x)Q_{k+1}(x)g_k^2(x)] > 0 \tag{9.11}$$

for all x in R^n.

$$\begin{aligned}(ii)x^T Q_k(x)x \geq x^T\{F^T(x)Q_{k+1}(x)F(x) + \sum_{i=1}^{l} r_i(G_k^i(x))^T Q_{k+1}(x)G_k^i(x)\}x \\ +F^T(x)Q_{k+1}(x)g_k^1(x)\phi_k^{-1}(x)g_k^{1T}(x)Q_{k+1}(x)F(x)\end{aligned} \tag{9.12}$$

for all x in R^n, where $r_i = E(\omega_k^i\omega_k^i)$.

$$(iii)Q_0(x) \leq \gamma^2 I, \quad \forall x \in R^n. \tag{9.13}$$

Then, the system (9.9) has l_2-gain$\leq \gamma$.

*(**b**) Assume the system (9.9) is dissipative with respect to the supply rate $\gamma^2||v||^2 - ||z||^2$ for some $\gamma > 0$, with a storage function $V_i(x) = x^T Q_i(x)x$, where $\{Q_i(x)\}_{i\geq 0}$ are symmetric with $Q_i(x) \geq 0$ for all $x \in R^n$. Then, the inequality (9.12) holds.*

Proof:

(**a**) Define $V_k(x)$ as $V_k(x) = x^T Q_k(x)x$, for all $x \in R^n$. Utilizing (9.12), it is straightforward to show that $\{V_k(x)\}_k$ satisfies (9.4), which together with (*iii*) imply, in view of the BRL, that the system (9.9) has l_2-gain$\leq \gamma$. In fact, in view of the particular structure of $V_k(x)$ we have

$$\begin{aligned}&\sup_v\{|z_k|^2 - \gamma^2|v|^2 + E_{\omega_k}[V_{k+1}\{f_k(x_k, v, \omega_k)\}]\}\\ &= \sup_v\{x^T H_k^T(x)H_k(x)x - \gamma^2|v|^2 + [F_k(x)x + g_k^1(x)v]^T Q_{k+1}(x)[F_k(x)x + g_k^1(x)v]\\ &+Tr[R^{\frac{1}{2}}G_k^T(x)Q_{k+1}G_k(x)R^{\frac{1}{2}}]\} + rv^T g_k^{2T}Q_{k+1}g_k^2 v\end{aligned} \tag{9.14}$$

where $R = diag\{[r_i]_{i=1}^l\}$.

From (9.11) it follows that there is a unique maximizing $v_k = v_k(x)$ given by

$$v_k = [\gamma^2 I - g_k^{1T}(x)Q_{k+1}(x)g_k^1(x) - rg_k^{2T}(x)Q_{k+1}(x)g_k^2(x)]^{-1}g_k^T(x)Q_{k+1}(x)F_k(x)x \tag{9.15}$$

Substituting (9.15) in (9.14) and using the following identity

$$Tr[R^{\frac{1}{2}}G_k^T(x)Q_{k+1}G_k(x)R^{\frac{1}{2}}] = x^T\{\sum_i r_i[G_k^i(x)]^T Q_{k+1}(x)G_k^i(x)\}x \tag{9.16}$$

in (9.14) we conclude, in view of (9.12), that the inequality (9.4) holds with $\{V_k(x)\}_k$ as defined above. Therefore, by the BRL the system has l_2-gain$\leq \gamma$.

(**b**) Since the sequence $\{x^T Q_k(x)x\}$ satisfies is a storage function for the system (9.9) it satisfies HJI of (9.4). Now evaluating the maximizing v, it yields the inequality (9.12). □

9.3 Norm-Bounded Uncertainty

We consider the infinite-time horizon case where the underlying stochastic system is time-invariant. The system considered is:

$$x_{k+1} = f(x_k)x_k + G(x_k)w_k + g_1(x_k)v_k + \bar{g}(x_k)u_k, \quad z_k = col\{h(x_k),\ \bar{D}u_k\} \tag{9.17}$$

where:

$$f(x) = (A + HF(x)E_1), \quad G(x) = (A_1 + HF(x)E_3)x,$$
$$g_1(x) = B_1, \quad \bar{g}(x) = B_2 + HF(x)E_2, \quad h(x) = \bar{C}_1 x$$

and where $\bar{D}$ is a constant matrix of appropriate dimensions. The sequence $\{w_k\}$ is a standard white noise scalar sequence and u_k is the control input, $G \in \mathcal{R}^{n\times n}$, $g \in \mathcal{R}^{n\times r}$, and $\bar{g} \in \mathcal{R}^{n\times m}$. The nonlinear part $F(x)$, which is in $\mathcal{R}^{q_1\times q_2}$, is assumed to be bounded. Namely,

$$F(x)^T F(x) \leq I. \tag{9.18}$$

In the sequel we view $F(x)$ as the norm-bounded uncertainty and we consider the case where $z_k(x) = C_1 x_k + Du_k$, where C_1 and D are constant matrices and $D^T D > 0$, namely we take $C_1 = \begin{bmatrix} \bar{C}_1 \\ 0 \end{bmatrix}$ and $D = \begin{bmatrix} 0 \\ \bar{D} \end{bmatrix}$, and we seek a state-feedback control $u_k = Kx_k$, where $K \in \mathcal{R}^{m\times n}$ is a constant gain matrix, that achieves a prescribed bound γ on the system's ℓ_2-gain.

The latter problem can be readily solved in the deterministic case by applying a linear approach. The extension of this approach to the stochastic case is, however, not immediate and one has to base his solution on the theory of the above sections in order to obtain a correct answer. Choosing $V_k(x_k) = x_k^T Q x_k$, where $0 < Q \in \mathcal{R}^{n\times n}$, it follows from Lemma 9.2.1 that the inequality which guarantees the required bound on the system's gain is given by:

$$
\begin{aligned}
&-Q+A_F^TQA_F+A_F^TQB_1(\gamma^2I-B_1^TQB_1)^{-1}B_1^TQA_F\\
&+(A_1+HF(x)E_3)^TQ(A_1+HF(x)E_3)+C_1^TC_1+K^TD^TDK<0,
\end{aligned}
\tag{9.19}
$$

for all $x \in \mathcal{R}^n$, where $A_F = A+B_2K+HF(x)(E_1+E_2K)$. The latter can be put in an equivalent matrix inequality applying Schur complement formula. The resulting inequality is:

$$
\begin{bmatrix}
-Q+C_1^TC_1 & A_F^TQ & 0 & (A_1+HF(x)E_3)^TQ & K^T\\
* & -Q & QB_1 & 0 & 0\\
* & * & -\gamma^2 I & 0 & 0\\
* & * & * & -Q & 0\\
* & * & * & * & -I
\end{bmatrix} \le 0. \tag{9.20}
$$

Multiplying the latter, from both sides, by $diag\{\bar{Q},\ \bar{Q},\ I,\ \bar{Q},\ I\}$ where $\bar{Q} \triangleq Q^{-1}$ and defining $Y = K\bar{Q}$ the following LMI is obtained.

$$
\begin{bmatrix}
-\bar{Q} & \bar{Q}A_F^T & 0 & \bar{Q}(A_1+HF(x)E_3)^T & Y^TD^T & \bar{Q}C_1^T\\
* & -\bar{Q} & B_1 & 0 & 0 & 0\\
* & * & -\gamma^2 I & 0 & 0 & 0\\
* & * & * & -\bar{Q} & 0 & 0\\
* & * & * & * & -I & 0\\
* & * & * & * & * & -I
\end{bmatrix} \le 0. \tag{9.21}
$$

The latter inequality can also be written as:

$$
\Gamma + \Phi_1F(x)\Phi_2^T + \Phi_2F^T(x)\Phi_1^T + \Phi_3F(x)\Phi_4^T + \Phi_4F^T(x)\Phi_3^T \le 0 \tag{9.22}
$$

where

$$
\begin{aligned}
&\Phi_1 = \begin{bmatrix}0 & H^T & 0 & 0 & 0 & 0\end{bmatrix}^T,\ \Phi_2 = \begin{bmatrix}E_1\bar{Q}+E_2Y & 0 & 0 & 0 & 0\end{bmatrix}^T,\\
&\Phi_3 = \begin{bmatrix}0 & 0 & 0 & H^T & 0 & 0\end{bmatrix}^T,\ \Phi_4 = \begin{bmatrix}E_3\bar{Q} & 0 & 0 & 0 & 0 & 0\end{bmatrix}^T
\end{aligned}
\tag{9.23}
$$

and where

$$
\Gamma = \begin{bmatrix}
-\bar{Q} & \bar{Q}A^T+Y^TB_2^T & 0 & \bar{Q}A_1^T & Y^TD^T & \bar{Q}C_1^T\\
* & -\bar{Q} & B_1 & 0 & 0 & 0\\
* & * & -\gamma^2 I & 0 & 0 & 0\\
* & * & * & -\bar{Q} & 0 & 0\\
* & * & * & * & -I & 0\\
* & * & * & * & * & -I
\end{bmatrix}. \tag{9.24}
$$

Using the fact that for any two matrices α and β of compatible dimensions and for any positive definite matrix $\bar{R}$ of compatible dimensions the following holds: $\alpha\beta^T + \beta\alpha^T \leq \alpha\bar{R}\alpha^T + \beta\bar{R}^{-1}\beta^T$, together with the bound of (9.18), we obtain the following:

$$\Gamma + \Phi_1\bar{R}_1\Phi_1^T + \Phi_2\bar{R}_1^{-1}\Phi_2^T + \Phi_3\bar{R}_2\Phi_3^T + \Phi_4\bar{R}_2^{-1}\Phi_4^T \leq 0 \tag{9.25}$$

for some positive definite matrices $\bar{R}_1$ and $\bar{R}_2$. Choosing the latter matrices to be scalar matrices we obtain the following.

Theorem 9.3. *Consider the system of (9.17). Given the scalar* $0 < \gamma$, *there exists a constant state-feedback controller* $u_k = Kx_k$ *that globally stabilizes the closed-loop exponentially, in the mean-square sense, and achieves* l_2*-gain*$\leq \gamma$ *if there exist* $0 < \bar{Q} \in R^{n\times n}$, $Y \in R^{p\times n}$ *and scalars* ρ_1 *and* ρ_2 *so that the following LMI is satisfied:*

$$\begin{bmatrix}
-\bar{Q} & \bar{Q}A^T{+}Y^TB_2^T & 0 & \bar{Q}A_1^T & Y^TD^T & \bar{Q}C_1^T & 0 & 0 & \bar{Q}E_1^T{+}Y^TE_2^T & \bar{Q}E_3^T \\
* & -\bar{Q} & B_1 & 0 & 0 & 0 & H\rho_1 & 0 & 0 & 0 \\
* & * & -\gamma^2 I & 0 & 0 & 0 & 0 & 0 & 0 & 0 \\
* & * & * & -\bar{Q} & 0 & 0 & 0 & H\rho_2 & 0 & 0 \\
* & * & * & * & -I & 0 & 0 & 0 & 0 & 0 \\
* & * & * & * & * & -I & 0 & 0 & 0 & 0 \\
* & * & * & * & * & * & -\rho_1 I & 0 & 0 & 0 \\
* & * & * & * & * & * & * & -\rho_2 I & 0 & 0 \\
* & * & * & * & * & * & * & * & -\rho_1 I & 0 \\
* & * & * & * & * & * & * & * & * & -\rho_2 I
\end{bmatrix} < 0. \tag{9.26}$$

If a solution to this LMI exists, the state-feedback gain is given by $K = Y\bar{Q}^{-1}$.

9.4 Conclusions

Discrete-time nonlinear systems with state-dependent noise are considered. Sufficient conditions for the robust stability of these systems and for achieving a given bound on their l_2-gain are obtained in the special case where the nonlinearity of the systems stems from the norm-bounded uncertainty model. These conditions are expressed in terms of LMIs. These LMIs are applied to achieve a required H_∞ performance via state-feedback control. An engineering-oriented control example of inverted pendulum state-feedback control is given in Chapter 13, Section 13.5.

10

H_∞ Output-Feedback Control of Discrete-Time Systems

We consider the following system

$$x_{k+1} = f_k(x_k, v_k, u_k, \omega_k) \tag{10.1}$$

together with a measurement process:

$$y_k = C(x_k) + D(x_k)v_k + G^2(x_k)w_k^2$$

where $\{x_k\}_{k\geq 0}$ is a solution to (10.1), with: an initial condition x_0, an exogenous disturbances $\{v_k\}_{k\geq 0}$, a control signal $\{u_k\}_{k\geq 0}$, and a white noise sequence $\omega = \{\omega_k\}_{k\geq 0}$ defined on a probability space (Ω, F, P). We assume that the assumptions of subsection 1.3.4 are all satisfied. We also consider the following controlled output

$$z(x, u) = col\{h(x),\ D_{12}u\} \tag{10.2}$$

where $h : R^n \to R^r$, and $D_{12}^T D_{12} = R > 0$.

In Chapter 9 (Th. 9.1) a BRL condition has been derived that guarantees a prescribed bound on the l_2−gain of the system 10.1. As an application of this BRL we consider now the time-invariant version of (10.1), that is:

$$x_{k+1} = f(x_k, v_k, u_k, \omega_k) \tag{10.3}$$

with the controlled output (10.2) and with time-invariant h. In this case one easily establishes, analogously to the time-varying case, a BRL which is a trivial corollary of Theorem 9.1. It yields a time-invariant algebraic HJI, the counterpart of the well known linear stochastic case. This BRL is given in the following Corollary.

Corollary 10.0.1 *Consider the system (10.3) with the controlled output (10.2), and let* $u_k = 0 \quad \forall k \geq 0$.

(a) *Let* $\gamma > 0$*, and let* V *be a positive real valued function:* $V : R^n \to R^+$ *satisfying the following Hamilton–Jacobi Algebraic Inequality (HJAI)*

E. Gershon & U. Shaked: *Advanced Topics in Control & Estimation,* LNCIS 439, pp. 165–176.
DOI: 10.1007/978-1-4471-5070-1_10

$$V(x) \geq \sup_{v \in R^n} \{||z(x)||^2 - \gamma^2 ||v||^2 + \mathcal{E}_{\omega_k}\{V[f(x, v, \omega_k)]\}\} \tag{10.4}$$

for some control signal $\{u_k\}_{k\geq 0}$, *and for all* $x \in R^n$. *Assume also:* $\mathcal{E}\{V(x_0)\} \leq \gamma^2 \mathcal{E}\{||x_0||^2\}$ *for all* F_0*-measurable random variables* x_0 *with* $\mathcal{E}\{||x_0||^2\} < \infty$. *Then, the system of (10.3) has* l_2*-gain*$\leq \gamma$.

(**b**) *Assume there is a control signal* $\{u_k\}_{k\geq 0}$ *which makes the system (10.3)* l_2*-gain*$\leq \gamma$ *for some* $\gamma > 0$ *and for all* F_0*-measurable* x_0 *with* $\mathcal{E}\{||x_0||^2\} < \infty$. *Then, there is a positive function* V *which satisfies the HJI (10.4).*

We remark that exponential mean square stability in the time invariant case is easily established under a proper adjustment of the conditions obtained for exponential mean square stability for the time varying case.

10.1 Sufficient Conditions for l_2-Gain$\leq \gamma$: A Special Case

In this section a special case of (10.1) is considered which characterizes a large class of systems (e.g. mechanical systems). For this system we establish a more tractable sufficient condition in terms of a HJI. We consider the following system:

$$x_{k+1} = f(x_k) + g^1(x_k)v_k + g^2(x_k)v_k\omega_k^2 + G(x_k)\omega_k^1, \quad z = h(x_k) \tag{10.5}$$

where $\{\omega_k^2\}_k$ is an R^l-valued white noise sequence with uncorrelated components defined on the probability space $(\Omega, \mathcal{F}, P)$, and with the covariance $\mathcal{E}\{\omega_k^1(\omega_k^1)^T\} = R = diag\{r_1, r_2, ..., r_l\}$. Also $\{\omega_k^1\}_k$ is an R^1-valued white noise process, with the covariance $r = \mathcal{E}\{(\omega_k^2)^2\}$. The exogenous disturbance $\{v_k\}_{k\geq 0}$ is assumed to be $\{\mathcal{F}_k\}_{k\geq 0}$-adapted, that is v_k is $\mathcal{F}_{k-1}$-measurable for all k, where $\mathcal{F}_k$ is the minimal σ-algebra generated by $\{\omega_i : i \leq k-1\}$. Moreover, it is assumed that $\sum_{k=1}^{\infty} ||v_k||^2 < \infty$. The functions $f_k : R^n \to R^n$, $g_k^1 : R^n \to R^{n\times m_1}$, $g_k^2 : R^n \to R^{n\times m_1}$, and $G_k : R^n \to R^{n\times l}$ are all assumed to be continuous. It is further assumed that the above functions are of the following structure. $G_k(x)$ is taken to be an $n \times l$ matrix, so that

$$G_k(x) = [G_k^1(x)x \quad G_k^2(x)x \cdots G_k^l(x)x] \tag{10.6}$$

where $G_k^i(x)$ are $n \times n$ matrices. It follows that$G_k(x)\omega_k = \sum_{i=1}^{l} G_k^i(x)x\omega_k^i$.

It is also assumed that: $f_k(x) = F_k(x)x$ and $h_k(x) = H_k(x)x$. Application of the BRL yields the following.

Lemma 10.1.1 *Consider the system (10.5), with* $f_k, g_k^1, g_k^2, G_k, h_k$ *as defined above*

(**a**) *Let* $\{Q_k(x)\}_{k\geq 0}$ *be a family of symmetric positive matrices for all* $x \in R^n$ *satisfying*

$$(i)\quad \phi_k(x) \triangleq [\gamma^2 I - g_k^{1T}(x)Q_{k+1}(x)g_k^1(x) - rg_k^{2T}(x)Q_{k+1}(x)g_k^2(x)] > 0 \quad (10.7)$$

for all x in R^n.
(ii) Assume also that there is a positive function $V : R^n \to R^+$ with $V(0) = 0$ which satisfies

$$\begin{aligned}&\mathcal{E}_{w_k}\{V_k[f_k(x)+g_k(x)u+g_k^1(x)v+G_k^1(x)w_k^1]\}-V_k(x)\\ &\leq\mathcal{E}_{w_k}\{[f_k(x)+g_k(x)u + g_k^1(x)v + G_k^1(x)w_k^1]^T Q_k(x)[f_k(x) + g_k(x)u + g_k^1(x)v\\ &+G_k^1(x)w_k^1]\} - x^T Q_k(x)x\end{aligned} \quad (10.8)$$

and

$$\begin{aligned}x^T Q_k(x)x > x^T\{&F_k^T(x)Q_{k+1}(x)F_k(x) + \sum_{i=1}^{l} r_i(G_k^i(x))^T Q_{k+1}(x)G_k^i(x)\\ &+F_k^T(x)Q_{k+1}(x)g_k^1(x)\phi_k^{-1}(x)g_k^{1T}(x)Q_{k+1}(x)F_k(x) + H_k^T(x)H_k(x)\}x\end{aligned} \quad (10.9)$$

for all x in R^n, where $r_i = \mathcal{E}(\omega_{ki}^1\omega_{ki}^1)$.

$$(iii)\qquad Q_0(x) \leq \gamma^2 I, \quad \forall x \in R^n. \quad (10.10)$$

Then, the system (10.5) has l_2-gain$\leq \gamma$.

(b) *Assume the system (10.5) is dissipative with respect to the supply rate $\gamma^2||v||^2 - ||z||^2$ for some $\gamma > 0$, with a storage function $V_i(x) = x^T Q_i(x)x$, where $\{Q_i(x)\}_{i\geq 0}$ are symmetric with $Q_i(x) \geq 0$ for all $x \in R^n$. Then, the inequality (10.9) holds.*

Remark 10.1. Assumption *(ii)* in the above lemma is trivially satisfied in the case where Q_k is independent of the state vector. It can also be applied in the case where the nonlinearities of (10.5) are of a polynomial type, for which the method of SOS (see, e.g [101]) can be used to obtain an equivalent problem with a quadratic storage function (namely, with an augmented Q_k that is independent of the states).

In the case where $\bar{g}_k^2$ in (10.9) is identically zero, we apply the matrix inversion lemma to the third term in the right side of (10.9) and obtain the following.

Corollary 10.1.1 *In the case where $\bar{g}_k^2$ is identically zero, the requirement of (10.9) becomes:*

$$\begin{aligned}&-\bar{Q}_k(x)+\bar{F}_k^T(x)[\bar{Q}_{k+1}^{-1}(x)-\gamma^{-2}\bar{g}_k^1(x)\bar{g}_k^{1T}(x)]^{-1}\bar{F}_k(x)+\sum_{i=1}^{l} r_i(\bar{G}_k^i(x))^T\bar{Q}_{k+1}(x)\bar{G}_k^i(x)\\ &+\bar{H}_k^T(x)\bar{H}_k(x)<0\end{aligned}$$

Consider now the time-invariant version of (10.5), that is:

$$x_{k+1} = f(x_k) + g^1(x_k)v_k + g^2(x_k)v_k\omega_k^2 + G(x_k)\omega_k^1, \quad z_k = h(x_k) \quad (10.11)$$

where the assumptions associated with (10.5) now read

$$G(x) = [G^1(x)x \quad G^2(x)x \cdots G^l(x)x] \tag{10.12}$$

so that $G(x)\omega_k = \sum_{i=1}^{l} G^i(x)x\omega_k^i$, $f(x) = F(x)x$ and $h(x) = H(x)x$.

In this context, Eq. (10.7) and the inequality (10.9) of Lemma 10.1.1 reduce to

$$\phi(x) \triangleq [\gamma^2 I - g^{1T}(x)Q(x)g^1(x) - rg^{2T}(x)Q(x)g^2(x)] > 0 \tag{10.13}$$

for all x in R^n. Assume also that there is a positive function $V : R^n \to R^+$ with $V(0) = 0$ which satisfies

$$\begin{aligned}&\mathcal{E}_{w_k}\{V[f(x)+g(x)u+g^1(x)v+G^1(x)w_k^1]\}-V(x)\\ &\leq \mathcal{E}_{w_k}\{[f(x)+g(x)u+g^1(x)v+G^1(x)w_k^1]^T Q(x)[f(x)+g(x)u+g^1(x)v\\ &+G^1(x)w_k^1]\}-x^T Q(x)x\end{aligned} \tag{10.14}$$

and

$$\begin{aligned}x^T Q(x)x > x^T\{&F^T(x)Q(x)F(x) + \sum_{i=1}^{l} r_i(G^i(x))^T Q(x)G^i(x)\\ &+F^T(x)Q(x)g^1(x)\phi^{-1}(x)g^{1T}(x)Q(x)F(x) + H^T(x)H(x)\}x\end{aligned} \tag{10.15}$$

for all x in R^n. In the case where $g^2 = 0$ the latter inequality is satisfied if, for all x in R^n,

$$\begin{aligned}&-Q(x)+F^T(x)[Q(x)^{-1}-\gamma^{-2}g^1(x)g^{1T}(x)]^{-1}F(x)+\sum_{i=1}^{l} r_i(G^i(x))^T Q(x)G^i(x)\\ &+H^T(x)H(x)<0\end{aligned} \tag{10.16}$$

The following example demonstrates the applicability of the algorithm of (10.13) and (10.15).

10.1.1 Example

We consider the following model for a single degree of freedom inverted pendulum with the multiplicative white noise ω:

$$mL^2\ddot{\theta} - mgL\sin(\theta) + (\varsigma + \omega)\dot{\theta} + \kappa\theta = u + 2v \tag{10.17}$$

where κ is the spring coefficient and ς is a damping coefficient. The signal v is a deterministic disturbance acting on the control input u and ω is a stochastic uncertainty in the damping. This uncertainty stems from the facts that damping is temperature dependent and that temperature possesses a component that can be modeled as a standard white-noise ($r_1 = 1$) . In this model, θ is the inclination angle of the pendulum, L and m are its length and mass, respectively, and g is the gravitation coefficient.

Denoting $x_1 = \theta$ and $x_2 = ml^2\dot{\theta}$, we apply Euler discretization to the model of (10.17) and obtain the following discrete-time state-space representation of the system.

$$\begin{bmatrix} x_1 \\ x_2 \end{bmatrix}_{k+1} = \begin{bmatrix} 1 & \frac{T}{ml^2} \\ Tmgl\frac{\sin(x_{1,k})}{x_{1,k}} - \kappa T & 1 - \frac{T\varsigma}{ml^2} \end{bmatrix} \begin{bmatrix} x_1 \\ x_2 \end{bmatrix}_k + \begin{bmatrix} 0 \\ T \end{bmatrix} u_k$$
$$+ \begin{bmatrix} 0 \\ 2T \end{bmatrix} \nu_k + \begin{bmatrix} 0 \\ -\sqrt{T}\frac{x_{2,k}}{ml^2} \end{bmatrix} \omega_k$$

where T is the sampling period. In this case $l = 1$ and we seek a constant state-feedback control gain matrix $K = \begin{bmatrix} k_1 & k_2 \end{bmatrix}$ such that the control law $u_k = Kx_k$ achieves l_2-gain $\leq \gamma$ for the closed-loop system that relates the disturbance ν_k to the angle θ_k with a weight of 0.1 on the control effort u_k, namely, $z_k = col\{x_{1k},\ 0.1u_k\}$.

Using the notation associated with (10.11) we have now $f(x) = F(x)x$ where

$$F(x) = \hat{F}(x) + \begin{bmatrix} 0 \\ T \end{bmatrix} K,\ \hat{F}(x) = \begin{bmatrix} 1 & \frac{T}{mL^2} \\ \alpha(x) - \kappa T & 1 - \frac{T\varsigma}{mL^2} \end{bmatrix},\ \alpha(x) = TmgL\frac{\sin(x_1)}{x_1},$$
$$g^1(x) = \begin{bmatrix} 0 \\ 2T \end{bmatrix};\quad G^1(x) = \begin{bmatrix} 0 & 0 \\ 0 & -\frac{\sqrt{T}}{mL^2} \end{bmatrix},\quad \text{and} \quad H = \begin{bmatrix} [1\ \ 0] \\ 0.1K \end{bmatrix}$$

We require a minimum γ in the case where: $m = 0.5kg$, $L = 0.7meter$, $\kappa = 0.5Newton/meter$ and $\varsigma = 0.25$ and where $T = 0.01sec$.

Considering a constant Q, the only function in (10.16) that depends on x is $F(x)$. Applying Schur complement formula, Eqs. (10.16) and (10.7) are then equivalent to the following matrix inequalities.

$$\begin{bmatrix} -Q & F^T(x)Q & 0 & H^T & G^{1T}Qr_1 \\ * & -Q & g^1 & 0 & 0 \\ * & * & -\gamma^2 I & 0 & 0 \\ * & * & * & -I & 0 \\ * & * & * & * & -r_1 Q \end{bmatrix} < 0, \quad \begin{bmatrix} \gamma^2 I & g^{1T}Q \\ * & Q \end{bmatrix} > 0. \tag{10.18}$$

The latter two inequalities should be satisfied by Q and K for all x in R^n. The first one, (10.18a), is bilinear in Q and K. Note that, in fact, (10.18b) is included in (10.18a) and thus only the latter should be solved for. We, therefore, define $P = Q^{-1}$ and $Y = KP$ and multiply, from both sides,

(10.18a), by $diag\{P, P, I, I, P\}$. We obtain the following LMI in the decision variables P and Y.

$$\begin{bmatrix} -P & P\hat{F}^T(\alpha) + Y^T[0 \;\; T] & 0 & P\begin{bmatrix}1\\0\end{bmatrix} & 0.1Y^T & PG^{1T}r_1 \\ * & -P & g^1 & 0 & 0 & 0 \\ * & * & -\gamma^2 I & 0 & 0 & 0 \\ * & * & * & -I & 0 & 0 \\ * & * & * & 0 & -I & 0 \\ * & * & * & * & * & -r_1 P \end{bmatrix} < 0 \quad (10.19)$$

where α is considered as an uncertain parameter that resides in the interval $[0 \;\; TmgL]$. Solving the latter LMIs at the two vertices ($\alpha = 0,\ TmgL$) a minimum attenuation level $\gamma = 0.2193$ for $K = -[22.5908 \quad 99.9143]$ (and $P = \begin{bmatrix} 8.2656 & -1.8945 \\ * & 8.7656 \end{bmatrix}$).

10.2 The OF Case

In this section we develop an output-feedback controller. the development is based on utilization of the bounded real lemma which has been established in the previous section. As only noisy partial state observations are available, we employ a nonlinear observer in the synthesis of the desired controller.

In what follows we consider a time invariant nonlinear system. This system, together with its observation process is given by

$$\begin{aligned} x_{k+1} &= f(x_k) + g^1(x_k)v_k + g(x_k)u_k + G^1(x_k)w_k^1 \\ y_k &= C(x_k) + D(x_k)v_k + G^2(x_k)w_k^2. \end{aligned} \quad (10.20)$$

In addition, the controlled output z of (10.2) is considered. Here, y is R^r-valued, w^1 and w^2 are independent white noise stochastic processes which are assumed to be one dimensional each (this assumption is made for the sake of convenience only). The matrices C, D and G^2 are taken to be of pertinent dimensions.

In order to establish the output-feedback case we need little more than the BRL as the central controller is to be utilized in the sequel rather than the one we have established in the preceding section. In fact we have the following theorem

Theorem 10.2.1 *Consider the the system (10.20) together with the controlled output (10.2) and let x_0 be an $\mathcal{F}_0$-measurable with $\mathcal{E}\{||x_0||^2\} < \infty$. Assume there is a positive matrix $Q(x) : Q(x) \geq \alpha I, \quad \forall x \in R^n$, for some $\alpha > 0$, so that*

$$\mathcal{E}\{x_0^T Q x_0\} \leq \gamma^2 \mathcal{E}\{||x_0||^2\}$$

for a prescribed positive scalar γ. *Let* $\phi_v(x) \triangleq \gamma^2 I - g^{1T}(x)Q(x)g^1(x)$ *and assume that* $\phi_v(x) > 0$ *for all* $x \in R^n$. *Define* $\phi_u(x) \triangleq R + g^T(x)\bar{Q}(x)g(x)$ *where* $\bar{Q}(x) = Q(x)g^1(x)\phi_v^{-1}(x)g^{1T}(x)Q(x) + Q(x)$, *and let:*
$v^*(x) = \phi_v^{-1}(x)g^{1T}(x)Q(x)[f(x) + g(x)u^*(x)]$
where
$u^*(x) = -\phi_u^{-1}(x)g^T(x)\bar{Q}(x)f(x)$.

Assume also that there is a positive function $V : R^n \rightarrow R^+$ *with* $V(0) = 0$ *which satisfies*

$$\begin{aligned}&\mathcal{E}_{w_k}\{V[f(x)+g(x)u+g^1(x)v+G^1(x)w_k^1]\}-V(x)\\&\leq\mathcal{E}_{w_k}\{[f(x)+g(x)u + g^1(x)v + G^1(x)w_k^1]^T Q(x)[f(x) + g(x)u + g^1(x)v\\&+G^1(x)w_k^1]\} - x^T Q(x)x\end{aligned} \tag{10.21}$$

Then, a sufficient condition for the closed-loop system to have l_2*-gain*$\leq \gamma$ *is the existence of the following HJI:*

$$\begin{aligned}&h^T(x)h(x)+f^T(x)[\bar{Q}(x)-\bar{Q}(x)g(x)\phi_u^{-1}(x)g^T(x)\bar{Q}(x)]f(x)-x^T Qx\\&+rx^T G^{1T}(x)Q(x)G^1(x)x \leq 0\end{aligned}$$

for all $x \in R^n$.

Proof: Applying completion to squares, the following is readily obtained.

$$\begin{aligned}&h^T(x)h(x) + ||D_{12}u||^2 - \gamma^2||v||^2 - x^T Q(x)x\\&+\mathcal{E}_{w_k}\{[f(x) + g(x)u + g^1(x)v + G^1(x)w_k^1]^T Q(x)[f(x) + g(x)u + g^1(x)v\\&+G^1(x)w_k^1]\} = h^T(x)h(x) + f^T(x)[\bar{Q}(x) - \bar{Q}(x)g(x)\phi_u^{-1}(x)g^T(x)\bar{Q}(x)]f(x)\\&-x^T Q(x)x + rx^T G^1(x)Q(x)G^1(x) + [u-u^*]^T\phi_u[u-u^*] - [v-v^*]^T\phi_v[v-v^*]\end{aligned} \tag{10.22}$$

where u^* and v^* are defined above. Letting now $u = u^*$ and $v = v^*$ and using the hypothesis of the theorem together with (10.21) yields

$$V(x)\geq \inf_{u\in R^n} \sup_{v\in R^m} \{||z||^2-\gamma^2||v||^2+\mathcal{E}_{w_k}\{V[f(x)+g^1(x)v+g(x)u+w_k]\}\},\ \forall x\in R^n \tag{10.23}$$

which implies the conclusion of the theorem. □

For the special case where $f(x) = F(x)x$ and $h(x) = H(x)x$ for some functions $F(x)$ and $H(x)$ one has the following corollary which establishes a nonlinear matrix inequality as a sufficient condition for the closed-loop with state-feedback to posses an l_2-gain$\leq \gamma$.

Corollary 10.2.1 *Suppose all the hypotheses of the last theorem hold, and let* $f(x) = F(x)x$ *and* $h(x) = H(x)x$. *Then, the system (10.20) has* l_2*-gain*$\leq \gamma$ *provided the following inequality.*

$$\begin{aligned} &H^T(x)H(x) + F^T(x)[\bar{Q}(x) - \bar{Q}(x)g(x)\phi_u^{-1}(x)g^T(x)\bar{Q}(x)]F(x) - Q \\ &+ rG^{1T}(x)Q(x)G^1(x) \leq 0 \end{aligned} \tag{10.24}$$

holds for all $x \in R^n$.

Remark 10.2. The hypothesis regarding the structure of $h(x)$ and $f(x)$ as assumed in the last corollary is not a severe restriction as it is well known that any differentiable function $f(x)$ with $f(0) = 0$ can be written as $f(x) = F(x)x$ where $F(x) = \int_0^1 f_x(tx)dt$.

Utilizing the last theorem we proceed now in the direction of establishing an output-feedback controller which guarantees an l_2-gain$\leq \gamma$ of the closed-loop system (10.20). As a full state observation is assumed to be unavailable we adopt a common practice for synthesizing the desired controller. In fact we assume a controller structure which is identical to the one established in Theorem 10.2.1 and which operates on the state-estimate that is given by

$$\hat{x}_{k+1} = f(\hat{x}_k)+g^1(\hat{x}_k)v^*(\hat{x}_k)+g(\hat{x}_k)u^*(\hat{x}_k)+\hat{K}(\hat{x}_k)[y_k-C(\hat{x}_k)-D(\hat{x}_k)v^*(\hat{x}_k)] \tag{10.25}$$

where $\hat{K}(\hat{x})$ is the estimator gain, and has to be determined so that the closed-loop system achieves an H_∞ performance. The controller is given by

$$\begin{aligned} u^*(\hat{x}) &= -\phi_u^{-1}(\hat{x})g^T(\hat{x})\bar{Q}(\hat{x})f(\hat{x}) \\ &= -\phi_u^{-1}(\hat{x})g^T(\hat{x})Q(\hat{x})[I - \gamma^{-2}g^{1T}(\hat{x})g^1(\hat{x})Q(\hat{x})]^{-1}f(\hat{x}), \end{aligned}$$

and $v^*(\hat{x}) = \phi_v^{-1}(\hat{x})g^{1T}(\hat{x})Q(\hat{x})[f(\hat{x}) + g(\hat{x})u^*(\hat{x})]$.

In order to facilitate the synthesis of the desired output-feedback closed-loop system, we consider the following augmented system.

$$x_{k+1}^e = f^e(x_k^e) + g^e(x_k^e)s_k + G^e(x_k^e)w_k^e \tag{10.26}$$

where

$$f^e(x_k^e)= col\{f(x_k)+g^1(x_k)v_k^*(\hat{x}_k)+g(x_k)u_k^*(\hat{x}_k),$$
$$f(\hat{x}_k)+g(\hat{x}_k)u_k^*(\hat{x}_k)+\hat{K}(\hat{x}_k)[C(x_k)-C(\hat{x}_k)] + \hat{K}(\hat{x})[D(x)-D(\hat{x})]v_k^*(\hat{x}_k)\}$$

$$x_k^e = \begin{bmatrix} x_k \\ \hat{x}_k \end{bmatrix},\ g^e(x_k^e) = \begin{bmatrix} g^1(x_k) \\ \hat{K}(\hat{x}_k)D(x_k) \end{bmatrix},\ G^e(x_k^e) = \begin{bmatrix} G^1(x_k) & 0 \\ 0 & \hat{K}(\hat{x}_k)G^2(x_k) \end{bmatrix},$$

$$w_k^e = \begin{bmatrix} w_k^1 \\ w_k^2 \end{bmatrix},\ s_k = v_k - v_k^*(\hat{x}_k).$$

Having introduced the above augmented system, we are now ready to state and prove the following theorem which establishes a solution to the output-feedback H_∞ control of the underlying system.

Theorem 10.2.2 *Consider the system (10.20) and (10.26). Let $\gamma > 0$ be fixed and assume that there exist a positive function $V(x)$ and a matrix Q :*

$R^n \to R^+$ for which the hypotheses of Theorem 10.2.1 hold. In addition, let $\hat{V}$ be a positive function defined on R^{2n}, i.e. $\hat{V} : R^{2n} \to R^+$ and let W be a positive matrix defined on R^{2n}, $W : R^{2n} \to R^+$ which satisfies the following: there is a positive number β so that $W(x^e) \geq \beta I, \quad \forall x^e \in R^{2n}$.

$$\begin{aligned}
&\mathcal{E}_{w_k^e}\{\hat{V}[f^e(x^e) + g^e(x^e)s + G^e(x^e)w_k^e]\} - \hat{V}(x^e) \\
&\leq \mathcal{E}_{w_k^e}\{[f^e(x^e) + g^e(x^e)s + G^e(x^e)w_k^e]^T W(x^e)[f^e(x^e) + g^e(x^e)s + G^e(x)w_k^e]\} \\
&-x^{eT}W(x^e)x^e.
\end{aligned} \tag{10.27}$$

Let $\phi_e(x^e) = \gamma^2 I - g^{eT}(x^e)W(x^e)g^e(x^e)$ and assume that there is a positive number, a so that $\phi_e(x^e) \geq aI, \quad \forall x^e \in R^{2n}$. Suppose there is a matrix $\hat{K}(\hat{x})$ of appropriate dimensions such that

$$\begin{aligned}
&f^{eT}(x^e)[W(x^e)g^e(x^e)\phi_e^{-1}(x^e)g^{eT}(x^e)W(x^e) + W(x^e)]f^e(x^e) + h^{eT}(x^e)h^e(x^e) \\
&-x^{eT}W(x^e)x^e + Tr\{G^{eT}(x^e)W(x^e)G^e(x^e)\} \leq 0, \quad \forall x^e \in R^{2n}
\end{aligned} \tag{10.28}$$

where $h^e(x^e) = u^*(x) - u^*(\hat{x})$.
Then, the closed-loop system has an l_2-gain$\leq \gamma$ and is internally stable in the mean square sense.

Proof: Let $S(x^e) = V(x) + \hat{V}(x^e)$. We shall show that $S(x^e)$ is a storage function for the system described by (10.20), together with the controller of Theorem 10.2.1, and that this closed-loop system is dissipative with respect to the supply rate function $\gamma^2||v||^2 - ||z||^2$ where $z = col\{h(x),\ D_{12}u(\hat{x})\}$. Having shown this, it readily follows that closed-loop system has l_2-gain$\leq \gamma$, that is

$$\mathcal{E}\{\sum_{i=0}^{\infty}||z_i||^2\} \leq \gamma^2\mathcal{E}\{||x_0||^2 + \sum_{i=1}^{\infty}||v_i||^2\}$$

for all stochastic processes $\{v_i\}_i$ with finite energy, i.e. $\sum_{i=1}^{\infty}\mathcal{E}\{||v_i||^2\} < \infty$.

It is clear that because of the nature of V and $\hat{V}$, $S(x^e)$ is a positive function and that $S(0) = 0$. To show now that the closed-loop system is dissipative in the sense mentioned above it is sufficient to show that there is a controller u such that

$$\begin{aligned}
&\mathcal{E}_{w_k}\{S(f^e(x^e) + g^e(x^e)s + G^e(x^e)w_k)\} - S(x^e) + ||z||^2 - \gamma^2||v||^2 \leq 0, \\
&\forall x^e \in R^{2n}, v \in R^m
\end{aligned} \tag{10.29}$$

which is equivalent to determining u so that

$$\begin{aligned}
&\mathcal{E}_{w_k^1}\{V(f(x) + g(x)u + g^1(x)v + G^1(x)w_k)\} - V(x) + ||z||^2 - \gamma^2||v||^2 \\
&+\mathcal{E}_{w_k^e}\{\hat{V}(f^e(x^e) + g^e(x^e)s + G^e(x^e)w_k^e)\} - \hat{V}(x^e) \leq 0, \quad \forall x^e \in R^{2n}, v \in R^m
\end{aligned}$$

where we recall that $s = v - v^*(\hat{x})$.

Since, by assumption, there is a positive matrix $Q(x)$ so that

$$\mathcal{E}_{w_k^1}\{V[f(x) + g(x)u + g^1(x)v + G^1(x)w_k^1]\} - V(x) \leq -x^T Q(x)x$$
$$+\mathcal{E}_{w_k^1}\{[f(x) + g(x)u + g^1(x)v + G^1(x)w_k^1]^T Q(x)[f(x) + g(x)u + g^1(x)v$$
$$+G^1(x)w_k^1]\}$$

it is concluded from (10.22) that
$\mathcal{E}_{w_k^1}\{V(f(x) + g(x)u + g^1(x)v + G^1(x)w_k)\} - V(x) + ||z||^2 - \gamma^2||v||^2$
$\leq [u - u^*]^T \phi_u[u - u^*] - [v - v^*]^T \phi_v[v - v^*]$
Consider now the augmented system

$$x_{k+1}^e = f^e(x_k^e) + g^e(x_k^e)s_k + G^e(x_k^e)w_k^e \tag{10.30}$$

with the supply rate

$$\gamma^2||s||^2_{\phi_v} - ||h^e(x^e)||^2_{\phi_u} = \gamma^2[v - v^*]^T\phi_v[v - v^*] - [u - u^*]^T\phi_u[u - u^*].$$

Using the BRL and (10.27), it follows that (10.28) renders the sufficient condition (10.29) for the system (10.30) to have l_2-gain$\leq \gamma$. As to the stability of the closed-loop system, assuming that $h(x)$ satisfies $||h(x)||^2 \geq \alpha||x||^2$ for some positive α, an application of Theorem 1.3.3 to (10.29) yields exponential mean square stability of the closed-loop system. □

Assuming now $f(x) = F(x)$, $h(x) = H(x)x$, and $C(x) = \bar{C}(x)x$, we have the following corollary which provides the desired sufficient condition for l_2-gain$\leq$ γ in terms of nonlinear matrix inequalities rather than scalar inequalities.

Corollary 10.2.2 *Assuming that all the hypotheses of Theorem 10.2.2 hold, together with above assumption regarding the structure of f, h and C. A sufficient condition for the closed-loop system to have l_2=gain$\leq \gamma$ is the following.*

$$\begin{aligned} &H^T(x)H(x) + F^T(x)[\bar{Q}(x) - \bar{Q}(x)g(x)\phi_u^{-1}(x)g^T(x)\bar{Q}(x)]F(x) - Q \\ &+rG^{1T}(x)Q(x)G^1(x) \leq 0 \end{aligned} \tag{10.31}$$

with the requirement that $\phi_v(x) > 0$ for all $x \in R^n$ and:

$$\begin{aligned} &F^{eT}(x^e)[W(x^e)g^e(x^e)\phi_e^{-1}(x^e)g^{eT}(x^e)W(x^e) + W(x^e)]F^e(x^e) + H^{eT}(x^e)H^e(x^e) \\ &-W(x^e) + [G^{eT}(x^e)W(x^e)G^e(x^e)] \leq 0, \end{aligned}$$

requiring

$$\phi_e(x^e) > 0,\ \forall x^e \in R^{2n}, \tag{10.32}$$

where $h^e(x^e) = H^e(x^e)x^e = u^*(x) - u^*(\hat{x})$.

10.2.1 Example

We consider example 10.1.1 of the previous section, this time with partial information, that is, we assume the availability of x_1 only. In fact, the measurements process is taken to be:

$$y_k = Cx_k + Dv_k = \begin{bmatrix}1 \ 0\end{bmatrix} x_k + v_k$$

We need to solve (10.31) and (10.32). Solving the first, for a prescribed γ, a state-feedback gain $K(x)$ will be derived which is different from the constant gain found in Example 1 using Theorem 1. This feedback gain and the corresponding matrix Q will be used in (10.28) to provide the filter gain $\hat{K}$. We obtained for $\gamma = 2.55$ that $K(x) = -[11.1570 \ \ 9.9434]\hat{F}(x)$ and

$$Q = \begin{bmatrix} 28.3112 & 12.3805 \\ 12.3805 & 11.0338 \end{bmatrix}.$$

The corresponding filter gain was found to be $\hat{K} = [0.0097 \ \ 0.2022]^T$ with

$$W = 10^4 \begin{bmatrix} 6.2922 & 0.0202 & -6.2098 & -0.0353 \\ 0.0202 & 1.3329 & 0.0151 & -1.3057 \\ -6.2098 & 0.0151 & 6.5412 & 0.1547 \\ -0.0353 & -1.3057 & 0.1547 & 1.4127 \end{bmatrix}.$$

Denoting $\Theta = I + g^1\phi_v^{-1}g^{1T}Q$ we obtain

$$\Theta = \begin{bmatrix} 1 & 0 \\ 0.0008 & 1.0007 \end{bmatrix}$$

and the resulting output-feedback controller that renders closed-loop l_2-gain of $\gamma = 2.55$ is described by $u_k = -[11.1570 \ \ 9.9434]\hat{F}(\hat{x}_k)\hat{x}_k$ where by (10.25) $\hat{x}_k$ satisfies:

$$\hat{x}_{k+1} = \Theta[F(\hat{x}_k) + gK(\hat{x}_k)]\hat{x}_k + \hat{K}y_k - \hat{K}[C + D\phi_v^{-1}g^{1T}Q(gK(\hat{x}_k) + F(\hat{x}_k))\hat{x}_k.$$

10.3 Conclusions

In this chapter we have developed some theory for H_∞ output-feedback controller design for nonlinear discrete-time stochastic systems. As a general theory for the discrete-time systems suffers from an inherent computational difficulties, we have resorted to a rather special case of quadratic storage functions which have facilitated the achievement of more complete results. Thus, we have established sufficient conditions for the underlying discrete-time nonlinear stochastic system to have an L_2-gain$\leq \gamma$ subject to an output feedback.

The resulting closed loop system has been shown to be exponentially mean square stable, provided the associated storage function satisfies certain conditions.

An explicit formula for the output feedback controller has been derived on the basis of two coupled HJI's. Finally we have demonstrated the applicability of the theory by means of an example, where LMIs methods have been employed in solving the pertinent HJI's.

11

H_∞ Control of Stochastic Switched Systems with Dwell Time

11.1 Introduction

In this chapter [62] we address the problems of stabilization and H_∞ state-feedback control of continuous-time, linear switched systems with state-multiplicative noise and dwell time constraint.

The stability analysis of deterministic linear switched systems with dwell time has been vastly investigated in the past [3], [139]. While the literature on switching of deterministic systems has been concentrated on the case where switching can occur immediately, the realistic case is where there is a minimum time period, the dwell time, during which no switching occurs. Switching, without dwell time, produces dynamics that is a combination of the dynamics of the subsystems. This leads to some switched systems that are unstable for switching without dwell time and stable with (short) dwell time [110]. In [90] it is shown that the worst case (most destabilizing) switching law does obey some dwell time constraint, at least for second order systems. Arbitrarily fast switching may also cause large state transients at the switching points. A dwell time may thus be required for these transients to subside. This is probably one of the reasons why the filed of switched systems with dwell time is becoming increasingly popular.

Switching of stochastic systems has been considered in the literature in the context of stochastic Markov jumps [23], [70], [84], [109] (see also [86] for the deterministic case and for stochastic stability of jump systems see [30]). Switching of stochastic state-multiplicative systems, and switching with dwell time constraint in particular, have hardly been investigated in the past.

In this chapter we apply a recently developed method which deals with the switching of deterministic systems and which has been shown to yield improved results in the uncertain case [3]. Based on the latter, we address the problem of stability of stochastic linear switched systems with dwell time and polytopic type parameter uncertainties. The best stability result attained so far for deterministic systems, that is both efficiently computable and yields a sufficient result, is the one of [43] (it is also used in [21]). This result is

E. Gershon & U. Shaked: *Advanced Topics in Control & Estimation*, LNCIS 439, pp. 177–191.
DOI: 10.1007/978-1-4471-5070-1_11

based on a quadratic LF. It is obtained for the case without uncertainty (namely for a nominal system), and it is derived using LMIs. These LMIs are not affine in the system matrices making it difficult to generalize to the uncertain case. In [14], a piecewise linear in time, quadratic form LF is used. Similarly to [3], we apply in the present work such a LF to stochastic switched systems with dwell time. The LF applied is non-increasing at the switching instants and is separately assigned to each subsystem. During the dwell time, this function varies piecewise linearly in time after switching occurs and it becomes time invariant afterwards. Our choice of the LF, allows derivation of sufficient conditions for robust stochastic stability of the switched system in terms of LMIs. In the deterministic case, these conditions have been shown to be less conservative than those obtained by bounding the matrix exponent of [43] by a scalar matrix.

This chapter is organized as follows: Following the problem formulation of Section 11.2, the stochastic stability is investigated in Section 11.3. The conditions obtained for the stability of the switched system are extended in Section 11.4 to find a robust bound on the L_2-gain of the system. The results are first derived there for the nominal case and they are then extended to cope with polytopic uncertainty. Section 11.4 also presents an alternative, less conservative, condition for the bound on the stochastic L_2-gain of the uncertain systems. The latter is applied, in Section 11.5, to the solution of the robust stochastic state-feedback control problem. The theory developed is applied in two examples, in Section 11.6 and Chapter 13, Section 13.3 . In the first example, a bound is obtained for the L_2-gain of a switched system with uncertainties for various values of dwell time. The second example (in Chapter 13) is one where a switched state-feedback controller is derived for robust H_∞control of the F4E fighter aircraft.

11.2 Problem Formulation

We consider the following linear stochastic state-multiplicative switched system [62]:

$$\begin{aligned} dx(t) &= A_{\sigma(t)}x(t)dt + B_{\sigma(t)}w(t)dt + F_{\sigma(t)}x(t)d\beta(t) + \bar{G}_{\sigma(t)}x(t)d\nu(t), \ x(0) = 0, \\ z(t) &= C_{\sigma(t)}x(t) + D_{\sigma(t)}w(t), \end{aligned} \tag{11.1}$$

which is defined for all $t \geq 0$, where $x(t) \in \mathcal{R}^n$ is the system state vector, $w(t) \in \mathcal{R}^q$ is the exogenous disturbance in $\tilde{L}^2_{\mathcal{F}_t}([0,\infty);\mathcal{R}^q)$ and $z(t) \in \mathcal{R}^z$ is the objective vector. The variables $\beta(t)$ and $\nu(t)$ are zero-mean real scalar Wiener processes that satisfy:

$$\mathcal{E}\{d\beta(t)^2\}=dt, \ \mathcal{E}\{d\nu(t)^2\}=dt, \ \mathcal{E}\{d\beta(t)d\nu(t)\} = 0. \tag{11.2}$$

The switching rule $\sigma(t)$, for each $t \geq 0$, is such that $A_{\sigma(t)} \in \{A_1, ..., A_M\}$, $F_{\sigma(t)} \in \{F_1, ..., F_M\}$, $B_{\sigma(t)} \in \{B_1, ..., B_M\}$, $\bar{G}_{\sigma(t)} \in \{\bar{G}_1, ..., \bar{G}_M\}$ and

$C_{\sigma(t)} \in \{C_1, ..., C_M\}$, $D_{\sigma(t)} \in \{D_1, ..., D_M\}$, where $A_i \in \mathcal{R}^{n\times n}, i = 1, ...M$ is a Hurwitz matrix. The matrices A_i, B_i, C_i and D_i are assumed to reside within the following polytope:

$$\Omega_i = \sum_{j=1}^{N_i} \eta^j(t)\Omega_i^{(j)}, \quad \sum_{j=1}^{N_i} \eta^j(t) = 1, \quad \eta^j(t) \geq 0 \tag{11.3}$$

where:

$$\Omega_i = \begin{bmatrix} A_i & B_i & F_i \\ C_i & D_i & \bar{G}_i \end{bmatrix} \quad \text{and} \quad \Omega_i^{(j)} = \begin{bmatrix} A_i^{(j)} & B_i^{(j)} & F_i^{(j)} \\ C_i^{(j)} & D_i^{(j)} & \bar{G}_i^{(j)} \end{bmatrix}, i = 1, ...M, \; j = 1, ...N. \tag{11.4}$$

The above model naturally imposes discontinuity in $A_{\sigma(t)}$ since this matrix jumps instantaneously from A_{i_1} to A_{i_2} for some $i_1 \neq i_2$ at the switching instant. The latter property also applies to all the system matrices.

For the stochastic state-feedback control problem, we consider the following system:

$$\begin{aligned} dx(t) &= [A_{\sigma(t)}x(t) + B_{\sigma(t)}w(t) + B_{2,\sigma(t)}u(t)]dt + F_{\sigma(t)}x(t)d\beta(t) \\ &\quad + G_{\sigma(t)}u(t)d\nu(t), \quad x(0) = 0, \\ z(t) &= C_{\sigma(t)}x(t) + D_{\sigma(t)}w(t) + D_{12,\sigma(t)}u(t), \end{aligned} \tag{11.5}$$

where $x(t)$, $w(t)$, $\beta(t)$, $\nu(t)$ and $\sigma(t)$ are defined above, $u(t) \in \mathcal{R}^l$ is the control signal, and A_i, B_i, C_i, $B_{2,i}$, D_i, and $D_{12,i}$ are assumed to reside in the following polytope:

$$\bar{\Omega}_i = \sum_{j=1}^{N_i} \bar{\eta}^j(t)\bar{\Omega}_i^{(j)}, \quad \sum_{j=1}^{N_i} \bar{\eta}^j(t) = 1, \quad \bar{\eta}^j(t) \geq 0 \tag{11.6}$$

where:

$$\bar{\Omega}_i = \begin{bmatrix} A_i & B_i & B_{2,i} \\ C_i & D_i & D_{12,i} \end{bmatrix} \quad \text{and} \bar{\Omega}_i^{(j)} = \begin{bmatrix} A_i^{(j)} & B_i^{(j)} & B_{2,i}^{(j)} \\ C_i^{(j)} & D_i^{(j)} & D_{12,i}^{(j)} \end{bmatrix}, i = 1, ...M, \; j = 1, ...N. \tag{11.7}$$

Note that in the system (11.5a) A_i is no longer required to be Hurwitz matrix and that in the polytope description of (11.6) and (11.7) we require that in each subsystem $i = 1, ...M$ the matrices F and G are constant matrices.

11.3 Stochastic Stability

We start by investigating the stability of the stochastic switched system (11.1a,b) where $w(t) \equiv 0$, under a constraint on the rate of allowed commutations (dwell time analysis), which means that if the dwell time is T, and the switching instants are $\tau_1, \tau_2, ...$, then $\tau_{h+1} - \tau_h \geq T, \quad \forall h \geq 1$.

We seek sufficient conditions for the system to be globally asymptotically stable in probability where the latter is defined as follows.

Definition 11.1 The solution $x(t)$ of (11.1a,b), where $w(t) \equiv 0$, is said to be globally asymptotically stable in probability if, for any scalar $\varepsilon > 0$, $\lim_{x \to 0} \mathcal{P}\{\sup_{t \geq 0} \|x(t)\| > \varepsilon\} = 0$, and if $\mathcal{P}\{\lim_{t \to \infty} x(t) = 0\} = 1$ for any initial state $x_0 \in R^n$.

A well known sufficient condition for global stability in probability is the following.

Lemma 11.3.1 *([9]) Assume that there exists a positive Lyapunov function $V(x,t) \in C^{n,1}$, with $V(0,t) = 0$. Let $\mathcal{L}(x,t)$ be the infinitesimal generator of the process x so that $\mathcal{E}\{(\mathcal{L}V)(x,t)\} < 0$ for all $x \in R^n$ and for all $t \geq 0$. Assume also that $\inf_{t>0} V(x,t) \to \infty$ as $\|x\| \to \infty$. Then, the system of (11.1a,b) is globally asymptotically stable in probability.*

We introduce a time-varying LF. For this purpose the following result, inspired by [14], is applied.

Lemma 11.3.2 *Assume that for some time interval $t \in [t_0, t_f]$, where $\delta = t_f - t_0$, there exist two symmetric matrices P_1 and P_2 of compatible dimensions that satisfy the following:*

$$
\begin{aligned}
&P_1, P_2 > 0, \quad \tfrac{P_2 - P_1}{\delta} + P_1 A + A^T P_1 + F^T P_1 F + \bar{G}^T P_1 \bar{G} < 0, \\
&\tfrac{P_2 - P_1}{\delta} + P_2 A + A^T P_2 + F^T P_2 F + \bar{G}^T P_2 \bar{G} < 0.
\end{aligned} \tag{11.8}
$$

Then, for the system $dx(t) = Ax(t)dt + Fx(t)d\beta + \bar{G}x(t)d\nu$ the expected value of the infinitesimal generator, denoted by $\mathcal{L}$ [9], of the following LF:

$$
V(t, x(t)) = x^T(t)P(t)x(t), \ \text{with} \ P(t) = P_1 + (P_2 - P_1)\frac{t - t_0}{\delta},
$$

is strictly negative over the time interval $t \in [t_0, t_f]$.

Proof: Applying *Itô* formula [9] to $V(t, x(t))$ and taking expectation we find that

$$
\begin{aligned}
\mathcal{E}\{\mathcal{L}V(t, x(t))\} &= \mathcal{E}\{x^T(t)\dot{P}(t)x(t) + 2x^T(t)P(t)Ax(t) + x^T(t)F^T P(t)Fx(t) \\
&\quad + x^T(t)\bar{G}^T P(t)\bar{G}x(t)\} \\
&= \mathcal{E}\{x^T(t)(\tfrac{P_2 - P_1}{\delta} + P(t)A + A^T P(t) + F^T P(t)F + \bar{G}^T P(t)\bar{G})x(t)\},
\end{aligned}
$$

where use was made of (11.2). Noting that

$$\frac{P_2-P_1}{\delta} + P(t)A + A^T P(t) + F^T P(t)F + \bar{G}^T P(t)\bar{G} = \lambda_1[\frac{P_2-P_1}{\delta} + P_1 A + A^T P_1$$

$$+F^T P_1 F + \bar{G}^T P_1 \bar{G}] + \lambda_2[\frac{P_2-P_1}{\delta} + P_2 A + A^T P_2 + F^T P_2 F + \bar{G}^T P_2 \bar{G}]$$

where $\lambda_1 = 1 - \frac{t-t_0}{\delta}$, $\lambda_2 = \frac{t-t_0}{\delta}$, it follows from (11.8b,c) that $\mathcal{E}\{\mathcal{L}V(t)\} < 0$ in the time interval specified above. □

The above proof is for a nominal system. The extension to the polytopic uncertainty case is, however, immediate choosing the same P_1 and P_2 for all the vertices of the uncertainty polytope Ω_i.

We next present sufficient conditions for the stability of nominal stochastic linear switched systems. These conditions are given in terms of LMIs which are affine in the systems matrices, and they can thus be easily extended to the polytopic uncertainty case. Choosing the integer K according to the allowed computational complexity, we obtain the following:

Theorem 11.1. *The nominal system (11.1a,b) with $B_{\sigma(t)} = 0$ is globally asymptotically stable for any switching law with dwell time greater than or equal to $T > 0$ if there exist a collection of positive definite matrices $P_{i,k}, i = 1,...M, k = 0,...K$ of compatible dimensions, such that, for all $i = 1,...M$ the following holds:*

$$K\frac{P_{i,k+1}-P_{i,k}}{T} + P_{i,k}A_i + A_i^T P_{i,k} + F_i^T P_{i,k} F_i + \bar{G}_i^T P_{i,k}\bar{G}_i < 0,$$
$$K\frac{P_{i,k+1}-P_{i,k}}{T} + P_{i,k+1}A_i + A_i^T P_{i,k+1} + F_i^T P_{i,k+1} F_i + \bar{G}_i^T P_{i,k+1}\bar{G}_i < 0,$$
$$k = 0,...K-1$$

$$P_{i,K}A_i + A_i^T P_{i,K} + F_i^T P_{i,K} F_i + \bar{G}_i^T P_{i,K}\bar{G}_i < 0, \quad \forall\, l = 1,...i-1, i+1,...M,$$
$$P_{i,K} - P_{l,0} \geq 0. \tag{11.9}$$

Proof: We define $\tau_{h,k} = \tau_h + k\frac{T}{K}$ and $\tau_{h,0} = \tau_h$ and note that the dwell time constraint implies $\tau_{h,K} \leq \tau_{h+1,0} = \tau_{h+1}$. We consider the LF, $V(t, x(t)) = x^T(t)P_{\sigma(t)}(t)x(t)$, where $P_{\sigma(t)}(t)$ is given by:

$$P_i(t) = \begin{cases} P_{i,k} + (P_{i,k+1} - P_{i,k})\frac{K(t-\tau_{h,k})}{T} & t \in [\tau_{h,k}, \tau_{h,k+1}) \\ P_{i,K} & t \in [\tau_{h,K}, \tau_{h+1,0}) \\ P_{i_0,K} & t \in [0, \tau_1) \end{cases} \tag{11.10}$$
$$, k = 0,...K-1,\ h = 1, 2, ...$$

and where $i = \sigma(t)$ is the index of the subsystem that is active at time t.

Assuming that at some switching instant τ_h the systems switches from the i-th subsystem to the l-th subsystem, we have $V(\tau_h^-, x(\tau_h)) = x(\tau_h)^T P_{i,K} x(\tau_h)$ and $V(\tau_h, x(\tau_h)) = x(\tau_h)^T P_{l,0} x(\tau_h)$. Therefore, for the LF to be non increasing at the switching instant we demand $P_{i,K} \geq P_{l,0}$ which is the condition (11.9d). After the dwell time, and before the next switching occurs, we have that $V(t,x(t)) = x(t)^T P_{i,K} x(t)$, where $dx(t) = A_i x(t)dt + F_i x(t)d\beta(t) + \bar{G}_i x(t)d\nu(t)$. Therefore,

$$\mathcal{E}\{(\mathcal{L}V)(t,x(t))\} = \mathcal{E}\{x(t)^T[P_{i,K}A_i + A_i^T P_{i,K} + F_i^T P_{i,K}F_i + \bar{G}_i^T P_{i,K}\bar{G}_i]x(t)\}$$

and (11.9c) guarantees then that this expression is negative for any $x(t) \neq 0$. During the dwell time we consider the time intervals $t \in [\tau_{h,k}, \tau_{h,k+1})$ where $\frac{T}{K} = \tau_{h,k+1} - \tau_{h,k}$. The matrix $P(t)$ changes then linearly from $P_{i,k}$ to $P_{i,k+1}$, and conditions (11.9a-c) guarantee that $V(t,x(t))$ is decreasing during this time interval, according to Lemma 11.3.2. □

We note that the LMIs in Theorem 11.1 are affine in the system matrices and thus they can be readily applied to the stability problem of polytopic uncertain systems. Robust stability in the case of uncertain subsystems is thus given by the following corollary where K is a prechosen integer:

Corollary 11.3.1 *Consider the system (11.1a,b) with $B_{\sigma(t)} = 0$ and (11.3). Assume that for some dwell time $T > 0$ there exist a collection of positive definite matrices $P_{i,k}, i = 1,...M, k = 0,...K$ of compatible dimensions, such that, for all $i = 1,...M$, and $j = 1,...N$ the following holds:*

$$\begin{aligned}
&K\tfrac{P_{i,k+1}-P_{i,k}}{T} + P_{i,k}A_i^{(j)} + A_i^{(j)^T}P_{i,k} + F_i^{(j)^T}P_{i,k}F_i^{(j)} + \bar{G}_i^{(j)^T}P_{i,k}\bar{G}_i^{(j)} < 0,\\
&K\tfrac{P_{i,k+1}-P_{i,k}}{T} + P_{i,k+1}A_i^{(j)} + A_i^{(j)^T}P_{i,k+1} + F_i^{(j)^T}P_{i,k+1}F_i^{(j)}\\
&+\bar{G}_i^{(j)^T}P_{i,k+1}\bar{G}_i^{(j)} < 0, \quad k = 0,...K-1,\\
&P_{i,K}A_i^{(j)} + A_i^{(j)^T}P_{i,K} + F_i^{(j)^T}P_{i,K}F_i^{(j)} + \bar{G}_i^{(j)^T}P_{i,K}\bar{G}_i^{(j)} < 0,\\
&P_{i,K} - P_{l,0} \geq 0, \quad \forall\, l = 1,...i-1, i+1,...M.
\end{aligned}$$

Then, the system (11.1a,b) is globally asymptotically stable for any switching law with dwell time greater than or equal to T.

11.4 Stochastic L_2-Gain

We seek a sufficient condition for the mean square stability of the system (11.1a-c) and for the following inequality to hold, given a prescribed scalar $\gamma > 0$,

$$J = \mathcal{E}\{\int_0^\infty (z^T z - \gamma^2 w^T w)dt\} \leq 0,\ \forall\, w \in \tilde{L}^2_{\mathcal{F}_t}([0,\infty);\mathcal{R}^q). \qquad (11.11)$$

Let $i_0 = \sigma(0)$, and let $\tau_1, \tau_2, ...$ be the switching instances, where $\tau_{h+1} - \tau_h \geq T,\ \ \forall h = 1,2,...$ We define $\tau_{h,k} = \tau_h + \frac{kT}{K}$ for $k \geq 1$, and $\tau_{h,0} = \tau_h$ and note that the dwell time constraint implies $\tau_{h,K} \leq \tau_{h+1,0} = \tau_{h+1}$. We choose the following LF.

$$V(t, x(t)) = x^T(t) P_\sigma(t) x(t), \tag{11.12}$$

with $P_i(t)$ as defined in (11.10) where K is an integer that is chosen a priori, according to the allowed computational complexity. We also denote:

$$\bar{J} = lim_{t\to\infty} \mathcal{E}\{V(t, x(t))\} + \mathcal{E}\{\int_0^t (z^T z - \gamma^2 w^T w) ds\}$$

Since $V(t, x(t)) \geq 0 \ \forall t$, we have that $J \leq \bar{J}$. Taking into account that $\mathcal{L}V(t, x(t))dt$ exists for all t, except for the switching instances, and that $x(0) = 0$, we write

$$lim_{t\to\infty} \mathcal{E}\{V(t, x(t))\} =$$

$$\mathcal{E}\{\sum_{h=0}^{\infty} \int_{\tau_h}^{\tau_{h+1}} \mathcal{L}V(t, x(t))dt + \sum_{h=1}^{\infty} (V(\tau_h,\ x(\tau_h)) - V(\tau_h^-,\ x(\tau_h^-)))\}$$

where $\tau_0 = 0$. If the conditions for non increasing $V(t, x(t))$ at the switching instances are satisfied we find that $V(\tau_h,\ x(\tau_h)) - V(\tau_h^-,\ x(\tau_h^-)) \leq 0 \ \forall\ h > 0$, and then:

$$lim_{t\to\infty} \mathcal{E}\{V(t, x(t))\} \leq \mathcal{E}\{\sum_{h=0}^{\infty} \int_{\tau_h}^{\tau_{h+1}} \mathcal{L}V(t, x(t))dt\}. \tag{11.13}$$

Denoting

$$\tilde{J} = \mathcal{E}\{\sum_{h=0}^{\infty} \int_{\tau_h}^{\tau_{h+1}} \mathcal{L}V(t, x(t))dt + \int_0^{\infty} (z^T z - \gamma^2 w^T w) dt\}, \tag{11.14}$$

we have from (11.13) that: $J \leq \bar{J} \leq \tilde{J}$. Consequently, if $\tilde{J} \leq 0$ and the above LF does not increase at the switching instants, the stochastic L_2-gain of the system will be less than γ. We apply below the following LMI condition of [53] for the L_2-gain of stochastic state-multiplicative linear systems (without switching):

$$P > 0, \quad \begin{bmatrix} \dot{P} + PA + A^T P & PB & C^T & F^T P & \bar{G}^T P \\ * & -\gamma^2 I_q & D^T & 0 & 0 \\ * & * & -I_z & 0 & 0 \\ * & * & * & -P & 0 \\ * & * & * & * & -P \end{bmatrix} < 0, \tag{11.15}$$

where A, B, C, F, $\bar{G}$ and D are as in (11.1a–c). Applying the method of [14] on the latter LMI for the switched system (11.1a–c), the following result is obtained for a prescribed integer K:

Theorem 11.2. *The stochastic L_2-gain of the nominal system (11.1a–c) is less than a prescribed $\gamma > 0$ for a dwell time of $\bar{T} \geq T$ if there exists a collection of positive definite matrices $P_{i,k}$, $i = 1, ...M$, $k = 0, ...K$ of compatible dimensions such that, for all $i = 1, ...M$ the following holds:*

$$\begin{bmatrix} K\frac{P_{i,k+1}-P_{i,k}}{T} + P_{i,k}A_i + A_i^T P_{i,k} & P_{i,k}B_i & C_i^T & F_i^T P_{i,k} & \bar{G}_i^T P_{i,k} \\ * & -\gamma^2 I_q & D_i^T & 0 & 0 \\ * & * & -I_z & 0 & 0 \\ * & * & * & -P_{i,k} & 0 \\ * & * & * & * & -P_{i,k} \end{bmatrix} < 0,$$

$$\begin{bmatrix} K\frac{P_{i,k+1}-P_{i,k}}{T} + P_{i,k+1}A_i + A_i^T P_{i,k+1} & P_{i,k+1}B_i & C_i^T & F_i^T P_{i,k+1} & \bar{G}_i^T P_{i,k+1} \\ * & -\gamma^2 I_q & D_i^T & 0 & 0 \\ * & * & -I_z & 0 & 0 \\ * & * & * & -P_{i,k+1} & 0 \\ * & * & * & * & -P_{i,k+1} \end{bmatrix} < 0,$$

$$k = 0, ...K-1$$

$$\begin{bmatrix} P_{i,K}A_i + A_i^T P_{i,K} & P_{i,K}B_i & C_i^T & F_i^T P_{i,K} & \bar{G}_i^T P_{i,K} \\ * & -\gamma^2 I_q & D_i^T & 0 & 0 \\ * & * & -I_z & 0 & 0 \\ * & * & * & -P_{i,K} & 0 \\ * & * & * & * & -P_{i,K} \end{bmatrix} < 0, \qquad (11.16)$$

$$P_{i,K} - P_{l,0} \geq 0, \ \forall \ l = 1, ...i-1, i+1, ...M.$$

Proof: The theorem is proved by applying the result of Lemma 11.3.2 and $P(t)$ of (11.10) to (11.15a,b). We first show that the above conditions guarantee the asymptotic stability of the system and that V does not increase at the switching points. We then turn to (11.14) showing that the integrand there is strictly negative. Condition (11.16d) means that once a switching occurs, $P(t)$ switches in such a way that it does not increase at the switching instant of time. From this time on, $P(t)$ becomes piecewise linear in time, where over the time interval $t \in [\tau_h + k\frac{T}{K}, \tau_h + (k+1)\frac{T}{K}]$, it changes linearly from $P_{i,k}$ to $P_{i,k+1}$, where i is the index of the active subsystem at $t \in [\tau_h, \tau_{h+1}]$.

Before the first switching occurs, $P(t)$ is constant and by (11.16c) it guarantees that $V(t)$ decreases while the system remains in the same subsystem. It follows then from (11.16a,b) that $V(t)$ is strictly decreasing during the dwell time, the LMIs (11.16c) guarantees that the LF is strictly decreasing for any $t \in [\tau_{h,K}, \tau_{h+1,0})$, and (11.16d) guarantees that the LF is non-increasing at the switching instances. Since the switching points are distinct, the system is asymptotically stable according to Lasalle's invariance principle [68]. The first diagonal block of the LMIs (11.16a-c) corresponds to $\mathcal{L}V(t, x(t))$ of (11.14). The second and the third columns and rows in these LMIs correspond to $w^T w$ and $z^T z$ there, respectively. □

The LMIs in Theorem 11.2 are all affine in the system matrices and the result there can thus be readily applied to the uncertain case where the system parameters lie in the polytope of Ω_i. One has then to replace A_i, B_i, C_i and D_i by A_i^j, B_i^j, C_i^j and D_i^j, respectively, and solve the resulting LMIs for all the vertices in Ω_i. The resulting LMIs are not suitable, however, for state-feedback control synthesis. For the latter we derive, in the sequel, a similar result which will be suitable for the state-feedback control.

Let $Q(t) = P^{-1}(t)$. We depart from the assumption of linearity in time of $P(t)$ and assign this linearity to $Q(t)$, namely

$$Q(t)=\begin{cases} Q_{i,k} + (Q_{i,k+1} - Q_{i,k})\frac{K(t-\tau_{h,k})}{T} & t \in [\tau_{h,k}, \tau_{h,k+1}) \\ Q_{i,K} & t \in [\tau_{h,K}, \tau_{h+1,0}) \\ Q_{i_0,K} & t \in [0, \tau_1) \end{cases}, \qquad h = 1, 2, ...$$

Multiplying (11.15b) by $diag\{Q, I_q, I_z, Q, Q\}$ from the left and the right and applying the method of Lemma 11.3.2 the following alternative LMI conditions are obtained for the L_2-gain bound of the system (11.1a-c):

$$Q_{i,k} > 0,$$

$$\begin{bmatrix} -K\frac{Q_{i,k+1}-Q_{i,k}}{T} + A_iQ_{i,k} + Q_{i,k}A_i^T & B_i & Q_{i,k}C_i^T & Q_{i,k}F_i^T & Q_{i,k}\bar{G}_i^T \\ * & -\gamma^2 I_q & D_i^T & 0 & 0 \\ * & * & -I_z & 0 & 0 \\ * & * & * & -Q_{i,k} & 0 \\ * & * & * & * & -Q_{i,k} \end{bmatrix} < 0,$$

$$\begin{bmatrix} \tilde{\Upsilon}_{i,k} & B_i & Q_{i,k+1}C_i^T & Q_{i,k+1}F_i^T & Q_{i,k+1}\bar{G}_i^T \\ * & -\gamma^2 I_q & D_i^T & 0 & 0 \\ * & * & -I_z & 0 & 0 \\ * & * & * & -Q_{i,k+1} & 0 \\ * & * & * & * & -Q_{i,k+1} \end{bmatrix} < 0,$$
$$\tilde{\Upsilon}_{i,k} = -K\frac{Q_{i,k+1}-Q_{i,k}}{T} + A_iQ_{i,k+1} + Q_{i,k+1}A_i^T, \quad k = 0, ...K-1,$$

$$\begin{bmatrix} A_iQ_{i,K} + Q_{i,K}A_i^T & B_i & Q_{i,K}C_i^T & Q_{i,K}F_i^T & Q_{i,K}\bar{G}_i^T \\ * & -\gamma^2 I_q & D_i^T & 0 & 0 \\ * & * & -I_z & 0 & 0 \\ * & * & * & -Q_{i,K} & 0 \\ * & * & * & * & -Q_{i,K} \end{bmatrix} < 0, \tag{11.17}$$

$$Q_{i,K} - Q_{l,0} \le 0, \ \forall\, l = 1, ...i-1, i+1, ...M.$$

The latter conditions can be readily applied to design a state-feedback controller for the switched systems. To reduce the conservatism entailed in the conditions of (11.17a–e) when applied to the uncertain case, the following result is obtained.

Corollary 11.4.1 *The stochastic L_2-gain of the system (11.1a–c) is less than γ for a dwell time $\bar{T} \geq T$ if there exists a collection of symmetric matrices $Q_{i,k} > 0$, $G_{i,k}$, $H_{i,k}$, $i = 1, ...M$, $k = 0, ..., K$ of compatible dimensions, such that, for all $i = 1, ...M$, the following holds:*

$$\begin{bmatrix} \tilde{\Psi}_{11} & Q_{i,k} - G_{i,k}^T + A_i H_{i,k} & G_{i,k}^T C_i^T & B_i & Q_{i,k} F_i^T & Q_{i,k} \bar{G}_i^T \\ * & -H_{i,k} - H_{i,k}^T & H_{i,k}^T C_i^T & 0 & 0 & 0 \\ * & * & -\gamma^2 I_q & D_i & 0 & 0 \\ * & * & * & -I_z & 0 & 0 \\ * & * & * & * & -Q_{i,k} & 0 \\ * & * & * & * & * & -Q_{i,k} \end{bmatrix} < 0,$$

$$\begin{bmatrix} \bar{\Psi}_{11} & Q_{i,k+1} - G_{i,k+1}^T + A_i H_{i,k+1} & G_{i,k+1}^T C_i^T & B_i & Q_{i,k+1} F_i^T & Q_{i,k+1} \bar{G}_i^T \\ * & -H_{i,k+1} - H_{i,k+1}^T & H_{i,k+1}^T C_i^T & 0 & 0 & 0 \\ * & * & -\gamma^2 I_q & D_i & 0 & 0 \\ * & * & * & -I_z & 0 & 0 \\ * & * & * & * & -Q_{i,k+1} & 0 \\ * & * & * & * & * & -Q_{i,k+1} \end{bmatrix} < 0,$$

$k = 0, ...K - 1,$

$\tilde{\Psi}_{11} = -K \frac{Q_{i,k+1} - Q_{i,k}}{T} + G_{i,k}^T A_i^T + A_i G_{i,k},$

$\bar{\Psi}_{11} = -K \frac{Q_{i,k+1} - Q_{i,k}}{T} + G_{i,k+1}^T A_i^T + A_i G_{i,k+1},$

$$\begin{bmatrix} G_{i,K}^T A_i^T + A_i G_{i,K} & Q_{i,K} - G_{i,K}^T + A_i H_{i,K} & G_{i,K}^T C_i^T & B_i & Q_{i,K} F_i^T & Q_{i,K} \bar{G}_i^T \\ * & -H_{i,K} - H_{i,K}^T & H_{i,K}^T C_i^T & 0 & 0 & 0 \\ * & * & -\gamma^2 I_q & D_i & 0 & 0 \\ * & * & * & -I_z & 0 & 0 \\ * & * & * & * & -Q_{i,K} & 0 \\ * & * & * & * & * & -Q_{i,K} \end{bmatrix} < 0,$$

$Q_{i,K} - Q_{l,0} \leq 0, \ \forall \ l = 1, ...i - 1, i + 1, ...M.$

(11.18)

Proof: Assuming that (11.18a–d) are feasible, we multiply (11.18a–c) by $\tilde{\Upsilon}^T$ from the left and by $\tilde{\Upsilon}$ from the right, where $\tilde{\Upsilon} \triangleq diag\{ \begin{bmatrix} I & 0 & 0 \\ A_i^T & I & C_i^T \\ 0 & 0 & I \end{bmatrix}, \ I, \ I, \ I\}$, and readily retrieve the corresponding LMIs of (11.17a–c). □

The result of Corollary 11.4.1 is readily extended to the uncertain case. Replacing the system matrices A_i, B_i, C_i and D_i by A_i^j, B_i^j, C_i^j, and D_i^j, respectively, and solving the resulting LMIs for all the vertices in Ω_i, a solution to the uncertain case is obtained.

11.5 H_∞ State-Feedback Control

Given the system (11.5a–c) with the uncertainty that is described in (11.6), a state-feedback controller is sought that stabilizes the system, and satisfies the performance criterion (11.11).

We distinguish between two cases: The first is characterized by a switching signal $\sigma(t)$ that is measured on-line and different state-feedback gains can therefore be applied to different subsystems. In the second case, it is assumed that the switching signal is unknown, and thus the same state-feedback gain should be applied to all subsystems.

We begin with the first case. Replacing the system matrices A_i, B_i, C_i, and D_i in (11.17) by $\bar{A}_i^j$, B_i^j, $\bar{C}_i^j$ and D_i^j of (11.5a,c) respectively, where $\bar{A}_i^j = A_i^j + B_{2,i}^j \bar{K}_{i,k}$, $\bar{C}_i^j = C_i^j + D_{12,i}^j \bar{K}_{i,k}$ and denoting $Y_{i,k} = \bar{K}_{i,k} Q_{i,k}$, we obtain the following result for a prescribed integer K:

Theorem 11.3. *There exists a state feedback gain $\bar{K}_{\sigma(t)}$ that stabilizes (11.5a-c), with uncertainty as in (11.6) and a dwell time of $\bar{T} \geq T$, and achieves closed-loop stochastic L_2-gain bound that is less than a prescribed positive scalar γ, if there exists a collection of matrices: $Y_{i,k}, Q_{i,k} > 0$, $i = 1, ...M$, $k = 0, ...K$ of compatible dimensions such that, for all $i = 1, ...M$, and $j = 1, ...N$ the following holds:*

$$\begin{bmatrix} \Upsilon_{11,k} & B_i^{(j)} & Q_{i,k} C_i^{(j),T} + Y_{i,k} D_{12,i}^{(j),T} & Q_{i,k} F_i^T & Y_{i,k}^T \bar{G}_i^T \\ * & -\gamma^2 I_q & D_i^{(j)} & 0 & 0 \\ * & * & -I_z & 0 & 0 \\ * & * & * & -Q_{i,k} & 0 \\ * & * & * & * & -Q_{i,k} \end{bmatrix} < 0,$$

where

$$\Upsilon_{11,k} = -K \frac{Q_{i,k+1} - Q_{i,k}}{T} + Q_{i,k} A_i^{(j),T} + A_i^{(j)} Q_{i,k} + B_{2,i}^{(j)} Y_{i,k} + Y_{i,k}^T B_{2,i}^{(j),T},$$

$$\begin{bmatrix} \bar{\Upsilon}_{11,k+1} & B_i^{(j)} & Q_{i,k+1} C_i^{(j),T} + Y_{i,k+1} D_{12,i}^{(j),T} & Q_{i,k+1} F_i^T & Y_{i,k+1}^T \bar{G}_i^T \\ * & -\gamma^2 I_q & D_i^{(j)} & 0 & 0 \\ * & * & -I_z & 0 & 0 \\ * & * & * & -Q_{i,k+1} & 0 \\ * & * & * & * & -Q_{i,k+1} \end{bmatrix} < 0,$$

$$k = 0, ...K-1,$$

where

$$\bar{\Upsilon}_{11,k+1} = -K \frac{Q_{i,k+1} - Q_{i,k}}{T} + Q_{i,k+1} A_i^{(j),T} + A_i^{(j)} Q_{i,k+1} + B_{2,i}^{(j)} Y_{i,k+1} + Y_{i,k+1}^T B_{2,i}^{(j),T},$$

$$\begin{bmatrix} \hat{\Upsilon}_{11,K} & B_i^{(j)} & Q_{i,K}C_i^{(j),T}+Y_{i,K}D_{12,i}^{(j),T} & Q_{i,K}F_i^T & Y_{i,K}^T\bar{G}_i^T \\ * & -\gamma^2 I_q & D_i^{(j)} & 0 & 0 \\ * & * & -I_z & 0 & 0 \\ * & * & * & -Q_{i,K} & 0 \\ * & * & * & * & -Q_{i,K} \end{bmatrix} < 0,$$

$$\hat{\Upsilon}_{11,K} = Q_{i,K}A_i^{(j),T} + A_i^{(j)}Q_{i,K} + B_{2,i}^{(j)}Y_{i,K} + Y_{i,K}^T B_{2,i}^{(j),T}$$

and (11.19)

$$Q_{i,K} - Q_{l,0} \le 0,$$
$$\forall\ l = 1, ...i-1, i+1, ...M.$$

The state-feedback gain is then given by:

$$\bar{K}_{\sigma(t)}(t) = \begin{cases} Y_\sigma(t)[Q_{\sigma,k} + K(Q_{\sigma,k+1} - Q_{\sigma,k})\frac{t-\tau_{h,k}}{T}]^{-1} & t \in [\tau_{h,k}, \tau_{h,k+1}), \\ Y_{\sigma,K}Q_{\sigma,K}^{-1} & t \in [\tau_{h,K}, \tau_{h+1,0}) \\ Y_{i_0,K}Q_{i_0,K}^{-1} & t \in [0, \tau_1). \end{cases}$$

where in the upper row $k = 0, 1, ..K-1$ *and* $Y_\sigma(t) = Y_{\sigma,k} + \frac{t-\tau_{h,k}}{T}K(Y_{\sigma,k+1} - Y_{\sigma,k})$.

Theorem 11.3 provides a time-varying state-feedback gain that may be difficult to implement, especially for large K. The solution there may also be quite conservative due to the fact that $Q_{i,k}$ are the same for all the points in $\bar{\Omega}_i$. This conservatism is reduced by applying the result of Corollary 11.4.1 which allows $Q_{i,k}$ to be vertex dependent. However, in order to remain in the linear arena, applying the latter result of Corollary 11.4.1 can be done only by taking $G_{\sigma(t)}(t) = 0$ in (11.5a). We thus obtain the following less conservative solution to the problem of state-feedback control of (11.5), with uncertainty as in (11.6), for a prescribed integer K and where we denote $Y_{i,k} = \bar{K}_{i,k}G_{i,k}$. Note that unlike the result of Corollary 11.4.1 which refers to the stochastic L_2-gain, in the design of the state-feedback controller, one is compelled to take $H_{i,k} = \eta G_{i,k}$, $\eta > 0$ with η as a tuning scalar parameter, in order to arrive at a feasible solution.

Corollary 11.5.1 *For a dwell time* $\bar{T} \ge T$, *there exists a state feedback gain* $\bar{K}_{\sigma(t)}$ *that stabilizes the system of (11.5a–c) and (11.6), and achieves, for a given scalar* $\eta > 0$, *closed-loop stochastic* L_2*-gain that is less than a prescribed positive scalar* γ, *if there exists a collection of matrices* $G_{i,k}$, $Y_{i,k}, Q_{i,k} > 0$, $i = 1, ...M$, $k = 0, ...K$ *of compatible dimensions such that, for all* $i = 1, ...M$, *and* $j = 1, ...N$ *the following holds.*

$$\begin{bmatrix} \bar{\Upsilon}_{1,k} & \bar{\Upsilon}_{2,k} & G_{i,k}^T C_i^{(j),T} + Y_{i,k}^T D_{12,i}^{(j)T} & B_i^{(j)} & Q_{i,k}^{(j)} F_i^T \\ * & -\eta(G_{i,k}+G_{i,k}^T) & \eta G_{i,k}^T C_i^{(j),T} + \eta Y_{i,k}^T D_{12,i}^{(j)T} & 0 & 0 \\ * & * & -\gamma^2 I_q & D_i^{(j)} & 0 \\ * & * & * & -I_z & 0 \\ * & * & * & * & -Q_{i,k}^{(j)} \end{bmatrix} < 0,$$

$$\bar{\Upsilon}_{1,k} = -K\frac{Q_{i,k+1}^{(j)}-Q_{i,k}^{(j)}}{T} + G_{i,k}^T A_i^{(j),T} + A_i^{(j)} G_{i,k} + B_{2,i}^{(j)} Y_{i,k} + Y_{i,k}^T B_{2,i}^{(j),T},$$

$$\bar{\Upsilon}_{2,k} = Q_{i,k}^{(j)} - G_{i,k}^T + \eta A_i^{(j)} G_{i,k} + \eta B_{2,i}^{(j)} Y_{i,k},$$

$$\begin{bmatrix} \bar{\Upsilon}_{1,k+1} & \bar{\Upsilon}_{2,k+1} & G_{i,k+1}^T C_i^{(j),T} + Y_{i,k+1}^T D_{12,i}^{(j)T} & B_i^{(j)} & Q_{i,k+1}^{(j)} F_i^T \\ * & -\eta(G_{i,k+1}+G_{i,k+1}^T) & \eta G_{i,k+1}^T C_i^{(j),T} + \eta Y_{i,k+1}^T D_{12,i}^{(j)T} & 0 & 0 \\ * & * & -\gamma^2 I_q & D_i^{(j)} & 0 \\ * & * & * & -I_z & 0 \\ * & * & * & * & -Q_{i,k+1}^{(j)} \end{bmatrix} < 0,$$

$$\bar{\Upsilon}_{1,k+1} = -K\frac{Q_{i,k+1}^{(j)}-Q_{i,k}^{(j)}}{T} + G_{i,k+1}^T A_i^{(j),T} + A_i^{(j)} G_{i,k+1} + B_{2,i}^{(j)} Y_{i,k+1} + Y_{i,k+1}^T B_{2,i}^{(j),T},$$

$k = 0, \dots K-1,$

$$\bar{\Upsilon}_{2,k+1} = Q_{i,k+1}^{(j)} - G_{i,k+1}^T + \eta A_i^{(j)} G_{i,k+1} + \eta B_{2,i}^{(j)} Y_{i,k+1},$$

$$\begin{bmatrix} \hat{\Upsilon}_{1,K} & \hat{\Upsilon}_{2,K} & G_{i,K}^T C_i^{(j),T} + Y_{i,K}^T D_{12,i}^{(j)T} & B_i^{(j)} & Q_{i,K}^{(j)} F_i^T \\ * & -\eta(G_{i,K}+G_{i,K}^T) & \eta G_{i,K}^T C_i^{(j),T} + \eta Y_{i,K}^T D_{12,i}^{(j)T} & 0 & 0 \\ * & * & -\gamma^2 I_q & D_i^{(j)} & 0 \\ * & * & * & -I_z & 0 \\ * & * & * & * & -Q_{i,K}^{(j)} \end{bmatrix} < 0,$$

$$\hat{\Upsilon}_{1,K} = G_{i,K}^T A_i^{(j),T} + A_i^{(j)} G_{i,K} + B_{2,i}^{(j)} Y_{i,K} + Y_{i,K}^T B_{2,i}^{(j),T},$$

$$\hat{\Upsilon}_{2,K} = Q_{i,K}^{(j)} - G_{i,K}^T + \eta A_i^{(j)} G_{i,K} + \eta B_{2,i}^{(j)} Y_{i,K},$$

$$Q_{i,K}^{(j)} - Q_{l,0}^{(j)} \le 0, \ \forall\ l = 1, \dots i-1, i+1, \dots M. \tag{11.20}$$

Remark 11.1. The result of Corollary 11.5.1 only improves the one of Theorem 11.3 where the switching strategy of $\sigma(t)$ is measured on line. The same method can also be applied to a case where time-invariant controller $\bar{K}_i$ is sought. One has then to choose the same G_i and Y_i for all the time intervals. One may also choose Y_i and $G_{i,k}$, $k = 1, \dots, K$ and obtain then a linearly dependent time-varying controller. In the case where the switching strategy is not known, one is obliged to choose the same constant G and Y for all the subsystems which leads to a constant controller gain $\bar{K}$.

11.6 Example – Stochastic L_2-Gain Bound

We consider the system of (11.1a–c) with the following matrices:

$$A_1 = \begin{bmatrix} 0 & 1 \\ -10(1+\delta_1) & -1 \end{bmatrix}, \; F_1 = \begin{bmatrix} 0 & 0.06 \\ 0 & 0 \end{bmatrix}, \; B_1 = \begin{bmatrix} 0 \\ 1 \end{bmatrix}, \; C_1 = \begin{bmatrix} 0.8715 & 0 \end{bmatrix},$$

$$A_2 = \begin{bmatrix} 0 & 1 \\ -0.1(1+\delta_1) & -0.5 \end{bmatrix}, \; F_2 = F_1, \, B_2 = B_1, \; C_2 = \begin{bmatrix} 0 & 0.335 \end{bmatrix},$$

$$D_1 = -0.8715, \; D_2 = 0.335, \; \bar{G}_1 = \begin{bmatrix} 0 & 0.02 \\ 0 & 0 \end{bmatrix}, \; \bar{G}_2 = \bar{G}_1$$

where $\delta_1 \in [-0.2 \;\; 0.2]$. The deterministic version of this system appeared in [21] without uncertainty. Table 11.1 below shows the minimal disturbance attenuation level γ that is achieved for various dwell times and two values of K by applying the robust version of Theorem 11.2 and the less conservative, robust version of Corollary 11.4.1. In Table 11.1, 'n.f' means 'non feasible'.

Table 11.1 Values of γ for the example

Dwell time (sec)	3.7	5	6	7.8	8.3	10	15	30
Theorem 11.2, $K = 10$	n.f	n.f	n.f	164.32	24.19	7.81	4.54	4.06
Theorem 11.2, $K = 6$	n.f	n.f	n.f	n.f	374.05	10.41	4.91	4.10
Corollary 11.4.1, $K = 10$	22.86	3.45	2.69	2.33	2.29	2.21	2.16	2.16

The results of the table clearly show the superiority of Corollary 11.4.1 over Theorem 11.2. They also demonstrate the role of K in achieving smaller bounds for the L_2-gain. We note that the Matlab codes of the results that are described in Table 11.1 are given in Appendix C. One can readily produce, with these codes, additional results by taking different values of T_d and K.

11.7 Conclusions

A new method for analyzing the mean square stability of stochastic state-multiplicative linear switched systems, using a switching dependent LF, is introduced. The method is applied to both nominal and uncertain polytopic-type systems. The stability result is extended to solve the stochastic L_2-gain problem via a set of LMIs. Based on the latter solution, the H_∞state-feedback control problem is solved where a time-varying switched state-feedback gain is derived. Constant and linear-in-time feedback gains are obtained then as special cases. In the case of polytopic type uncertainties, the results of the present work can also be generalized, to situations where the uncertainty polytopes of the subsystems have different number of vertices. Other immediate extensions are to stochastic systems with a different dwell time for each subsystem, and systems whose switching is limited, in the sense that each subsystem can only be switched to a given subset of subsystems (a case which is encountered in modeling large uncertainties). The theory presented can also be easily extended to include norm-bounded uncertainties. The example in Section 11.6 and in Chapter 13, Section 13.3 clearly demonstrate the tractability and applicability of the present work to real engineering systems. In both examples, a significant reduction of the bound on the attenuation level was obtained by applying the less conservative results of Corollaries 11.4.1 and 11.5.1.

12

Robust L_∞-Induced Control and Filtering

12.1 Introduction

In the this chapter we address the problems of induced L_∞ state-feedback control and estimation of continuous-time, state-multiplicative noise linear systems. The design and stability analysis of state-feedback controllers for systems with stochastic uncertainties have drawn a great deal of attention in the past (see [53] and the references therein). Recent articles on this subject and on related problems present results such as solutions to stochastic control and filtering problems which guarantee a worst case performance bound in the H_∞ style (see, e.g. [53]- [56])

The H_∞ approach to control has motivated, yet another approach which views a stable linear system driven by an L_∞ disturbance signal as an operator defined on a Banach space L_∞ which is supported by the observation that often, in practice, the disturbances are not of an L_2 type (finite energy signals), but rather, they are only bounded, or equivalently they are of L_∞ type. In this case the induced operator norm is the L_1 norm of the underlying system's impulse response. Within this framework, Scherer ([103]) has formulated what is called now L_1 or peak-to-peak nominal performance. He established conditions, in terms of certain LMI's for the guarantee of a prescribed bound on the induced L_1 norm of the underlying system. Abedor et. al [1] have synthesized a state feedback controller and a filter based on the peak-to-peak approach. This has been established by means of a solution to a certain LMI. In this work the authors leave the fundamental question of convexity open (they only conjecture convexity on the grounds of numerical results). We note here that both articles consider deterministic systems.

It is also noted that Wilson [123] has derived an exact expression for the peak-to-peak norm of a system. However, the result there is suitable for analysis and not for synthesis. In order to synthesize a controller (or a filter) for the nominal case, and obviously for the uncertain case, Wilson's result cannot be used; instead, we use the LMI approach. Obviously, the latter approach renders sufficiency only.

E. Gershon & U. Shaked: *Advanced Topics in Control & Estimation,* LNCIS 439, pp. 193–202.
DOI: 10.1007/978-1-4471-5070-1_12

In the present paper we extend the theory developed in [1] and [103] to include stochastic systems. As our approach differs fundamentally from the one presented in [1], it is more closely related to the Scherer's approach as presented in [103]. In fact, we consider the underlying system as one which operates between two L_∞ Banach spaces of stochastic functions, so that its induced operator norm is an L_∞-norm. Thus, we consider what is called P2P (peak-to-peak) control and estimation problem for systems with a multiplicative white noise. As has been mentioned above, the mathematical setup for stochastic systems is motivated by [103] where the framework for the deterministic counterpart is laid out.

First we address the stability issue of the stochastic state multiplicative noise systems and we then formulate and find the conditions for the induced $L_\infty-$ norm of the system to be less than or equal to a given bound γ. We also establish (see Section 3) a lower bound for the minimal gain γ by means of the minimum achievable disturbance attenuation level $\bar{\gamma}$ of the corresponding H_∞ problem, which is new in the stochastic setup. We then solve the state-feedback and the filtering problem and demonstrate the applicability and tractability of the results regarding the filtering, via a numerical example. This example clearly indicates an out-performance of the stochastic P2P filter compared to the Kalman and the stochastic H_∞ filters.

12.2 Problem Formulation and Preliminaries

We consider the following system:

$$\begin{aligned} dx &= [Ax(t)+B_1w(t)]dt+Gx(t)d\beta(t)+B_2u(t)dt, \ x(0)=0,\\ dy(t) &= [C_2x(t) + D_{21}w(t)]dt + Fx(t)d\zeta(t),\\ z(t) &= C_1x(t) + D_{12}u(t) + D_{11}w(t) \end{aligned} \tag{12.1}$$

where $x(t) \in R^n$ is the state vector, and $w(t)$ is an exogenous disturbance which is assumed to be a stochastic process in $L^\infty_{\mathcal{F}_t}(\mathcal{R}^k)$, $y(t) \in R^m$ is the measurement vector, $z(t) \in R^m$ is the objective vector and $u(t) \in R^\ell$ is the control input signal. The zero-mean real scalar Wiener processes $\beta(t)$, $\zeta(t)$ satisfy:

$$E\{\zeta(t)\zeta(s)\} = min(t,s), \ \ E\{\beta(t)\beta(s)\} = min(t,s),$$
$$E\{\beta(t)\zeta(s)\} = \tilde{\alpha}\cdot min(t,s), \ |\tilde{\alpha}| \le 1.$$

We treat the following two problems:

i) Induced L_∞ state-feedback control of systems with state-multiplicative noise:
We consider the system of (12.1a,c) and the following performance index:

$$J_E \stackrel{\Delta}{=} ||z||_\infty - \gamma||w||_\infty. \tag{12.2}$$

Our objective is to find a state-feedback control law $u(t) = Kx(t)$ that achieves $J_E \leq 0$, for all the process disturbances $w(t) \in L^{\infty}_{\mathcal{F}_t}(\mathcal{R}^k)$ and for the prescribed scalar $\gamma > 0$.

ii) Induced L_∞ filtering of systems with state-multiplicative noise: We consider the system of (12.1a–c) where $B_2 = 0,\ D_{12} = 0$ and assume an estimator of the following general form:

$$d\hat{x}(t) = A_f \hat{x}(t)dt + B_f dy,\ \ \hat{x}(0) = 0, \qquad \hat{z} = C_f \hat{x}(t), \tag{12.3}$$

where A_f is a stability matrix.

We use the following notation

$$e(t) = x(t) - \hat{x}(t), \quad \text{and} \quad \bar{z}(t) = z(t) - \hat{z}(t) \tag{12.4}$$

and consider the cost function:

$$J_F \stackrel{\Delta}{=} ||\bar{z}||_\infty - \gamma ||w||_\infty. \tag{12.5}$$

Given $\gamma > 0$, we seek an estimate $C_f \hat{x}(t)$ of $C_1 x(t)$ over the infinite time horizon $[0, \infty)$ such that J_F given by (12.5) is non-positive for all $w(t) \in L^{\infty}_{\mathcal{F}_t}(\mathcal{R}^k)$.

Remark 12.1. We note that the fact that, say, J_E is non-positive which implies that the ratio between the peak value of $E\{||z(t)||^2\}$ as expressed by $\|z\|_\infty$ and the peak value of $E\{||w(t)||^2\}$ where $w(t)$ is the input signal which produces $z(t)$, is bounded by γ. This is why the induced L_∞ norm of the system (12.1a,c) is sometimes called the peak-to-peak (P2P) norm of the system.

12.3 Stability and P2P Norm Bound of Multiplicative Noisy Systems

We consider the system (12.1a) with $B_2 = 0$, $w \equiv 0$, as we are interested in the internal stability of this system, and the following positive definite function $V(x) = \frac{1}{2} x^T Q x$, where $Q > 0$ is a constant matrix. An application of Theorem 7.1 of [67] yields the following well known sufficient condition for the mean square internal stability of (12.1a).

Theorem 12.1. *A sufficient condition for the internal exponential stability of (12.1a) in the mean square stability sense is the following*

$$QA + A^T Q + G^T Q G < 0 \tag{12.6}$$

Using Theorem 1 we establish now the following result for stochastic systems, which is an extension of the deterministic counterpart that was introduced in [103].

Lemma 12.3.1 *The system (12.1a,c) with $B_2 = 0$ and $D_{12} = 0$ is exponentially stable in mean square and J_E of (12.2) is negative for all nonzero $w(t) \in L^\infty_{\mathcal{F}_t}(\mathcal{R}^k)$ if there exist $Q > 0$, μ and λ that satisfy the following two LMIs:*

$$\Gamma_1 \triangleq \begin{bmatrix} A^TQ + QA + \lambda Q + G^TQG & QB_1 \\ B_1^TQ & -\mu I \end{bmatrix} < 0 \tag{12.7}$$

and

$$\Gamma_2 \triangleq \begin{bmatrix} \lambda Q & 0 & C_1^T \\ 0 & (\gamma-\mu)I & D_{11}^T \\ C_1 & D_{11} & \gamma I \end{bmatrix} > 0. \tag{12.8}$$

Proof: We first notice that the stability follows from the first diagonal block in (12.7) and (12.6).

Defining $\xi = col\{x,\ w\}$ and denoting $Z(t) = E\{V(x(t)\}$ we obtain from (12.7) that $E\{\xi^T\Gamma_1\xi\} \le 0$ which implies that $\dot{Z}(t) + \lambda Z(t) - \mu E\{\|w(t)\|^2\} \le 0, \forall t \ge 0, \quad V(x(0)) = 0$. It is readily found now that $Z(t) \le \bar{Z}(t)$ where $\bar{Z}(t)$ satisfies the equation:

$$\dot{\bar{Z}} + \lambda\bar{Z} - \mu E\{\|w(t)\|^2\} = 0, \quad \bar{Z}(0) = 0 \tag{12.9}$$

Namely,

$$Z \le \bar{Z} = \mu\int_0^t \exp\{-\lambda(t-s)\}E\{||w(s)||^2\}ds \le \lambda^{-1}\mu||w||^2_\infty \tag{12.10}$$

It follows, on the other hand, from (12.8) that

$$\begin{bmatrix} \lambda Q & 0 \\ 0 & (\gamma-\mu)I \end{bmatrix} - \gamma^{-1}\begin{bmatrix} C_1^T \\ D_{11}^T \end{bmatrix}\begin{bmatrix} C_1 & D_{11} \end{bmatrix} > 0$$

Multiplying the two sides of the latter inequality by $\xi(t)^T$ and $\xi(t)$, from the left and the right, respectively, and taking expectation, we find that

$$\lambda Z(t) + (\gamma-\mu)E\{\|w(t)\|^2\} > \gamma^{-1}E\{\|z(t)\|^2\} \quad \forall t \ge 0, \quad w \in L^\infty_{\mathcal{F}_t}(\mathcal{R}^k). \tag{12.11}$$

Using (12.10) in (12.11) yields

$$\mu||w||^2_\infty + (\gamma-\mu)E\{\|w(t)\|^2\} > \gamma^{-1}E\{\|z(t)\|^2\} \quad \forall t \ge 0, \quad w \in L^\infty_{\mathcal{F}_t}(\mathcal{R}^k) \tag{12.12}$$

which implies that $||z||_\infty \le \gamma||w||_\infty$.

Remark 12.2. The source of conservatism for the filtering and control results in the following sections is due to (12.10), which may yield a moderate conservatism in many cases. We stress that there is no way to estimate quantitatively the conservatism.

Remark 12.3. In the case where $D_{11} = 0$, it clearly follows from (12.8) that the optimal value of μ in the solution of (12.7) and (12.8) is $\mu = \gamma$. It is then obtained from (12.11) that $E\{||z(t)||^2\} \le \gamma\lambda\bar{Z}(t) \quad \forall t \ge 0$.

The scalar function $\bar{Z}(t)$ satisfies (12.9) and it can thus be considered as an output of a first order time invariant low-pass filter driven by the input function $E\{||w(t)||^2\}$. The transfer function of this 'filter' is $\frac{\gamma}{s+\lambda}$; it is thus concluded that, in the case where $D_{11} = 0$, the function $E\{||z(t)||^2\}$ is bounded, from above, by the output of a low-pass filter with transfer function $\frac{\gamma\lambda}{s+\lambda}$. The pole of the latter filter resides in the open interval $(0,\ min\{\frac{1}{2}eigA\})$.

The role of Lemma 1 in the analysis and design of P2P control and filtering is similar to the role that the stochastic bounded real lemma [69] plays in the H_∞ analysis and design. We apply Lemma 1 in the next section to derive the state-feedback control law that solves problem **i** of Section 2 and the filter that solves problem **ii** of the same section.

There is some similarity between the LMIs of Lemma 1 and the LMI of the stochastic Bounded Real Lemma (BRL)[53]. Considering the system (12.1a,c) with $B_2 = 0$ and $D_{12} = 0$. The H_∞norm of the system is less or equal to $\bar{\gamma} > 0$, namely:

$$E\{\int_0^\infty ||z(t)||^2 dt - \bar{\gamma}^2 \int_0^\infty ||w(t)||^2 dt\} \le 0 \tag{12.13}$$

for all $w(t) \in \tilde{L}^2_{\mathcal{F}_t}([0,\infty);\mathcal{R}^q)$ if there exists a matrix $\tilde{P} > 0$ that satisfies the following LMI:

$$\begin{bmatrix} A^T\tilde{P} + \tilde{P}A + C_1^T C_1 + G^T\tilde{P}G & \tilde{P}B_1 + C_1^T D_{11} \\ * & -\bar{\gamma}^2 I_p + D_{11}^T D_{11} \end{bmatrix} < 0. \tag{12.14}$$

Comparing (12.14) and the result of Lemma 1 the following is obtained.

Lemma 12.3.2 *The H_∞-norm of the system (12.1a,c), with $B_2 = 0$ and $D_{12} = 0$, sets a lower bound to the minimum achievable P2P ratio γ that is achieved in Lemma 1.*

Proof: Using Schur complement formula we have from (12.7) that

$$\begin{bmatrix} -\lambda P + \gamma^{-1}C_1^T C_1 & \gamma^{-1}C_1^T D_{11} \\ \gamma^{-1}D_{11}^T C_1 & (\mu-\gamma)I_p + \gamma^{-1}D_{11}^T D_{11} \end{bmatrix} < 0$$

Adding the latter to (12.8) we obtain:

$$\begin{bmatrix} A^T P + PA + \gamma^{-1}C_1^T C_1 + G^T PG & PB_1 + \gamma^{-1}C_1^T D_{11} \\ * & -\gamma I_p + \gamma^{-1}D_{11}^T D_{11} \end{bmatrix} < 0.$$

Multiplying the later LMI by γ it is readily found that (12.14) is satisfied for $\tilde{P} = \gamma P$ and for $\bar{\gamma} = \gamma$.

12.4 P2P State-Feedback Control

Seeking a SF control law $u = Kx$ that achieves $J_E \le 0$, for the system (12.1a,c), we use Schur complement formula [15] and obtain the following requirement.

$$\bar{\Gamma}_1 \triangleq \begin{bmatrix} A^TQ + QA + \lambda Q & QB_1 & G^TQ \\ B_1^TQ & -\mu I & 0 \\ QG & 0 & -Q \end{bmatrix} < 0.$$

Defining then $P \triangleq Q^{-1}$, multiplying $\bar{\Gamma}_1$ and Γ_2, from both sides, by $diag\{P, I, P\}$, and $diag\{P, I, I\}$, respectively, we find that the following LMIs are equivalent to the LMIs in Lemma 1.

$$\begin{bmatrix} PA^T + AP + \lambda P & B_1 & PG^T \\ B_1^T & -\mu I & 0 \\ GP & 0 & -P \end{bmatrix} < 0, \quad P > 0 \quad \text{and} \quad \begin{bmatrix} \lambda P & 0 & PC_1^T \\ 0 & (\gamma - \mu)I & D_{11}^T \\ C_1P & D_{11} & \gamma I \end{bmatrix} > 0.$$

Replacing in the latter LMIs A and C_1 by $A + B_2K$ and $C_1 + D_{12}K$, respectively and denoting $KP = Y$ we obtain the following result for the P2P state-feedback control.

Theorem 12.2. *Given a $\gamma > 0$. Then, the system (12.1a,c) with the state feedback law $u = Kx$ is exponentially stable in mean square and J_E of (12.2) is non-positive for all nonzero $w(t) \in L^\infty_{\mathcal{F}_t}(\mathcal{R}^k)$ if there exist $P > 0$, Y, μ and λ so that $\gamma - \mu > 0$ and so that the following two LMIs are satisfied.*

$$\begin{bmatrix} PA^T + AP + \lambda P + B_2Y + Y^TB_2^T & B_1 & PG^T \\ B_1^T & -\mu I & 0 \\ GP & 0 & -P \end{bmatrix} < 0, \quad P > 0 \qquad (12.15)$$

and

$$\begin{bmatrix} \lambda P & 0 & PC_1^T + Y^TD_{21}^T \\ 0 & (\gamma - \mu)I & D_{11}^T \\ C_1P + D_{21}Y & D_{11} & \gamma I \end{bmatrix} > 0. \qquad (12.16)$$

If the latter LMIs have a solution the state feedback gain is given by $K = YP^{-1}$.

The result of Theorem 1 can be extended to the case where the parameters in the state space model (12.1) are uncertain. Assuming that the system parameters lie within the polytope $\Omega \triangleq \left[\, A \ G \ B_1 \ B_2 \ C_1 \ D_{11} \,\right]$ which is described by its vertices. That is, for

$$\Omega_i \triangleq \left[\, A_i \ G_i \ B_{1i} \ B_{2i} \ C_{1i} \ D_{11i} \,\right] \qquad (12.17)$$

we have $\Omega = Co\{\Omega_1, \Omega_2, ..., \Omega_N\}$ where N is the number of vertices. In other words:

$$\Omega = \sum_{i=1}^{N} \Omega_i f_i \quad , \sum_{i=1}^{N} f_i = 1 \quad , f_i \geq 0. \qquad (12.18)$$

The affinity of (12.15) and (12.16) in the system parameters readily leads to the following condition for robust stabilization and P2P control design.

Corollary 12.4.1 *Given that the parameters of the system (12.1a,c) reside in the polytope (12.17). There exists a state feedback law $u = Kx$ that exponentially stabilize the system, in mean square sense, and achieves a non-positive J_E of (12.2) for all nonzero $w(t) \in L^{\infty}_{\mathcal{F}_t}(\mathcal{R}^k)$ over Ω if there exist $P > 0$, Y, μ and λ that satisfy the following LMIs:*

$$\begin{bmatrix} PA_i^T + A_iP + \lambda P + B_{2i}Y + Y^T B_{2i}^T & B_{1i} & PG_i^T \\ B_{1i}^T & -\mu I & 0 \\ G_iP & 0 & -P \end{bmatrix} < 0, \quad i = 1, 2, ...N, \quad P > 0 \tag{12.19}$$

and

$$\begin{bmatrix} \lambda P & 0 & PC_{1i}^T + Y^T D_{21i}^T \\ 0 & (\gamma - \mu)I & D_{11i}^T \\ C_{1i}P + D_{21i}Y & D_{11i} & \gamma I \end{bmatrix} > 0. \tag{12.20}$$

If the latter LMIs have a solution the state feedback gain is given by $K = YP^{-1}$.

12.5 P2P Filtering

In this section we address the filtering problem of the state-multiplicative noise stable system. We consider the system described by (12.1a-c) with $B_2 = 0$, $D_{12} = 0$ and $D_{11} = 0$ and the general type filter of (12.3). Denoting $\xi^T(t) \stackrel{\Delta}{=} [x(t)^T \quad \hat{x}(t)^T]$, $\bar{w}^T(t) \stackrel{\Delta}{=} [w(t)^T \quad n(t)^T]$ we obtain the following augmented system:

$$\begin{aligned} d\xi(t) &= [\tilde{A}_0\xi(t) + \tilde{B}\bar{w}(t)]dt + \tilde{G}\xi(t)d\beta(t) + \tilde{F}\xi(t)d\zeta(t), \ \xi(0) = 0, \\ \tilde{z}(t) &= \tilde{C}\xi(t), \end{aligned} \tag{12.21}$$

where

$$\tilde{A}_0 = \begin{bmatrix} A_0 & 0 \\ B_fC_2 & A_f \end{bmatrix}, \ \tilde{B} = \begin{bmatrix} B_1 & 0 \\ 0 & B_fD_{21} \end{bmatrix}, \ \tilde{G} = \begin{bmatrix} G & 0 \\ 0 & 0 \end{bmatrix}, \ \tilde{F} = \begin{bmatrix} 0 & 0 \\ B_fF & 0 \end{bmatrix}, \ \tilde{C} = [C_1 \ \ -C_f]. \tag{12.22}$$

We use the result of Lemma 1, and in addition, we take $\tilde{\alpha} = 0$ (for simplicity). This yields the following.

Theorem 12.3. *Consider the system of (12.1a-c) with $A-$ a Hurwitz matrix, $B_2 = 0$, $D_{12} = 0$, $D_{11} = 0$ and the general type filter of (12.3). Given $\gamma > 0$, the following holds:*

i) *The system (12.21) is stable and J_F is negative for all nonzero $w(t) \in L^{\infty}_{\mathcal{F}_t}(\mathcal{R}^k)$, if there exist $R = R^T \in \mathcal{R}^{n \times n}$, $W = W^T \in \mathcal{R}^{n \times n}$, $Z \in \mathcal{R}^{n \times r}$, $S \in \mathcal{R}^{n \times n}$, $T \in \mathcal{R}^{m \times n}$, and a positive tuning scalar λ such that*

$$\Sigma_1(R, W, Z, S) < 0, \quad \Sigma_2(R, W, T) > 0 \tag{12.23}$$

where

$$\Sigma_1 \triangleq \begin{bmatrix} RA + A^T R + \lambda R & * & * & * & * & * \\ WA + ZC + S & -S - S^T + \lambda W & * & * & * & * \\ RG & 0 & -R & * & * & * \\ WG & 0 & 0 & -W & * & * \\ ZF & 0 & 0 & 0 & -W & * \\ B_1^T R & B_1^T W + D_{21}^T Z^T & 0 & 0 & 0 & -\gamma I_q \end{bmatrix},$$

$$\Sigma_2 \triangleq \begin{bmatrix} \lambda R & 0 & C_1^T - T^T \\ * & \lambda W & T^T \\ * & * & \gamma I \end{bmatrix}.$$

ii) *If (12.23) is satisfied, a mean square stabilizing filter in the form of (12.3) that achieves $J_F < 0$ is given by:*

$$A_f = -W^{-1}S, \quad B_f = -W^{-1}Z \quad \text{and} \quad C_f = T. \qquad (12.24)$$

Proof: The assertions that J_F is negative for all nonzero $w(t) \in L^\infty_{\mathcal{F}_t}(\mathcal{R}^k)$ and that the filter of (12.3) stabilizes the filtering error (namely, $\{\tilde{A}, \tilde{G}, \tilde{F}\}$ defines a mean square stable evolution) are equivalent, by Lemma 1 to the solvability of the following inequalities.

$$\begin{bmatrix} \tilde{A}^T Q + Q\tilde{A} + \lambda Q & \tilde{G}^T Q & \tilde{F}^T Q & Q\tilde{B} \\ * & -I & 0 & 0 \\ * & * & -Q & 0 \\ * & * & * & -\mu I_q \end{bmatrix} < 0, \ \gamma \geq \mu, \text{ and } \lambda Q > \gamma^{-1}\tilde{C}_1^T \tilde{C}_1. \quad (12.25)$$

where, due to the fact that $\tilde{D}_{11}$ in the augmented system (12.21) is zero, $\mu = \gamma$ is the largest possible μ that solves (12.25a)

Following [42], we partition Q and Q^{-1} as: $Q \triangleq \begin{bmatrix} X & M \\ M^T & U \end{bmatrix}$, $\quad Q^{-1} \triangleq \begin{bmatrix} Y & N \\ N^T & V \end{bmatrix}$, where we require that $X > Y^{-1}$.

Defining : $J \triangleq \begin{bmatrix} Y & I_n \\ N^T & 0 \end{bmatrix}$ and $\tilde{J} \triangleq diag \left[J, I_m, J, J, I_q \right]$, we pre- and post-multiply (12.25a) by $\tilde{J}^T$ and $\tilde{J}$, and (12.25c) by J^T and J, respectively. Substituting:

$$Z \triangleq MB_f, \quad \tilde{Z} \triangleq C_f N^T, \text{ and } \hat{Z} \triangleq MA_f N^T \qquad (12.26)$$

in (12.25a,c) and carrying out the various multiplications there we obtain:

$$\begin{bmatrix} AY+YA^T+\lambda Y & A+YA^TX+YC_2^TZ^T+\hat{Z}^T+\lambda I & YG^T & YG^TX & 0 & YF^TZ^T & B_1 \\ * & XA+A^TX+\lambda X+ZC_2+C_2^TZ^T & G^T & YG^TX & 0 & YF^TZ^T & B_1 \\ * & * & -Y & -I & 0 & 0 & 0 \\ * & * & * & -X & 0 & 0 & 0 \\ * & * & * & * & -Y & -I & 0 \\ * & * & * & * & * & -X & 0 \\ * & * & * & * & * & * & -\gamma I_q \end{bmatrix} < 0,$$

$$\begin{bmatrix} \lambda Y & \lambda I & YC_1^T - \tilde{Z}^T \\ * & \lambda X & C_1^T \\ * & * & \gamma I \end{bmatrix} > 0, \quad X > Y^{-1} > 0 \tag{12.27}$$

where

$$Z \triangleq MB_f, \quad \tilde{Z} \triangleq C_f N^T \quad \text{and} \quad \hat{Z} \triangleq MA_f N^T. \tag{12.28}$$

Pre- and post-multiplying (12.27a) by Υ and Υ^T, and (12.27b) by $\bar{\Upsilon}$ and $\bar{\Upsilon}^T$, respectively, (12.23a) and (12.23b) are obtained, where we define the following

$$\Upsilon \triangleq diag\{\begin{bmatrix} R & 0 \\ -R & I_n \end{bmatrix}, I_m, \begin{bmatrix} R & 0 \\ -R & I_n \end{bmatrix}, \begin{bmatrix} R & 0 \\ -R & I_n \end{bmatrix}, I_q\}, \text{ and } \bar{\Upsilon} \triangleq \begin{bmatrix} R & 0 & 0 \\ -R & I_n & 0 \\ 0 & 0 & I_q \end{bmatrix}$$

$$S \triangleq \hat{Z}R, \quad T \triangleq \tilde{Z}R, \quad R \triangleq Y^{-1}, \quad W = X - R. \tag{12.29}$$

ii) If a solution to (12.23) exists, we obtain from (12.28) that

$$A_f = M^{-1}\hat{Z}N^{-T}, \quad B_f = M^{-1}Z \quad \text{and} \quad C_f = \tilde{Z}N^{-T}. \tag{12.30}$$

Denoting the transfer function matrix of the filter of (12.3) by $H_{\hat{z}y}(s)$, we find from (12.30) that:

$$H_{\hat{z}y}(s) = \tilde{Z}N^{-T}(sI_n - M^{-1}\hat{Z}N^{-T})^{-1}M^{-1}Z$$

which leads to:

$$H_{\hat{z}y}(s) = \tilde{Z}(sMN^T - \hat{Z})^{-1}Z = \tilde{Z}[s(I_n - XY) - \hat{Z}]^{-1}Z.$$

The result of (12.24) follows using (12.29).

We note that the requirement (12.27c) is satisfied by the fact that $W > 0$.

Remark 12.4. It follows from the first diagonal block in (12.27a) that $(A + \frac{\lambda}{2}I)Y + Y(A + \frac{\lambda}{2}I)^T < 0$. Hence, the positive tuning parameter λ in Theorem 12.3 is bounded by $-2max(real(eig\{A\}))$.

12.6 Conclusions

The stochastic $P2P$ control and estimation problems have been considered. The resulting designs are applicable to cases where, say, the tracking and/or the estimation errors should not necessarily be minimized; they should be bounded, instead, by prescribed bound values. It is demonstrated in the example that the peak-to-peak design may provide better results compared to those obtained by the L_2 and the H_∞design methods.

The results of this chapter first assure a $P2P$ performance for stochastic systems with known state space matrices. Since these results are based on solving LMIs that are affine in the latter matrices, they are readily extended to the case where the matrices of the state space model of the system are uncertain but they are known to reside in a given polytope.

13

Applications

13.1 Reduced-Order Control

In this example we treat a stochastic version of the practical example taken from [132], where we apply the theory of Chapter 3. In the stochastic version, the extent of the delay is accentuated in order to demonstrate the effect of the delay on the attenuation level.

Consider a modified version of the pitch control of F4E [2] described by

$$dx(t) = \begin{bmatrix} a_{11} & a_{12} & a_{12} & b_1 \\ a_{21} & a_{22} & a_{23} & 0 \\ 0 & 0 & -30 & 30 \\ 0 & 0 & 0 & -10^4 \end{bmatrix} x(t)dt + A_1 x(t-\tau(t))dt + \begin{bmatrix} 0 \\ 0 \\ 0 \\ 10^4 \end{bmatrix} u(t)dt$$

$$+ \begin{bmatrix} 1\,0\,0 \\ 0\,1\,0 \\ 0\,0\,1 \\ 0\,0\,0 \end{bmatrix} w(t)dt + \begin{bmatrix} 0 \\ 0 \\ 3\delta_e \\ 0 \end{bmatrix} d\beta(t), \quad y(t) = \begin{bmatrix} 1\,0\,0\,0 \\ 0\,1\,0\,0 \end{bmatrix} x(t),$$

$$z(t) = \begin{bmatrix} 1\,0\,0\,0 \\ 0\,1\,0\,0 \\ 0\,0\,1\,0 \end{bmatrix} x(t) + \begin{bmatrix} 0 \\ 0 \\ 1 \end{bmatrix} u(t), \; x(t) \stackrel{\Delta}{=} \begin{bmatrix} N_z \\ q \\ \delta_e \\ \tilde{\delta}_e \end{bmatrix},$$

where $H = 0, \; A_1 = diag\{0, 0, 1.5, 0\}$ and $h = 0.8, \; d = 0$. The state-vector consists of the aerodynamic components which are the load-factor N_z, the pitch-rate q and the servo components which are the elevon angle δ_e , and an additional state $\tilde{\delta}_e$. The state where δ_e relates to the elevon command u via a first-order model of the mechanical lag of the servo of a bandwidth of $30 rad/sec$, whereas $\tilde{\delta}_e$ is the electrical lag of the servo. We note that the stochastic term in the system may be caused by a flutter of the elevon which is a phenomenon related to the aero-elasticity of the elevon. The parameters

E. Gershon & U. Shaked: *Advanced Topics in Control & Estimation*, LNCIS 439, pp. 203–217.
DOI: 10.1007/978-1-4471-5070-1_13

Table 13.1 The parameters of the four operating points

Operating point	1	2	3	4
Mach number	.5	.9	.85	1.5
Altitude (ft)	5000	35000	5000	35000
a_{11}	-.9896	-.6607	-1.702	-.5162
a_{12}	17.41	18.11	50.72	29.96
a_{13}	96.15	84.34	263.5	178.9
a_{21}	.2648	.08201	.2201	-.6896
a_{22}	-.8512	-.6587	-1.418	-1.225
a_{23}	-11.39	-10.81	-31.99	-30.38
b_1	-97.78	-272.2	-85.09	-175.6

Table 13.2 The minimum achievable γ for different orders in comparison with the state-feedback.

order r	0	1	2	3	4 (Full-order)	State-feedback
γ	7.94	4.16	4.10	4.09	4.08	3.91

$a_{i,j}, i = 1,2; j = 1,2,3$ and b_1 are given in [2] in the following four operating points which are assumed here as vertices of a polytopic plant We first designed a zero-order output-feedback controller which achieved a minimal $\gamma = 7.94$ with the feedback gain matrix $K = \begin{bmatrix} 0.1103 \; 1.0471 \end{bmatrix}$. We next aim at reducing γ by adding dynamics to the controller. To this end, we applied the design approach of Remark 3.5, with a fixed $C_c = B_2^T F$. For each order $r \in \{1,2,3,4\}$ we have chosen $F \in \mathcal{R}^{n \times r}$ from 30 random trials. The results are described in Table 13.2. Comparing the result of $r = 1$ with the results of higher orders, a substantial improvement in the attenuation level is observed going from zero-order controller to the higher ones. The resulting 1st order controller turns out to be almost as good as the full order controller. Obviously, the latter two results achieve slightly higher attenuation levels compared to the state-feedback result. Note that when $h = 0$ and $H = 0$, $G = 0$ (i.e. for the deterministic, delay free counterpart system [132]), the same phenomenon appears, namely, the 1st order controller performs similarly to the full-order one. We note that the case of full-order output feedback control of this example can not be solved by the theory of [59], since the latter can cope only with norm-bounded uncertainties.

13.2 Terrain Following Control

In this example we apply the theory of Chapter 4 to a modified version of the closed-loop terrain following control problem of [53] where a time-varying delay is encountered in the thrust dynamics. Our system is described by (4.1a,b) where:

$$A = \begin{bmatrix} -1.00 & -0.15 & -0.017 & 0 & 0 & 0 & -0.011 \\ 1.00 & 0 & 0 & 0 & 0 & 0 & 0 \\ 0 & 1.00 & 0 & 0 & 0 & 0 & 0 \\ 0 & -0.00 & 0 & -0.001 & 0 & 0 & 0 \\ 0 & 0 & 0 & 0 & -1.00 & 0 & 0 \\ -0.088 & 0.034 & 0 & 0 & 1.00 & -0.0032 & 0 \\ 0 & 0 & 0.05 & 0 & 0 & 0 & -10^{-5} \end{bmatrix}, \quad B_1 = B_3 = \begin{bmatrix} 0 \\ 0 \\ 0 \\ 0 \\ 0 \\ 0 \\ -\bar{\beta} \end{bmatrix},$$

$$A_1 = \begin{bmatrix} 0&0&0&0&0&0&0 \\ 0&0&0&0&0&0&0 \\ 0&0&0&0&0&0&0 \\ 0&0&0&0&0&0&0 \\ 0&0&0&1&0&0&0 \\ 0&0&0&0&0&0&0 \\ 0&0&0&0&0&0&0 \end{bmatrix}, \quad B_2 = \begin{bmatrix} 1 & 0 \\ 0 & 0 \\ 0 & 0 \\ 0 & 0.09 \\ 0 & 0 \\ 0 & 0 \\ 0 & 0 \end{bmatrix}, \quad F = \begin{bmatrix} 0&0&-0.68&0&0&0&0 \\ 0&0&0&0&0&0&0 \\ 0&0&0&0&0&0&0 \\ 0&0&0&0&0&0&0 \\ 0&0&0&1&0&0&0 \\ 0&0&0&0&0&0&0 \\ 0&0&0&0&0&0&0 \end{bmatrix},$$

and

$$D_{12} = \begin{bmatrix} 0&0&1&0&0&0&0 \\ 0&1&0&0&0&0&0 \\ 0&0&0&0&0&0&1 \end{bmatrix}, \quad C_1 = \begin{bmatrix} 0&0&0&0&0&0&1 \\ 0&0&0&0&0&2.23&0 \\ 0&0&0&0&0&0&0 \\ 0&0&0&0&0&0&0 \end{bmatrix},$$

where $\bar{\beta} = 1/20[rad/sec]$ and $D_{13} = [0.1, 0, 0, 0]^T$. The time varying delay $\tau(t)$ appears in the 5th element of the state vector (the thrust) and satisfies $|\tau(t)| \leq \bar{\tau} = 0.1 sec$ and $|\dot{\tau}(t)| \leq 0.2$.

In order to tackle the delay, we assume a full access to the system states, namely to the vertical acceleration, height rate, height, thrust command, thrust, air speed and height error integral (see [53], page 187 for the detailed description of the dynamics involved). We apply the result of Corollary 4.4.2 where we solve the LMI of (4.21) for $\epsilon_2 = 50$, $\epsilon_m = 0.035$ and $\epsilon_r = 1$ in line with Remark 4.2. We obtain a near minimum attenuation level of $\gamma = 2.78$. The control law is given in Corollary 4.4.2 (or by (4.23b), taking $\bar{K}_\Theta = 0$) where:

$$\bar{K}_x = \begin{bmatrix} -71.6172 & -167.6193 & -142.2041 & 3.0649 & -2.6471 & -4.1063 & -986.7272 \\ 2.7585 & 5.0545 & 5.9221 & -26.7665 & -13.8189 & -15.0261 & 41.2898 \end{bmatrix},$$

$\bar{K}_r = \begin{bmatrix} -0.0577 \\ -0.1826 \end{bmatrix}$, $\bar{K}_\theta = 0$. The resulting 7 closed-loop poles are:

$$[-70.2744,\ -2.3986,\ -1.1027,\ -0.6257 \pm 0.4951,\ -0.0032,\ -1.0000].$$

We note that for $\bar{\tau} = 0$, namely when there is no delay in the thrust dynamics, a minimum attenuation level of $\gamma = 1.87$ is achieved.

13.3 State-Feedback Control of Switched Systems

In this example, we apply the theory of Chapter 11. We consider the problem of stabilizing and attenuating disturbances acting on the longitudinal short period mode of the F4E fighter aircraft with additional canards, taken from [98].

The state space model for this aircraft is described by (11.5) and is given by:

$$d\begin{bmatrix} N_z \\ q \\ \delta_e \end{bmatrix} = [\begin{bmatrix} a_{11} & a_{12} & a_{13} \\ a_{21} & a_{22} & a_{23} \\ 0 & 0 & -30 \end{bmatrix}\begin{bmatrix} N_z \\ q \\ \delta_e \end{bmatrix} + \begin{bmatrix} b_1 \\ 0 \\ 30 \end{bmatrix} u + I_3 w]dt + \begin{bmatrix} 0 \\ 0 \\ 4.5\delta_e \end{bmatrix} d\beta,$$

where $\bar{G} = 0$ and where the objective and measurement vectors are given by: $z = [N_z\ q\ u]'$ and $y = [N_z\ q]'$, respectively. In this model, N_z is the *normal acceleration*, q is the *pitch rate* and δ_e is the *elevator angle*. The parameters of the model for 4 Operating Points (O.Ps) are described in the following table:

O.P.	Mach	Altitude (ft)	a_{11}	a_{12}	a_{13}	a_{21}	a_{22}	a_{23}	b_1
1	.5	5000	−.9896	17.41	96.15	.2648	−.8512	−11.39	−97.78
2	.9	35000	−.6607	18.11	84.34	.08201	−.6587	−10.81	−272.2
3	0.85	5000	−1.702	50.72	263.5	.2201	−1.418	−31.99	−85.09
4	1.5	35000	−.5162	29.96	178.9	−.6896	−1.225	−30.38	−175.6

and it is assumed that between the O.Ps the system parameters are a convex combination of the four parameter sets of the table. A state-feedback controller is sought that achieves a minimum bound on the L_2-gain of the closed-loop system.

The standard way of solving the latter feedback problem is to apply the method of [53] and to find a solution over the polytope that has 4 O.Ps as vertices. The latter polytope is quite large, and the resulting L_2 bound, if exists, may be too large. An alternative way of solving the problem is to split the parameters polytope into 4 regions as described in Figure 13.1 The four regions overlap and it is assumed that there is an online indication which is the current subsystem. We solve the problem by considering it to be a switched

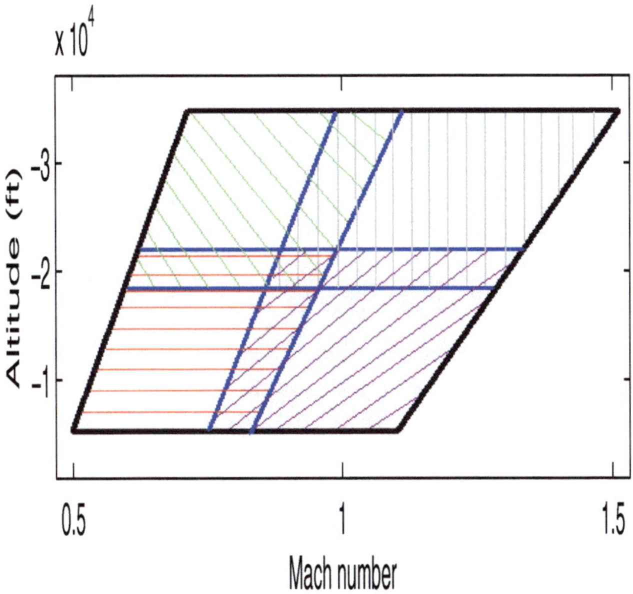

Fig. 13.1 Division into overlapping subpolytopes

system with dwell time that is dictated by the system time constant and the width of the overlap as in Figure 13.1.

This method is known in the literature.

We use the theory of Theorem 11.3 and Corollary 11.5.1, with various constraints, to design robust state-feedback controllers, assuming that the switching signal is measured online, and a dwell time of $T = 3$ secs. We shall compare our results with those that are achieved by standard polytopic state-feedback design and by gain scheduling [4]. The results are given in Table 13.3, where the minimum values of γ are obtained for different switched controllers (for either $K = 1$ or $K = 5$) and for gain scheduled controllers. In Table 13.3, 't.v gain' stands for time-varying gain, 'n.f' stands for 'not feasible', 'quad.' represents a quadratic stability solution and 'GS' stands for 'gain scheduling'. The first two results in Table 13.3 are obtained using Theorem 11.3 (the quadratic solution). In the second we took in (11.19) Q_i instead of $Q_{i,k}$. The next three results are obtained using Corollary 11.5.1. The result of row no. 4 is obtained by taking in (11.20) G_i, Y_i instead of $G_{i,k}$, $Y_{i,k}$ where in the result of row 5, the latter two matrices are replaced by G, Y that produce a controller that is independent of the measurement of $\sigma(t)$. The last two results are obtained by applying GS control design [4]. The result of row 7 is obtained by the least conservative version of the GS method which is obtained by using a modified version of Corollary 11.4.1. The constant feedback gains that correspond to the result in row no. 4 are: $\bar{K}_1 = [0.33\ 2.57\ -0.79]$, $\bar{K}_2 = [0.16\ 0.75\ \ -0.11]$, $\bar{K}_3 = [0.37\ 3.43\ \ -0.49]$, $\bar{K}_4 = [0.17\ 1.08\ \ -0.72]$.

Table 13.3 The minimum achievable values of γ

Result no.	Controller type	K=1	K=5
1	Theorem 11.3 - t.v gain	3.16	3.12
2	Theorem 11.3 - const. gain	n.f	n.f.
3	Corollary 11.5.1 - t.v gain	2.22	2.18
4	Corollary 11.5.1 - const. gain I	2.25	2.23
5	Corollary 11.5.1 - const. gain II	3.37	3.36
6	GS - quad.	n.f	n.f
7	GS - improved version	3.32	3.32

The results of Table 13.3 clearly demonstrate the adventage of the method of Corollary 11.5.1. It is less conservative than Theorem 11.3 and its result, which is based on a partial knowledge of $\sigma(t)$, is even better than the one obtained by the less conservative version of the GS design method (row 7) that requires online knowledge of $\sigma(t)$. Applying Corollary 11.5.1, the best attenuation is obtained by allowing t.v gain. However, the constant switched gain controller of row 7 is almost as good but is much simpler to implement. We note that the Matlab code for the result achieved by applying Theorem 11.3 (first two rows in Table 13.3) is given in Appendix C. This code can be readily modified to produce the rest of the results in Table 13.3.

13.4 Non Linear Systems: Measurement Output-Feedback Control

In this example we make use of the theory of Chapter 8. We consider the following model for a single degree of freedom inverted pendulum with the multiplicative white noise ω:

$$ml^2\ddot{\theta} - mgl\sin(\theta) + (\varsigma + \omega)\dot{\theta} + k\theta = u + 2v \tag{13.1}$$

where k is the spring coefficient and ς is damping coefficient. The signal v is a deterministic disturbance acting on the control input u and ω is the stochastic uncertainty in the damping. This uncertainty stems from the facts that damping is temperature dependent and that temperature possesses a component that can be modeled as white-noise. In this model, θ is the inclination angle of the pendulum, l and m are its length and mass, respectively, and g is the gravitation coefficient. Assume the available information is a noisy measurements of the angle θ, that is:

$$y(t) = \theta(t) + D_{21}v_1(t)$$

where $v_1(t)$ is a scalar bounded energy measurement noise.

The state-space representation of this model is given by:

$$\begin{bmatrix} \dot{x}_1 \\ \dot{x}_2 \end{bmatrix} = \begin{bmatrix} 0 & \frac{1}{ml^2} \\ mgl\frac{\sin(x_1)}{x_1} - k & -\frac{\varsigma}{ml^2} \end{bmatrix} \begin{bmatrix} x_1 \\ x_2 \end{bmatrix} + \begin{bmatrix} 0 \\ 1 \end{bmatrix} u + \begin{bmatrix} 0 \\ 2 \end{bmatrix} v + \begin{bmatrix} 0 \\ -\frac{x_2}{ml^2} \end{bmatrix} \omega$$

where

$$y = \begin{bmatrix} 1 & 0 \end{bmatrix} \begin{bmatrix} x_1 \\ x_2 \end{bmatrix} + 0.1 v_1.$$

The Ito formalism of this representation is:

$$\begin{aligned} dx &= [A(x) + HF(x)E_1]xdt + \begin{bmatrix} 0 \\ 2 \end{bmatrix} vdt + \begin{bmatrix} 0 \\ -\frac{x_2}{ml^2} \end{bmatrix} dW_t + \begin{bmatrix} 0 \\ 1 \end{bmatrix} udt \\ y &= Cx + D_{21} v_1 \end{aligned} \tag{13.2}$$

where $H_1 = \begin{bmatrix} 0 & mgl \end{bmatrix}^T$, $E_1 = \begin{bmatrix} 1 & 0 \end{bmatrix}$, $F(x) = \frac{sin(x_1)}{x_1}$, $A = \begin{bmatrix} 0 & \frac{1}{ml^2} \\ -k & -\frac{\varsigma}{ml^2} \end{bmatrix}$ and $C_2 = \begin{bmatrix} 1 & 0 \end{bmatrix}$.

We have:

$$\begin{aligned} & f = Ax, \; g = \begin{bmatrix} 0 \\ 1 \end{bmatrix}, \; g_1 = \begin{bmatrix} 0 \\ 2 \end{bmatrix}, \; g_2 = 0, \; \bar{g} = 0, \; G = \begin{bmatrix} 0 & 0 \\ 0 & -\frac{1}{ml^2} \end{bmatrix} x, \\ & h = C_1 x, \; H_2 = 0. \end{aligned}$$

We choose:

$$z = 10x_1 + u$$

so that in the notations of Section 8.1 we have $h = \begin{bmatrix} 10 & 0 \end{bmatrix} x$. Applying Lemma 8.3.1 we solve for $0 < P$ and $\bar{\varepsilon}$ that satisfy the following inequality:

$$\begin{bmatrix} PA + A^T P - Pgg^T P & G^T P & Pg_1 & C_1^T & 2PH_1 & \frac{\bar{\varepsilon}}{2} E_1^T \\ * & -P & 0 & 0 & 0 & 0 \\ * & * & -\gamma^2 I & 0 & 0 & 0 \\ * & * & * & -I & 0 & 0 \\ * & * & * & * & -\bar{\varepsilon} I & 0 \\ * & * & * & * & * & -\bar{\varepsilon} I \end{bmatrix} < 0.$$

Multiplying the latter inequality, from both sides, by $diag\{Q, I, I, I, I, I\}$, where $Q = P^{-1}$, and denoting $\varepsilon = \frac{4}{\bar{\varepsilon}}$, the following LMI is obtained:

$$\begin{bmatrix} AQ + QA^T - gg^T & QG^T & g_1 & QC_1^T & \varepsilon H_1 & QE_1^T \\ * & -Q & 0 & 0 & 0 & 0 \\ * & * & -\gamma^2 I & 0 & 0 & 0 \\ * & * & * & -I & 0 & 0 \\ * & * & * & * & -\varepsilon I & 0 \\ * & * & * & * & * & -\varepsilon I \end{bmatrix} < 0.$$

Taking $m = 0.5kg$, $l = 0.7meter$, $k = 0.5Newton/meter$ and $\varsigma = 0.25$, the latter LMI has a solution for $2 < \gamma$. For, say, $\gamma = 2.2$ the solution is:

$$Q = \begin{bmatrix} 0.0036 & -0.0011 \\ -0.0011 & 0.0052 \end{bmatrix} \text{ and } \quad \rho = 0.0062. \tag{13.3}$$

The latter leads to $g^T P = \begin{bmatrix} 66.077 \; 208.587 \end{bmatrix}$. The corresponding function $V = x^T Q^{-1} x$ will then satisfy the condition of Lemma 8.3.1.

Choosing $W = \begin{bmatrix} x^T \; \hat{x}^T \end{bmatrix} \bar{Q} \begin{bmatrix} x \\ \hat{x} \end{bmatrix}$ the following inequality is obtained from (8.30) in the decision variables $\bar{Q}$, K and $\hat{\varepsilon}$.

$$\begin{bmatrix} \Xi_{11} & \bar{Q}\begin{bmatrix} g_1 \\ KD_{21} \end{bmatrix} & \begin{bmatrix} I \\ -I \end{bmatrix} Pg & \begin{bmatrix} G^T & 0 \\ 0 & 0 \end{bmatrix}\bar{Q} & 2\bar{Q}\begin{bmatrix} H_1 \\ 0 \end{bmatrix} & \frac{\hat{\varepsilon}}{2}\begin{bmatrix} E_1^T \\ 0 \end{bmatrix} \\ * & -\gamma^2 I & 0 & 0 & 0 & 0 \\ * & * & -I & 0 & 0 & 0 \\ * & * & * & -\bar{Q} & 0 & 0 \\ * & * & * & * & -\hat{\varepsilon} I & 0 \\ * & * & * & * & * & -\hat{\varepsilon} I \end{bmatrix} < 0$$

where

$$\Xi_{11} = \bar{Q} diag\{\tilde{A}, \, \tilde{A}\} + diag\{\tilde{A}^T, \, \tilde{A}^T\}\bar{Q} + \bar{Q}\begin{bmatrix} 0 \\ I \end{bmatrix} KC_2 \begin{bmatrix} I & -I \end{bmatrix}$$
$$+ \begin{bmatrix} I \\ -I \end{bmatrix} C_2^T K^T \begin{bmatrix} 0 & I \end{bmatrix} \bar{Q}$$

and $\tilde{A} = A + (\gamma^{-2} g_1 g_1^T - g g^T)P$.

We choose now the following structure for $\bar{Q}$:

$$\bar{Q} = \begin{bmatrix} Q_1 & \alpha Q_2 \\ \alpha Q_2 & Q_2 \end{bmatrix},$$

and denote $\tilde{\varepsilon} = \frac{\hat{\varepsilon}}{4}$ and $Y = Q_2 K$ and obtain the following LMI in the decision variables Q_1, Q_2, Y and $\tilde{\varepsilon}$.

$$\begin{bmatrix} \tilde{\Xi}_{11} & \begin{bmatrix} Q_1 g_1 + \alpha Y D_{21} \\ \alpha Q_2 g_1 + Y D_{21} \end{bmatrix} & \begin{bmatrix} I \\ -I \end{bmatrix} Pg & \begin{bmatrix} G^T Q_1 & \alpha G^T Q_2 \\ 0 & 0 \end{bmatrix} & \begin{bmatrix} Q_1 H_1 \\ \alpha Q_2 H_1 \end{bmatrix} & \tilde{\varepsilon}\begin{bmatrix} E_1^T \\ 0 \end{bmatrix} \\ * & -\gamma^2 I & 0 & 0 & 0 & 0 \\ * & * & -I & 0 & 0 & 0 \\ * & * & * & -\begin{bmatrix} Q_1 & \alpha Q_2 \\ \alpha Q_2 & Q_2 \end{bmatrix} & 0 & 0 \\ * & * & * & * & -\tilde{\varepsilon} I & 0 \\ * & * & * & * & * & -\tilde{\varepsilon} I \end{bmatrix} < 0$$

where

$$\tilde{\Xi}_{11} = \begin{bmatrix} Q_1\tilde{A} & \alpha Q_2\tilde{A} \\ \alpha Q_2\tilde{A} & Q_2\tilde{A} \end{bmatrix} + \begin{bmatrix} \tilde{A}^T Q_1 & \alpha\tilde{A}^T Q_2 \\ \alpha\tilde{A}^T Q_2 & \tilde{A}^T Q_2 \end{bmatrix} + \begin{bmatrix} \alpha Y \\ Y \end{bmatrix} C_2 \begin{bmatrix} I & -I \end{bmatrix}$$
$$+ \begin{bmatrix} I \\ -I \end{bmatrix} C_2^T \begin{bmatrix} \alpha Y^T & Y^T \end{bmatrix}.$$

In the latter, α is a tuning scalar parameter. If a solution to the latter is obtained such that $0 < \bar{Q}$, then the filter gain matrix K is given by $K = Q_2^{-1}Y$.

Substituting for P in the latter LMI the observer gain $K = 10^6 col\{0.194,\ 4.831\}$ is obtained, using $\alpha = 8$. Since, in this example, the system matrices are all constant in time the transfer function of the controller is given by:

$$\begin{aligned}\bar{G}(s) &= -g^t P(sI - A + gg^T P - \gamma^{-2} g_1 g_1^T P + KC_2)^{-1} K \\ &= -1020.5(s + 1.27)/(s^2 + 37.41s + 75.8).\end{aligned}$$

13.5 Discrete-Time Non Linear Systems: l_2-Gain

In this example, we make use of the theory of Chapter 9. We consider the following model for a single degree of freedom inverted pendulum with the multiplicative white noise ω:

$$ml^2\ddot{\theta} - mgl\sin(\theta) + (\varsigma + \omega)\dot{\theta} + \kappa\theta = u + 2v \tag{13.4}$$

where κ is the spring coefficient and ς is a damping coefficient. The signal v is a deterministic disturbance acting on the control input u and ω is a stochastic uncertainty in the damping. In this model, θ is the inclination angle of the pendulum, l and m are its length and mass, respectively, and g is the gravitation coefficient.

Denoting $x_1 = \theta$ and $x_2 = ml^2\dot{\theta}$, we discretize the model description of (13.4) and obtain the following discrete-time state-space representation of the system.

$$\begin{bmatrix} x_1 \\ x_2 \end{bmatrix}_{k+1} = \begin{bmatrix} 1 & \frac{T}{ml^2} \\ Tmgl\frac{\sin(x_{1,k})}{x_{1,k}} - \kappa T & 1 - \frac{T\varsigma}{ml^2} \end{bmatrix} \begin{bmatrix} x_1 \\ x_2 \end{bmatrix}_k + \begin{bmatrix} 0 \\ T \end{bmatrix} u_k$$
$$+ \begin{bmatrix} 0 \\ 2T \end{bmatrix} v_k + \begin{bmatrix} 0 \\ -\sqrt{T}\frac{x_{2,k}}{ml^2} \end{bmatrix} \omega_k$$

where T is the sampling period. Regarding now the notation associated with (9.17) we have now

$$A = \begin{bmatrix} 1 & \frac{T}{ml^2} \\ -\kappa T & 1 - \frac{\varsigma T}{ml^2} \end{bmatrix}; \quad H = \begin{bmatrix} 0 \\ Tmgl \end{bmatrix}; \quad E_1 = \begin{bmatrix} 1 & 0 \end{bmatrix}; \quad F(x) = \frac{sin(x)}{x}$$

$$A_1 = \begin{bmatrix} 0 & 0 \\ 0 & -\frac{1}{ml^2} \end{bmatrix}; \quad B_1 = \begin{bmatrix} 0 \\ 2T \end{bmatrix}; \quad B_2 = \begin{bmatrix} 0 \\ T \end{bmatrix}$$

For $m = 0.5kg,\ l = 0.7meter,\ \kappa = 0.5Newton/meter$ and $\varsigma = 0.25$ we choose: $T = 0.01sec$,

$$C_1 = \begin{bmatrix} 1 & 0 \\ 0 & 0 \end{bmatrix}, \quad D = \begin{bmatrix} 0 \\ 0.1 \end{bmatrix},$$

and we require $\gamma = 0.23$.

Applying the LMI of Theorem 9.3 (for $\rho_2 \to 0$) we obtain a feasible solution with

$$\bar{Q} = 10^3 \begin{bmatrix} 3.4 & -1.8 \\ -1.8 & 9.3 \end{bmatrix}, \quad Y = \begin{bmatrix} 0.0033 & -0.8124 \end{bmatrix} \quad \text{and} \quad \rho = 0.3067. \quad (13.5)$$

The latter leads to $K = -\begin{bmatrix} 51.1802 & 97.5356 \end{bmatrix}$. The corresponding function $V = x^T \bar{Q}^{-1} x$ then satisfies the HJI of (9.19).

13.6 L_∞ Control and Estimation

In order to demonstrate the application of the theory developed in this Chapter 12, we consider the problem of altitude estimation with measurements from a RADAR altimeter and a baro altimeter. This problem was partially treated in [53]. The barometric altitude measurement is based on a static pressure measurement. As a result of various sources of error, (e.g. initial reference error, static pressure measurement bias, or temperature measurement errors) the baro altimeter is corrupted with a bias error (see page 32 [87]) up to 1000 ft together with a small white noise component. Denoting the true altitude above ground by h, we have the following approximate model for the altitude hold loop which is commanded by the altitude command $w(t)$:

$$dh = -1/\bar{\tau}(h - w)dt, \quad h_{baro} = h + b + \zeta_1 \quad (13.6)$$

where $\bar{\tau}$ is the time constant of the command response, b represents the baro altitude measurement bias and ζ_1 is a standard zero-mean white noise with intensity R_1, that is: $\mathcal{E}\{\zeta_1(t)\zeta_1(\tau)\} = R_1\delta(t-\tau)$.

The RADAR altimeter measures the height above ground level without bias, however, its output is corrupted by a broad band measurement noise, the intensity of which increases with height due to a lower SNR (signal to noise ratio) effect at higher altitude (see [83], page 196, for various models depending on the SNR levels).

A reasonable model for this effect is:

$$\begin{aligned} h_{radar} &= h(1+\eta) + \zeta_2, \\ \mathcal{E}\{\zeta_2(t)\zeta_2(\tau)\} &= R_2\delta(t-\tau), \quad \mathcal{E}\{\eta(t)\eta(\tau)\} = R_3\delta(t-\tau). \end{aligned} \quad (13.7)$$

In this model, η represents the altitude-dependent broad-band measurement noise and ζ_2 represents the altitude-independent white noise. The overall model is completed by: $\dot{b} = -0.09b + \sqrt{2}\bar{w}$, where $\bar{w}$ is a standard white noise signal which interferes with the measurement of the bias. The later signal is not correlated with the other white noise signals in the system.

To obtain a state-space description we denote: $x \triangleq [h \quad b]^T$, and $y \triangleq [h_{baro} \quad h_{radar}]^T$ where x is the state vector and y is the measurement vector. Using the measurements $Y(t) \triangleq \{y(\tau), \quad \tau < t\}$, we want to estimate $z(t) = [h(t) \quad b(t)]^T$.

Combining the dynamical descriptions in (13.6), (13.7) and that of $\dot{b}$, we obtain:

$$dx = (Ax + B_1\tilde{w})dt, \quad y = (Cx + D_{21}\tilde{w}) + Fxd\eta, \quad z = Lx \tag{13.8}$$

where

$$\tilde{w} = \begin{bmatrix} w \\ \bar{w} \\ \zeta_1 \\ \zeta_2 \end{bmatrix} \quad A = \begin{bmatrix} -1/\bar{\tau} & 0 \\ 0 & -0.09 \end{bmatrix}, \quad B_1 = \begin{bmatrix} 1/\bar{\tau} & 0 & 0 & 0 \\ 0 & \sqrt{2} & 0 & 0 \end{bmatrix}, \quad C = \begin{bmatrix} 1 & 1 \\ 1 & 0 \end{bmatrix},$$

$$F = \begin{bmatrix} 0 & 0 \\ \sqrt{R_3} & 0 \end{bmatrix}, \quad D_{21} = \begin{bmatrix} 0 & 0 & \sqrt{R_1} & 0 \\ 0 & 0 & 0 & \sqrt{R_2} \end{bmatrix}, \quad L = \begin{bmatrix} 1 & 0 \\ 0 & 0 \end{bmatrix}.$$

We note here that $\tilde{w}$ in (13.8) is assumed to comprise of components each of which is a stochastic processes with wide band spectrum so that each is of a finite anergy. As a result of this (13.8) agrees with the underlying setup of this paper as introduced by eq.(12.1). We also remark that $\bar{w}$, ζ_1 and ζ_2 are in general white noise processes, however, since they are acted upon by a system of a limited bandwidth they may be considered as stochastic processes in $L^{\infty}_{\mathcal{F}_t}(\mathcal{R}^k)$.

In our problem, the altitude hold-loop dynamics is characterized by a time-constant of $\bar{\tau} = 30sec$. We consider a step maneuver in w at a time $5.6sec$ from an initial condition of $w = h(0) = 500ft$ to the value of $w = 5000ft$. We also take an initial bias of $b = 0ft$ and we solved the case where $R_1 = R_2 = 0.4$ and $R_3 = 0.0016$.

In order to demonstrate the P2P performance of the new filter, we compare it with the P2P performance of three other popular filters, since they are not designed to optimize the P2P performance of the resulting estimator process, we tuned these filters to improve their P2P behavior. We consider the four filters:

1) **Kalman filter**: This filter ignores the multiplicative noise $\eta(t)$. For the design of the filter we assume that w is an unbiased white-noise signal with: $\mathcal{E}\{w(t)w(\tau)\} = Q\delta(t-\tau)$ where Q should be determined to achieve the best P2P result for the filter (with a tuned Q).

2) H_∞ **stochastic filter**: In order to compare the results with those obtained by the Kalman filter we multiply B by $\sqrt{Q}$. The measurement equation is changed to $y = Cx + \sqrt{R}\tilde{\zeta}$ where $\tilde{\zeta} = [\zeta_1 \quad \zeta_2]^T$ and where $R = diag\{R_1, R_2\}$. The term $F\eta$ is taken into consideration using the results of [53].
3) $P2P$ **filter**: The multiplicative noise $\eta(t)$ is neglected and a standard $P2P$ design is used. We apply the same changes in the state equations as we did in the H_∞filter.
4) **Stochastic** $P2P$ **filter**: The term $F\eta$ is taken into consideration using the results of Theorem 12.3. Since the best Q-tuned Kalman filter is obtained for $Q = 1000$, we designed the four filters for this value of Q. We obtained the following:

- The Kalman filter: The following transfer function matrix for the tuned Kalman filter relates the two measurements of h_{radar} and h_{baro} to the estimate of the height h:

$$G_{Kalman} = (s^2 + 4.25s + 3.73)^{-1}[0.674s + 0.061 \quad 1.49s + 3.6].$$

  The behavior of the resulting L_2 estimate is depicted in Fig. 13.2.1b, where the true trajectory of h is described in Fig. 13.2.1a. The Kalman estimate achieves a mean square error of $34060ft^2$.
- The stochastic H_∞ filter: This filter is calculated for $\gamma = 0.80$, a value slightly higher than the minimun achievable value of γ . The corresponding transfer function matrix is given by $G_{sto-hinf} = (s^2 + 5.98s + 7.59)^{-1}[1.58s + 0.067 \quad 1.86s + 7.49]$.
  The time behavior of the estimate that is achieved by this filter is given in Fig. 13.3.1c. The corresponding mean square estimation error is $35991ft^2$.
- The $P2P$ filter : This filter is calculated for $F = 0$, $\lambda = 0.0665 \cdot 10^{-6}$ and for the minimum achievable value $\gamma = 3.028$. The corresponding transfer function matrix is given by: $G_{p2p} = (s^2 + 4.31s + 3.87)^{-1}[0.675s + 0.061 \quad 1.52s + 3.74]$.
  The time behavior of the estimate that is achieved by this filter is given in Fig. 13.3.1d. The corresponding mean square estimation error is $35133ft^2$.
- The stochastic $P2P$ filter: This filter is calculated for $\lambda = 0.034$ and a near minimum value $\gamma = 5.595$. The corresponding transfer function matrix is given by: $G_{sto-p2p} = \frac{1}{(s^2+3.23s+1.18)}[0.893s + 0.08 \quad 0.384s + 1.05]$. The corresponding state space representation is given by: $A_f = -\begin{bmatrix} 1.314 & 0.907 \\ 1.468 & 1.913 \end{bmatrix}$, $B_f = \begin{bmatrix} 0.907 & 0.39 \\ 1.823 & -0.355 \end{bmatrix}$ and $C_f = \begin{bmatrix} 0.9845 & 0 \\ 0 & 0 \end{bmatrix}$. The time behavior of the estimate that is achieved by this filter is given in Fig. 13.4.1e. This filter is seen to be the best of all the four filters considered. Its estimate is closest to the actual trajectory of h and its mean square error is $28654ft^2$.

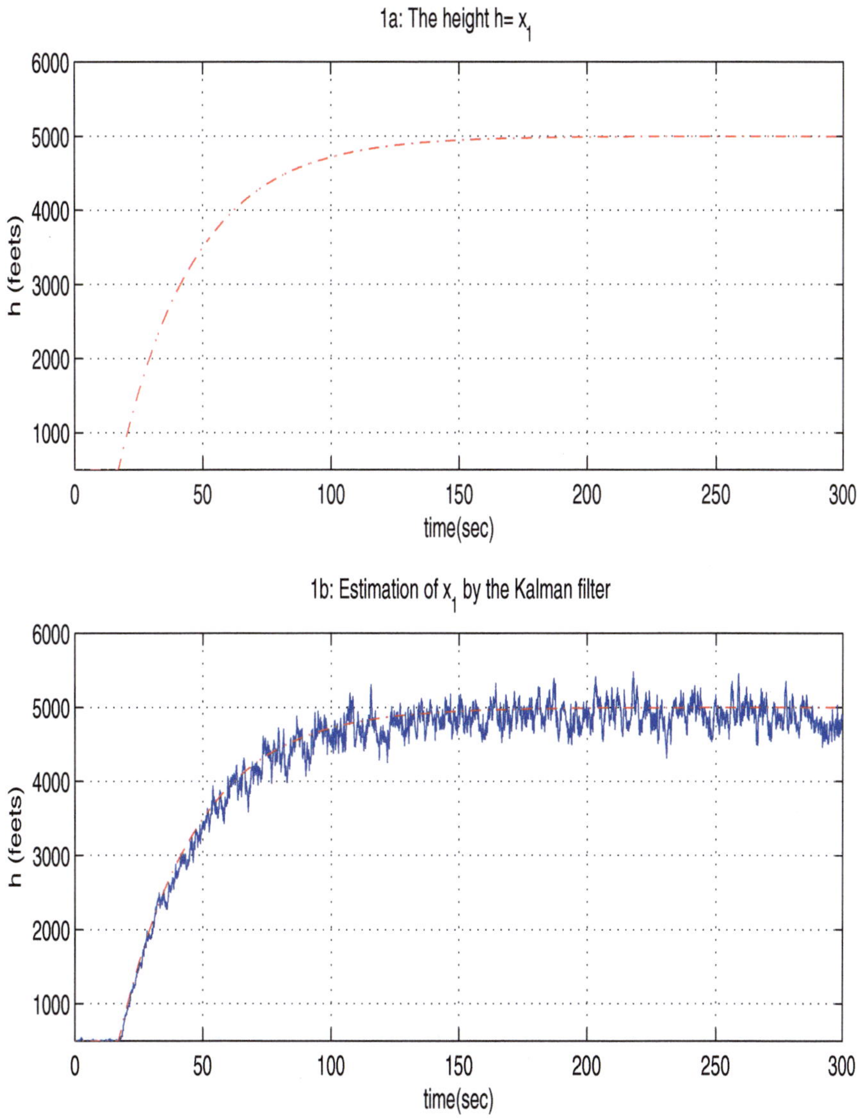

Fig. 13.2 Estimation by Kalman Filter

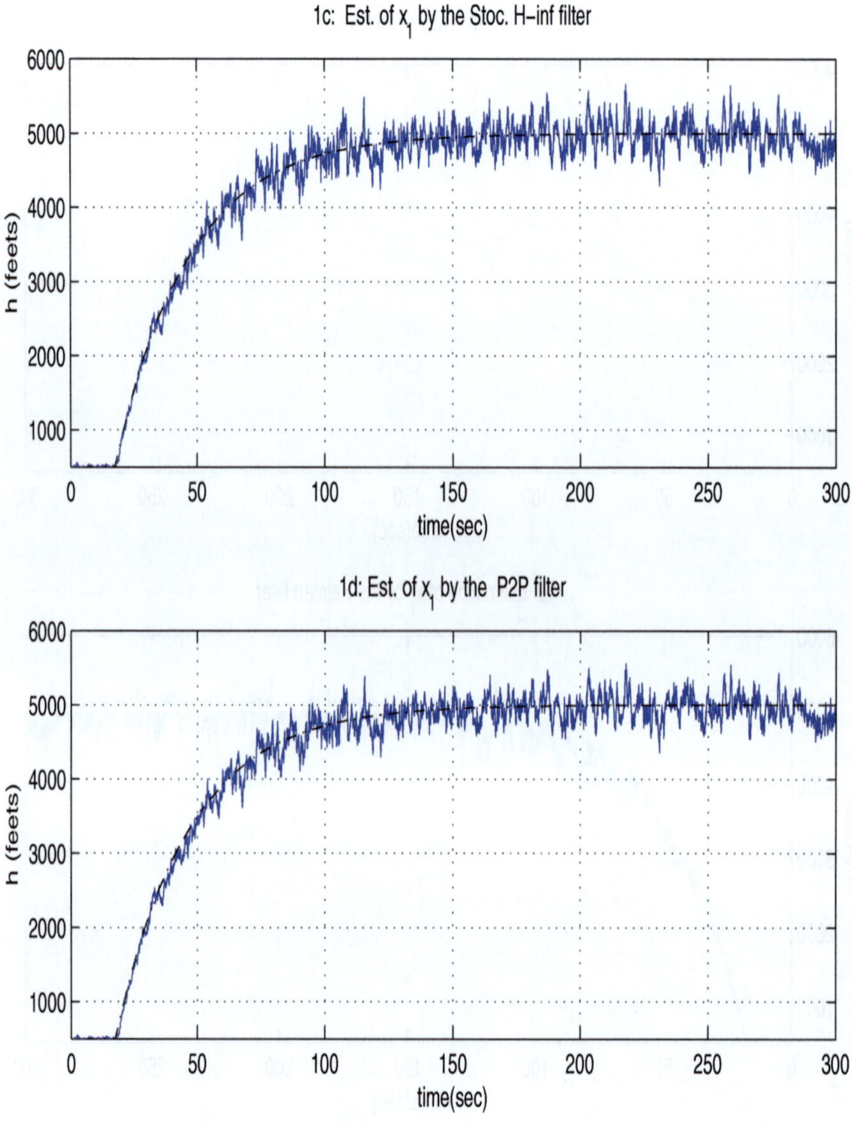

Fig. 13.3 Estimation by the Stochastic H_∞ and the standard P2P filters

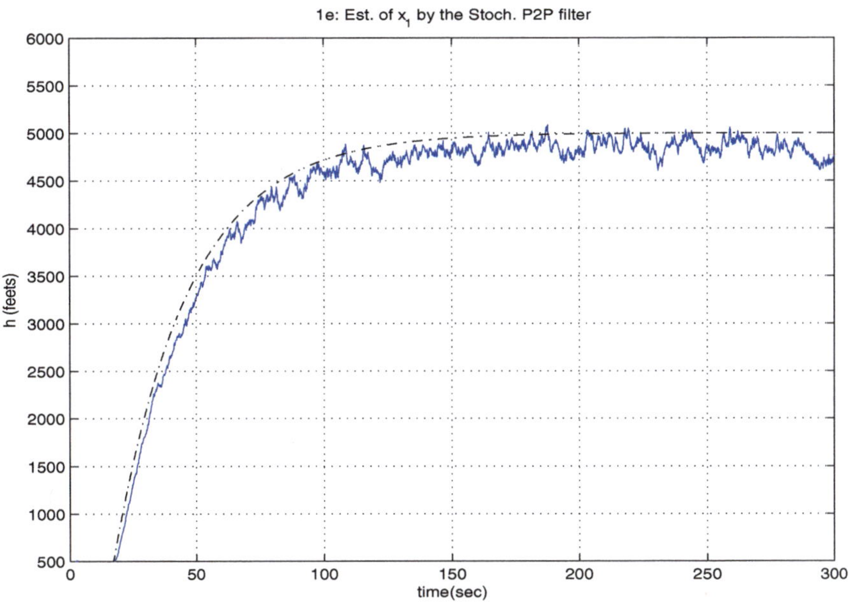

Fig. 13.4 Estimation by the Stochastic P2P filter

It would now be appropriate to verify whether the ratio between the L_∞ norms of the estimation error and the input w is indeed less than the designed bound of 0.88. It follows from the results that were depicted in Fig. 13.4.1e for the stochastic P2P filter that the latter ratio is 1.017, compared to the $\gamma = 5.5949$. This however does not indicate how close this γ is to the true operator norm, since the input signal which is considered in our simulation is a particular one (a step). The conservatism issue is yet an open question, and it hasn't even been discussed in the deterministic case literature (see[103], [1]).

Remark 13.1. As seen from Fig 13.4.1e the P2P estimation error may entail a nonzero steady state estimation error. As long as this steady state value is less than the obtained γ, this error is still consistent with the P2P objective.

Remark 13.2. The low DC gain achieved for the P2P stochastic filter may seem to be an excessive overdesign in view of the particular step input we have employed in the simulation of the example. However, this is misleading, as we recall that the filter is designed for receiving all input signals in L_∞, which include also those that possess a high frequency content.

A

Appendix: Stochastic Control Processes – Basic Concepts

A.1 Basics of σ-Algebra

We repeat here a short description of the σ-algebra already brought in our first book in this field due to its importance. The proper use of σ-algebra calculus is inherent in this monograph to the study of systems either delayed or non-delayed with state-multiplicative noise in the H_∞ context, given the nature of the multiplicative term (be it in the dynamics, the disturbance or the input matrices). This term is a white noise sequence (in the discrete-time case), or a Wiener process (in the continuous-time case). In the sequel we bring the description of a system with state-multiplicative noise and the signals involved in both the continuous- and the discrete-time setups.

A.1.1 The Continuous-Time Case

Given the following system with state-multiplicative noise:

$$\begin{aligned} dx &= [A(t)x(t) + B_1(t)w(t) + B_2(t)u(t)]dt + D(t)x(t)d\beta(t), \; x(0) = x_0, \\ y(t) &= C_2(t)x(t) + D_{21}(t)u(t), \\ z(t) &= C_1(t)x(t) + D_{12}(t)u(t), \end{aligned} \tag{A.1}$$

where, for simplicity, we consider a single, zero-mean, Wiener process, denoted by $\beta(t)$. We note that all the stochastic differential equations in this monograph are interpreted to be of the Ito type.

In order to elaborate on the nature of the signals involved in the system i.e: $x(t)$, $w(t)$, $u(t)$, $y(t)$ and $z(t)$ we introduce here the notations used for linear the continuous-time systems considered in this monograph.

In the continuous-time setup we provide all spaces $\mathcal{R}^k$, $k \geq 1$ with the usual inner product $< \cdot, \cdot >$, and with $|| \cdot ||$ we denote the standard Euclidean norm.

It is assumed here that the Wiener process β_t is defined on a probability space $(\Omega, \mathcal{F}, P)$. We then write $(\Omega, \mathcal{F}, \{\mathcal{F}_t\}_{t\geq 0}, P)$ to denote the filtered probability space generated by the Wiener process β_t, that is, the increasing family of σ-algebras $\{\mathcal{F}_t\}_{t\geq 0}$ is generated by β_t in the usual sense. The control signal u_t is taken to be a function of the observations, namely, $u_t = K(\mathcal{Y}_t)$, where $\mathcal{Y}_t = \{y(s) : s \leq t\}$. Thus, u_t is $\mathcal{F}_t$-measurable for all $t \in [0, T]$, and it is also non-anticipative process. Before characterizing the disturbance process w_t, we introduce the Hilbert space $\tilde{L}^2([0, T]; \mathcal{R}^k)$ which consists of, by definition, the the family of all non-anticipative stochastic processes, say f(t), defined on the filtered probability space $(\Omega, \mathcal{F}, \{\mathcal{F}_t\}_{t\geq 0}, P)$, and satisfy

$$||f||^2_{\tilde{L}^2} = E\{\int_0^T ||f(t)||^2 dt\} = \int_0^T E\{||f(t)||^2\}dt < \infty.$$

Thus, we assign for the system of (A.1) the above notations where $x(t) \in \tilde{L}^2([0, T]; \mathcal{R}^n$ is the state vector, $w(t) \in \tilde{L}^2([0, T]; \mathcal{R}^p)$ is an exogenous disturbance, $z(t) \in \tilde{L}^2([0, T]; \mathcal{R}^m$ is the objective vector and $u(t) \in \tilde{L}^2([0, T]; \mathcal{R}^l$ is the control input signal and $y(t) \in \tilde{L}^2([0, T]; \mathcal{R}^z$ is the measurement signal.

By writing $w(t) \in \tilde{L}^2([0, T]; \mathcal{R}^p)$ we thus mean that $w(t)$ is a finite (mean square) energy signal, measurable on $\mathcal{F}_t$, and consequently that $w(t)$ is, for example, a nonanticipative feedback strategy of the type $w(t) = f(x(\tau), \tau \leq t$.

A.1.2 The Discrete-Time Case

Similarly to the continuous-time case, we consider the following system with state-multiplicative noise:

$$\begin{aligned} x_{k+1} &= (A_k + D_k\beta_k)x_k + B_{2,k}u_k + B_{1,k}w_k, \qquad k = 0, 1, ..., N-1, \\ y_k &= C_{2,k}x_k + D_{21,k}u_k, \\ z_k &= C_{1,k}x_k + D_{12,k}u_k, \end{aligned} \tag{A.2}$$

where, for simplicity, we consider a single zero-mean white noise sequence, denoted by β_k. In order to describe the nature of the signals involved in the system, i.e: x_k, w_k, u_k, y_k and z_k, we bring below the notations used for the discrete-time systems of Chapter 5.

Denoting by $\mathcal{N}$ the set of the positive integers in $[1 \ N]$, we denote by $l^2(\Omega, \mathcal{R}^n)$ the space of square-integrable $\mathcal{R}^n-$ valued functions on the probability space $(\Omega, \mathcal{F}, \mathcal{P})$, where Ω is the sample space, $\mathcal{F}$ is a $\sigma-$algebra of subsets of Ω called events, and $\mathcal{P}$ is the probability measure on $\mathcal{F}$. By $(\mathcal{F}_k)_{k\in\mathcal{N}}$ we denote an increasing family of $\sigma-$algebras which is generated by β_j, $0 \leq j \leq k-1$. $\mathcal{F}_k \subset \mathcal{F}$. We also denote by $\tilde{l}^2(\mathcal{N}; \mathcal{R}^n)$ the space of n-dimensional nonanticipative stochastic processes $\{f_k\}_{k\in\mathcal{N}}$ with respect to $(\mathcal{F}_k)_{k\in\mathcal{N}}$ where $f_k \in L^2(\Omega, \mathcal{R}^n)$. On the latter space the following l^2-norm is defined:

$$||\{f_k\}||^2_{\tilde{l}_2} = \mathcal{E}\{\sum_0^{N-1} ||f_k||^2\} = \sum_0^{N-1} \mathcal{E}\{||f_k||^2\} < \infty, \quad \{f_k\} \in \tilde{l}_2(\mathcal{N}; \mathcal{R}^n) \tag{A.3}$$

where $|| \cdot ||$ is the standard Euclidean norm. Let $\phi : R^n \times R^m \to R$ be a continuous function and let $z \in R^n$ and W be an R^m valued random variable defined on $(\Omega, \mathcal{F}, P)$. Define $E_W\{\phi(z, W)\} = \int_\Omega \phi(z, W(\omega))dP_W(\omega)$, where P_W is the measure induced by W (which is the restriction of P to the the $\sigma-$algebra generated by W). Note that it readily follows from this definition that $E_W\{\phi(z, W)\} = E[\phi(Z, W)/Z = z]$ where Z is any random variable defined on $(\Omega, \mathcal{F}, P)$, which is independent on W.

Thus, we assign for the system (A.2) the above notations where $\{x_k\} \in \tilde{l}^2([0, N]; \mathcal{R}^n)$ is the state vector, $\{w_k\} \in \tilde{l}^2([0, N-1]; \mathcal{R}^p)$ is an exogenous disturbance, $\{z_k\} \in \tilde{l}^2([0, N]; \mathcal{R}^m)$ is the objective vector and $\{u_k\} \in \tilde{l}^2([0, N-1]; \mathcal{R}^l)$ is the control input vector and $\{y_k\} \in \tilde{l}^2([0, N]; \mathcal{R}^z)$ is the measurement vector.

A.2 Application of Ito Lemma

Ito lemma is useful in evaluating the covariance of state multiplicative processes considered in the present monograph (see [108] for comprehensive coverage). Consider

$$dx = Axdt + Dxd\xi + Bd\beta$$

where ξ is a scalar valued standard Wiener process and where β is also a scalar Wiener process independent of ξ so that $\mathcal{E}\{d\beta d\beta^T\} = Qdt$. We define $w = col\{\beta,\ \xi\}$. The intensity of w is given by $\tilde{Q} \triangleq \begin{bmatrix} Q & 0 \\ 0 & 1 \end{bmatrix}$. We also define $\tilde{G} = \begin{bmatrix} B & Dx \end{bmatrix}$

Defining $\varphi(x(t)) = x_i(t)x_j(t)$, where x_i is the i-th component of x, we get (see [75]), using Ito lemma, that

$$d\varphi = x_i(t)dx_j(t) + x_j(t)dx_i(t) + \frac{1}{2}Tr\{\tilde{G}\tilde{Q}\tilde{G}^T \Sigma\},$$

where the only non zero entries in Σ are at locations i, j and j, i. Consequently, we have that

$$d(xx^T) = (xdx^T + dxx^T + \tilde{G}\tilde{Q}\tilde{G}^T)dt.$$

Taking the expectation of both sides in the latter, and defining $P(t) = \mathcal{E}\{x(t)x^T(t)\}$, the following result is obtained.

$$\dot{P} = AP + PA^T + BQB^T + DPD^T.$$

A.3 Stochastic Stability

The issue of Stochastic stability is central to all the various topics treated in this monograph, as one can expect. Consider the following stochastic system

$$dx = f(x(t), t)dt + g(x(t), t)d\beta(t) \tag{A.4}$$

for $x \in \mathcal{R}^n$ and the corresponding Ito formula:

$$d\varphi = \varphi_t dt + \varphi_x dx + \frac{1}{2}Tr\{gQg^T\varphi_{xx}\}$$

$$= \varphi_t dt + \varphi_x(f + gd\beta) + \frac{1}{2}Tr\{gQg^T\varphi_{xx}\}$$

$$= \varphi_t dt + \varphi_x f + \frac{1}{2}Tr\{gQg^T\varphi_{xx}\} + \varphi_x gd\beta = L\varphi + \varphi_x gd\beta$$

where

$$L \triangleq \frac{\partial}{\partial t} + f^T\frac{\partial}{\partial x} + \frac{1}{2}\Sigma_{i,j=1}^{n}(gg^T)_{i,j}\frac{\partial^2}{\partial x_i \partial x_j}$$

is referred to as the infinitesimal generator of the system (A.4). Since

$$d\varphi = L\varphi + Md\beta$$

where $M \triangleq \varphi_x g$, then if $\varphi(x) = V(x) > 0$ for all x and $LV \leq 0$, then V is said to be Lyapunov function.

From the above we have:

$$V(x(t)) = V(x_0) + \int_0^t LV(x(t))dt + \int_0^t Md\beta \leq V(x_0) + \int_0^t Md\beta. \tag{A.5}$$

Using results about Martingales, which are beyond the scope of the present monograph, it is shown in [105] that (A.5) implies that all trajectories with x_0 sufficiently close to the origin remain, at all times, in the origin's neighborhood, except for a set of trajectories with arbitrarily small probability. The latter result, which corresponds to almost sure stability, can be understood also intuitively: if $V(x)$ is reduced so that $x(t)$ gets closer to the origin, then its gradient V_x becomes small and consequently $M = V_x g$ tends to zero. Therefore, $V(x)$ is reduced in a manner which is similar to the deterministic case. In fact, when $g(x,t) \equiv 0$ the above discussion reduces to the stability analysis of the deterministic system

$$\dot{x} = f(x(t), t).$$

In this case, $LV \leq 0$ reduces to $\frac{dV}{dt} \leq 0$. Furthermore, if $g(x,t) = x$, then the term

$$M = V_x g = V_x x$$

tends to zero if $x(t) \to 0$, thus allowing the effect of $\beta(t)$ to vanish, ensuring the convergence of $x(t)$ to zero. If in addition, $lim_{t\to\infty}x(t) = 0$ with probability 1, then the origin is stochastically asymptotically stable. Furthermore, if the above results hold for all x_0 rather than those that are in a small neighborhood, then the origin is globally asymptotically stable. It turns out that when $LV(x) < 0$ holds for all nonzero x then the origin is asymptotically stable in probability. If also $LV(x) < -kV(x)$ for all nonzero x, for some $k > 0$, then also mean square global stability is obtained.

Consider the class of linear systems with state-multiplicative noise that is treated in the present monograph where

$$f(x) = Ax, \text{ and } g(x) = Dx,$$

and where all eigenvalues of A are in the open left half of the complex plane. Choosing

$$V(x) = x^T Px,$$

with $P > 0$ and requiring that $LV < -kV$, for $k > 0$, (which generally means small enough $||D||$), imply square stability, in addition to the global asymptotic stability in probability.

It may be interesting to see a case [130] where the different concepts of stability may lead to different conclusions.

Consider the so-called Geometrical Brownian Motion (GBM) (see also [105]) which evolves according to the following stochastic differential equation:

$$dx(t) = ax(t)dt + bx(t)d\beta(t), \quad x(0) = x_0 \tag{A.6}$$

where $\beta(t)$ is again a standard Wiener process (or Brownian Motion), namely $\mathcal{E}\{d\beta^2(t)\} = dt$ and where $a - b^2/2 < 0$ but $a > 0$ so that the deterministic case with $b = 0$ would be unstable.

Define $dz(t) = bd\beta(t), z(0) = 0$ (meaning that $z(t) = b\beta(t)$) and consider $\varphi(z) = e^{(a-b^2/2)t+z(t)}$. Applying Ito lemma to evaluate $d\varphi$ we obtain, since z satisfies (A.4) with $f(z) = 0$, and $g(z) = b$, that

$$d\varphi = \varphi_t dt + \varphi_z dz + \tfrac{1}{2}\varphi_{zz}dt = [(a - b^2/2)\varphi + \varphi dz + \tfrac{b^2}{2}\varphi]dt = a\varphi dt + b\varphi dz$$
$$= a\varphi dt + b\varphi d\beta.$$

Since φ satisfies (A.6) we arrive at the conclusion (see [105]) that the solution (A.6) is

$$x(t) = e^{(a-b^2/2)t+b\beta(t)} = e^{[(a-b^2/2)+b\beta(t)/t]t}$$

Since $\beta(t)/t$ tends to zero as t tends to infinity ($\beta(t)$ is of the order of $\sqrt{t}$), $x(t)$ tends to zero as t tends to infinity (see [105]). Also, the corresponding Lyapunov exponential is negative [130]:

$$\lim_{t\to\infty} \lambda(t) = \lim_{t\to\infty} [log(x(t))]/t = a - b^2/2 < 0.$$

Consider now $V(x) = px^2$ where $p > 0$. Then,

$$LV = (2a + b^2)px^2 > 0$$

and also $P(t) = \mathcal{E}\{x^2(t)\}$ is shown to satisfy

$$dP/dt = (2a + b^2)P$$

which means that $P(t) = x_0^2 e^{(2a+b^2)t}$. The latter tends to infinity as t tends to infinity. Namely, (A.6) can not be shown to be stochastically stable in probability but it is clearly unstable in the mean square sense. Considering a non quadratic Lyapunov function $V(x) = p|x|^\alpha$ where $p > 0$, then

$$LV = [a + \frac{1}{2}b^2(\alpha - 1)]\alpha p|x|^\alpha,$$

which, by choosing $0 < \alpha < 1 - \frac{2a}{b^2}$, leads to $LV < 0$. Therefore, the system (A.6) is stochastically stable, in probability, but not mean square stable.

If both $a < 0$ and $a - b^2/2 < 0$ are satisfied, then for $b^2 < -2a$ (in this case b plays the role of D in $dx = Axdt + Dxd\beta$), the above discussion shows that (A.6) is mean square stable.

A.4 Simulation of Stochastic Differential Equations

Consider

$$dx = Axdt + Dxd\xi + Bd\beta.$$

Consider now the discrete-time stochastic process

$$\frac{x(k+1) - x(k)}{h} = Ax(k) + Dx(K)\bar{\xi}(k)/\sqrt{h} + D\bar{\beta}(k)/\sqrt{h},$$

where $\bar{\xi}(k)$ is a normally distributed random sequence of zero mean and unit variance and $\bar{\beta}(k)$ is a random vector sequence of zero mean and covariance Q. Notice that $x(k)$, in fact, satisfies

$$x(k+1) = Fx(k) + Gw(k),$$

where $F = I + Ah$ and $G = \sqrt{h}\,[\,B\ Dx(k)\,]$ and where $w(k) = col\{\bar{\beta}(K),\ \bar{\xi}(K)\}$. Defining $S(k) = \mathcal{E}\{x(k)x^T(k)\}$ we see, by the independence of $x(k)$ on $w(k)$, that

$$P(k+1) = FP(k)F^T + \mathcal{E}\{GQG^T\} = FP(k)F^T + hBQB^T + hDP(k)D^T.$$

Substituting for F, neglecting terms of the order of h^2 and dividing both sides of the resulting equation by h we get

$$\frac{P(k+1) - P(k)}{h} = AP(k) + P(k)A^T + BQB^T + DP(k)D^T,$$

meaning that the discrete-time process $x(k)$ correctly represents the continuous-time process $x(t)$, in the sense that the first two moments are identical in the limit where the integration step h tends to zero. Furthermore, we note that the increments $\xi((k+1)h) - \xi(kh)$ are represented in the discrete-time model by $\bar{\xi}(k)\sqrt{h}$, which is zero mean and of variance h thus approximating the property of Wiener process increments $\mathcal{E}\{d\xi^2(t)\} = dt$.

B

The LMI Optimization Method

The following is a short description of the LMI technique, which will be used extensively in the proposed research for stationary systems.

A LMI is any constraint of the form

$$A(x) \equiv A_0 + x_1 A_1 + ... + x_N A_N < 0 \tag{B.1}$$

where $x = (x_1, ..., x_N)$ is a vector of unknown scalars and $A_0, ..., A_N$ are given symmetric matrices. The above LMI is a convex constraint on x since $A(y) < 0$ and $A(z) < 0$ imply that $A((y+z)/2)) < 0$. As a result its solution set, if it exists, is called the feasible set and it is a convex subset of R^N. Within the feasible set one may choose to solve optimization problems which are still convex such as to minimize $c_0 + c_1 x_1 + ... + c_n x_n$. It should be noted that often cost functions for state-multiplicative noise often lead, as in the deterministic case, to quadratic inequalities in the search variables vector $x = (x_1, ..., x_N)$ of the following form :

$$Q(x) - S(x)R(x)^{-1}S^T(x) > 0, R(x) > 0$$

where $Q(x) = Q^T(x)$ and $R(x) = R^T(x)$ and $S(x)$ depend affinely on x. In such cases, the application of the Schur complement (see e.g. [15]) shows that an equivalent LMI is :

$$\begin{bmatrix} Q(x) & S(x) \\ S^T(x) & R(x) \end{bmatrix} > 0$$

where in many cases, repeated application of Schur complement is needed to achieve an LMI which is affine not only in the search variables but also in some of the parameters (e.g. matrices A, B_2, etc.).

The LMI technique for solving convex optimization problems has led to a major development in the field of H_∞ control and estimation in the past few years [15], [40] motivated by the efficiency of the interior point algorithms to solve LMIs. In fact, all problems which are convex in the optimizing parameters in the field of system theory are amenable to this technique,and, therefore, since any matrix inequality of the form $L(X_1, X_2, ...) < 0$ being affine in matrix valued arguments $X_1, X_2, ...$ can be written at the form of (B.1). Consider, for example, the following Lyapunov inequality:

$$A^T P + PA < 0.$$

Defining

$$P = \begin{bmatrix} x_1 & x_2 \\ x_2^T & x_3 \end{bmatrix}, \quad A = \begin{bmatrix} A_{11} & A_{12} \\ A_{21} & A_{22} \end{bmatrix},$$

The above inequality can be readily expressed in the form of B.1. Note also that in the case where the dynamic matrix A is uncertain say, $A = \Sigma A_i f_i,\ f_i \in [0\ 1],\ \Sigma f_i = 1$, the resulting inequality should also be affine in A in order to be formulated as (B.1).

In a case where the system is linear and stationary, the LMI technique is easily applied to the solution of the H_∞ control problem, either with perfect state-measurement or with noisy one (and thereby also to the filtering problem). It is important to note that unlike the Riccati equation approach, which supplies the central solution, the LMI approach produces a solution which may be far from the central solution [66]. The inequality-based solution using the latter approach can be easily extended to other additional requirements such as the minimum upper-bound on variance, in the case of nominal H_∞ estimation problem (the "mixed H_2/H_∞" problem). It is also important to note that in the case where the system parameters are modelled in a polytopic framework, the LMI technique can be easily applied to allow solutions for both the control and the estimation problems (see, for example, [46] for the solution of the H_∞ filtering problem, in the presence of polytopic uncertainties) .

C

Stochastic Switching with Dwell Time – Matlab Scripts

In this appendix we bring three scripts of the Matlab codes of Example 11.6 in Chapter 11 and Example 13.3 in Chapter 13 - both are taken from the field of stochastic state-multiplicative switching systems. These codes were run on Matlab 7.6.0 (R2008a), using the YALMIP package with the SDPT3 solver, part of the CVX optimization toolbox. The codes produce the results of Table 11.1 in Chapter 11 (two scripts) and one result of Table 13.3, Chapter 13 (one script). Each script is self-contained and maybe run independently of the others. One can change the different parameters such as the dwell time T_d and the integer K (according to the allowed computational complexity).

C.1 Example 11.6

C.1.1 Code for the Result of Theorem 11.2 in Table 11.1

```
 %— Stochastic Switching
%— Exmp1_L2_gain_Theorem_11_2.m
%— taken from Chapter 11.
%— This file  produces the result of Example 11.6,
%— first and second   rows of Table 11.1
%— using Theorem   11.2
%— Run on Matlab 7.6.0 (R2008a), using the YALMIP
%— package and the SDPT3
%— solver , part of the CVX optimization toolbox

clear
close all
%
del=0.2;
%——— First subsystem
```

```
A{1,1}=[0 1;-10*(1+del) -1];
A{1,2}=[0 1;-10*(1-del) -1];
Bw{1,1}=[0;1];
Bw{1,2}=[0;1];
Cz{1,1}=[0.8715 0];
Cz{1,2}=[0.8715 0];
Dzw{1,1}=[-0.8715];
Dzw{1,2}=[-0.8715];

%
spar=0.2;
F{1,1}=spar*[0 0.3;0 0];
F{1,2}=spar*[0 0.3;0 0];
G{1,1}=spar*[0 0.1;0 0];
G{1,2}=spar*[0 0.1;0 0];
%
A{2,1}=[0 1;-0.1*(1+del) -0.5];
A{2,2}=[0 1;-0.1*(1-del) -0.5];
Bw{2,1}=[0;1];
Bw{2,2}=[0;1];
Cz{2,1}=[0 0.335];
Cz{2,2}=[0 0.335];
Dzw{2,1}=[0.335];
Dzw{2,2}=[0.335];
%
F{2,1}=spar*[0 0.3;0 0];
F{2,2}=spar*[0 0.3;0 0];
G{2,1}=spar*[0 0.1;0 0];
G{2,2}=spar*[0 0.1;0 0];


noss=2; %number of sub systems;
n=2; %dim x
nw=1; %dim w
nu=1; %dim u
nz=1; %dim z
ny=1; %dim y
pv=2; %------Polytop vertices
i_sys=menu('Choose the following', ...
'Stochastic L2-gain, Theorem 11.2, ...
Example 11.6, Robust , K=10, td=10 sec' ...
,'Stochastic L2-gain, Theorem 11.2, ...
Example 11.6 - choose parameters');
%
if i_sys==1
K=10
td=10
else;end;
if i_sys==2
```

```
td=input(' enter td (the dwell time)=')
K=input(' enter K = ')
%— K is the number of breaks
%— in the function.
%— 1 is the minimum and
%— denotes a single straight line.
else;end;
%
BS=K+1;
zr=0; %zero margin - to be used with penbmi
%— to ensure feasibility of the solution.
tbs=td/K; %time between sections
constraints=[];
for i=1:noss
%vs=num2str(v);
for k=1:BS
%bs=num2str(b);
p{i,k}=sdpvar(n);
%eval(['p',vs,'_',bs,'=sdpvar(n);'])
constraints=[constraints,p{i,k}>0];
end
end
gamsq=sdpvar(1);

for i=1:noss
for k=1:K
for v=1:pv
Ac=A{i,v}; %A Current
Bc=Bw{i,v};
Cc=Cz{i,v};
Dc=Dzw{i,v};
lm1{i,k,v}=[(p{i,k+1}-p{i,k})/tbs+p{i,k}* ...
Ac+Ac'*p{i,k} , p{i,k}*Bc, ...
Cc' , F{i,v}'*p{i,k} , G{i,v}'*p{i,k}; ...
Bc'*p{i,k}, -gamsq*eye(nw), Dc', zeros(nw,n) ...
,zeros(nw,n); ...
Cc,Dc,-eye(nz),zeros(nz,n),zeros(nz,n); ...
p{i,k}*F{i,v},zeros(n,nw),zeros(n,nz), ...
-p{i,k},zeros(n); ...
p{i,k}*G{i,v} , zeros(n,nw), zeros(n,nz), ...
zeros(n) , -p{i,k}]; ...
lm2{i,k,v}=[(p{i,k+1}-p{i,k})/tbs+p{i,k+1}*Ac ...
+Ac'*p{i,k+1} , p{i,k+1}*Bc, ...
Cc', F{i,v}'*p{i,k+1} , G{i,v}'*p{i,k+1}; ...
Bc'*p{i,k+1},-gamsq*eye(nw),Dc', zeros(nw,n), ...
zeros(nw,n); ...
Cc,Dc,-eye(nz),zeros(nz,n),zeros(nz,n); ...
p{i,k+1}*F{i,v},zeros(n,nw),zeros(n,nz), ...
-p{i,k+1} , zeros(n); ...
```

```
p{i,k+1}*G{i,v} , zeros(n,nw), ...
zeros(n,nz) , zeros(n) , -p{i,k+1}];
constraints=[constraints,lm1{i,k,v}<0,lm2{i,k,v}<0];
end
end
k=K+1;
for v=1:pv
Ac=A{i,v}; %A Current
Bc=Bw{i,v};
Cc=Cz{i,v};
Dc=Dzw{i,v};
lm1{i,k,v}=[p{i,k}*Ac+Ac'*p{i,k},p{i,k}*Bc, ...
Cc',F{i,v}'*p{i,k}, G{i,v}'*p{i,k}; ...
Bc'*p{i,k},-gamsq*eye(nw),Dc',zeros(nw,n), ...
zeros(nw,n);Cc,Dc,-eye(nz),zeros(nz,n) ...
,zeros(nz,n); p{i,k}*F{i,v},zeros(n,nw), ...
zeros(n,nz) , -p{i,k} , zeros(n); ...
p{i,k}*G{i,v}, zeros(n,nw), ...
zeros(n,nz) , zeros(n) , -p{i,k}];
constraints=[constraints,lm1{i,k,v}<0];
end
for i2=[1:i-1,i+1:noss]
lmsi{i,i2}=[p{i,K+1}-p{i2,1}];
%LM Switching Instances
constraints=[constraints,lmsi{i,i2}>0];
end
end
ops = sdpsettings('solver','sdpt3');
solvesdp(constraints,gamsq,ops);
checkset(constraints)
Stoc_L2_Gain=all(checkset(constraints)>0);
if Stoc_L2_Gain == 0
display(' No feasible solution ')
else
gamma=sqrt(double(gamsq))
end
```

C.1.2 Code for the Result of Corollary 11.4.1 in Table 11.1

```
%— Stochastic Switching
%— Exmp1_L2_gain_Corollary_11_4_1.m
%— taken from Chapter 11.
%— This file produces the result of Example 11.6,
%— third row of Table 11.1
%— using Corollary 11.4.1.
%— Run on Matlab 7.6.0 (R2008a),
%— using the YALMIP package and the SDPT3
%— solver, part of the CVX optimization toolbox
```

```
clear
close all
%
%— Note: A_ij , i is the subsystem , j is the vertex
%— del is the uncertainty interval
disp(' del = The uncertainty Interval ')
del=0.2
%———— Subsystem 1
A{1,1}=[0 1;-10*(1+del) -1];
A{1,2}=[0 1;-10*(1-del) -1];
Bw{1,1}=[0;1];
Bw{1,2}=[0;1];
Cz{1,1}=[0.8715 0];
Cz{1,2}=[0.8715 0];
Dzw{1,1}=[-0.8715];
Dzw{1,2}=[-0.8715];
%———— Subsystem 2
A{2,1}=[0 1;-0.1*(1+del) -0.5];
A{2,2}=[0 1;-0.1*(1-del) -0.5];
Bw{2,1}=[0;1];
Bw{2,2}=[0;1];
Cz{2,1}=[0 0.335];
Cz{2,2}=[0 0.335];
Dzw{2,1}=[0.335];
Dzw{2,2}=[0.335];
%———— Stochastic parameters ,
%———— Subsystems 1 and 2
spar=0.2;
F{1}=spar*[0 0.3;0 0];
F{2}=spar*[0 0.3;0 0];
Gb{1}=spar*[0 0.1;0 0];
Gb{2}=spar*[0 0.1;0 0];
noss=2;
%————Number of sub systems;
n=2; %dim x
nw=1; %dim w
nu=1; %dim u
nz=1; %dim z
ny=1; %dim y
pv=2; %——Polytop vertices

i_sys=menu('Choose the following ', ...
'Stochastic L2-gain , ...
Corollary 11.4.1 - the ...
chapter results [Example 11.6 , ...
td=3.7 sec , K=10] ' ...
,'Stochastic L2-gain , Corollary 11.4.1 ...
-choose dwell time (td) and K');
%—GH=i_sys -1;
```

```
%The chapter result for K=10, Example 11.6, Table 11.1
if i_sys==1
K=10
td=3.7
else;end;
if i_sys==2
td=input(' enter td  (the dwell time)  = ')
K=input(' enter K = ')
%— K is the  number of breaks in the function
else;end;
BS=K+1;
zr=0;
%——zero margin – to be used with penbmi
%——to ensure feasibility of the solution.

%— Stochastic L2 gain
tbs=td/K; %time between sections

constraints=[];
for i=1:noss
%vs=num2str(v);
for k=1:BS
%bs=num2str(b);
g{i,k}=sdpvar(n,n,'full');
h{i,k}=sdpvar(n,n,'full');
%eval(['p',vs,'_',bs,'=sdpvar(n);'])
for v=1:pv
q{i,k,v}=sdpvar(n);
constraints=[constraints,q{i,k,v}>0];
end
end
end
gamsq=sdpvar(1);

for i=1:noss
for k=1:K
for v=1:pv
Ac=A{i,v}; %A Current
Bc=Bw{i,v};
Cc=Cz{i,v};
Dc=Dzw{i,v};
lm1{i,k,v}=[-(q{i,k+1,v}-q{i,k,v})/tbs+ ...
g{i,k}'*Ac'+Ac*g{i,k}, ...
q{i,k,v}-g{i,k}'+Ac*h{i,k}, ...
g{i,k}'*Cc' , Bc , q{i,k,v}*F{i}', ...
q{i,k,v}*Gb{i}';
q{i,k,v}-g{i,k}+h{i,k}'*Ac',-h{i,k} ...
-h{i,k}' , h{i,k}'*Cc', ...
zeros(n,nw),zeros(n),zeros(n); ...
```

```
Cc*g{i,k} , Cc*h{i,k} , -gamsq*eye(nz), ...
Dc,zeros(nz,n), zeros(nz,n); ...
Bc' , zeros(nw,n) , Dc' , -eye(nw), ...
zeros(nw,n) , zeros(nw,n); ...
F{i}*q{i,k,v} , zeros(n,n) , zeros(n,nz), ...
zeros(n,nw) , -q{i,k,v}, zeros(n); ...
Gb{i}*q{i,k,v} , zeros(n) , zeros(n,nz), ...
zeros(n,nw) , zeros(n), -q{i,k,v}]; ...
lm2{i,k,v}=[-(q{i,k+1,v}-q{i,k,v})/tbs+ ...
g{i,k+1}'*Ac'+Ac*g{i,k+1}, ...
q{i,k+1,v}-g{i,k+1}' ...
+Ac*h{i,k+1} , g{i,k+1}'*Cc', Bc,q{i,k+1,v}*F{i}' ...
,q{i,k+1,v}*Gb{i}'; ...
q{i,k+1,v}-g{i,k+1}+h{i,k+1}'*Ac',-h{i,k+1} ...
-h{i,k+1}' ,h{i,k+1}'*Cc' , zeros(n,nw) ...
,zeros(n) , zeros(n); ...
Cc*g{i,k+1} , Cc*h{i,k+1}, -gamsq*eye(nz), ...
Dc , zeros(nz,n) , zeros(nz,n); ...
Bc' , zeros(nw,n), Dc',-eye(nw), ...
zeros(nw,n) , zeros(nw,n); ...
F{i}*q{i,k+1,v} , zeros(n,n) , zeros(n,nz), ...
zeros(n,nw) , -q{i,k+1,v} , zeros(n); ...
Gb{i}*q{i,k+1,v} , zeros(n) , zeros(n,nz), ...
zeros(n,nw) , zeros(n) , -q{i,k+1,v}];
constraints=[constraints,lm1{i,k,v}<0,lm2{i,k,v}<0];
end
end
k=K+1;
for v=1:pv
Ac=A{i,v}; %A Current
Bc=Bw{i,v};
Cc=Cz{i,v};
Dc=Dzw{i,v};
lm1{i,k,v}=[g{i,k}'*Ac'+Ac*g{i,k}, ...
q{i,k,v}-g{i,k}'+Ac*h{i,k}, ...
g{i,k}'*Cc' , Bc , q{i,k,v}*F{i}', ...
q{i,k,v}*Gb{i}';
q{i,k,v}-g{i,k}+h{i,k}'*Ac', ...
-h{i,k}-h{i,k}' , h{i,k}'*Cc' ...
,zeros(n,nw) , zeros(n) , zeros(n); ...
Cc*g{i,k} , Cc*h{i,k} , -gamsq*eye(nz), ...
Dc , zeros(nz,n) , zeros(nz,n); ...
Bc' , zeros(nw,n) , Dc' , -eye(nw), ...
zeros(nw,n) , zeros(nw,n); ...
F{i}*q{i,k,v} , zeros(n,n) , zeros(n,nz), ...
zeros(n,nw) , -q{i,k,v} , zeros(n); ...
Gb{i}*q{i,k,v} , zeros(n) , zeros(n,nz), ...
zeros(n,nw) , zeros(n) , -q{i,k,v}];
constraints=[constraints,lm1{i,k,v}<0];
```

```
end
for i2=[1:i-1,i+1:noss]
lmsi{i,i2}=[q{i,K+1}-q{i2,1}];
%——LM Switching Instances
constraints=[constraints,lmsi{i,i2}<0];
end
end
ops = sdpsettings('solver','sdpt3');
solvesdp(constraints,gamsq,ops);
checkset(constraints)
Stoc_L2_Gain=all(checkset(constraints)>0);
if Stoc_L2_Gain == 0
display(' No feasible solution ')
else

[A{1,1} A{1,2} F{1}  Gb{1}];
[A{2,1} A{2,2} F{2} Gb{2}];
gamma=sqrt(double(gamsq))
end
```

C.2 Example 13.3

C.2.1 Code for the Result of Theorem 11.3 in Table 13.3

```
%— Stochastic Switching
%— Exmp_13_3_SF_Theorem_11_3_tv_or_constant_gain.m
%— Stochastic State-feedback  Switching - Chapter 13,
%— Example 13.3,  (Theory of Chapter 11)
%— This file  produces the results of Example 13.3,
%— first and second rows  of
%— Table 13.3 for K=1 or K=5, using the theory of
%— Theorem 11.3  with time-varying
%— gain ( q is  q{i,k}), or constant gain.
%— Run on Matlab 7.6.0 (R2008a), using the YALMIP
%— package and the SDPT3
%— solver, part of the CVX optimization toolbox

% In this example we assume that the switching can
%-be measured online, and
% therefore LMIs are used to design the controller
% The dynamics of the F4 are used, where it's assumed
% that the F4 is flying
% in 5000ft fo altitude (vertices 1 and 3 of the
% original dynamics), and
% that the mach number is between 0.5 and 0.7 for
% the first subsystem, and
% between 0.65 and 0.85 for the second subsystem.
```

```
clear
close all
%—clc

noss=4; %number of sub systems;
n=3; %dim x
nw=5; %dim w
nu=1; %dim u
nz=3; %dim z
ny=2; %dim y
pv=4; %Polytop vertices

%the 4 vertices of the original system are:
%0.5 Mach, 5000 ft
oA{1}=[-0.9896 17.41 96.15;0.2648 -0.8512 -11.39;0 0 -30];
oBu{1}=[-97.78 ; 0 ; 30];
%0.9 Mach, 35000 ft
oA{2}=[-0.6607 18.11 84.34;0.08201 -0.6587 -10.81;0 0 -30];
oBu{2}=[-272.2 ; 0 ; 30];
%0.85 Mach, 5000 ft
oA{3}=[-1.702 50.72 263.5;0.2201 -1.418 -31.99;0 0 -30];
oBu{3}=[-85.09 ; 0 ; 30];
%1.5 Mach, 35000 ft
oA{4}=[-0.5162 29.96 178.9;-0.6896 -1.225 -30.38;0 0 -30];
oBu{4}=[-175.6 ; 0 ; 30];


% A{1,1}=oA1;
% A{1,2}=(3/7)*oA1+(4/7)*oA2;
% A{1,3}=(3/7)*oA1+(4/7)*oA3;
% A{1,4}=(3/7)*oA1+(4/7)*oA4;
% A{3,1}=(4/7)*oA1+(3/7)*oA3;
for i=1:noss
    for j=1:pv
        A{i,j}=(3/7)*oA{i}+(4/7)*oA{j};
        Bu{i,j}=(3/7)*oBu{i}+(4/7)*oBu{j};
        Bw{i,j}=[eye(n),zeros(n,nw-n)];
        Cz{i,j}=[1 0 0 ; 0 1 0 ; 0 0 0];
        Dzu{i,j}=[0 ; 0 ; 1];
        Dzw{i,j}=zeros(nz,nw);
    end
end
td=3
%
zr=0; %zero margin - to be used with penbmi
%— to ensure feasibility of the solution.
elm=-eye(n+nw+nz)*zr;
elm=zeros(14,14);
```

```
en=eye(n)*zr;

%% Synthesis


spar=15;
F{1}=spar*[0 0 0 ; 0 0 0; 0 0 0.3 ];
F{2}=spar*[0 0 0 ; 0 0 0; 0 0 0.3 ];
F{3}=spar*[0 0 0 ; 0 0 0; 0 0 0.3 ];
F{4}=spar*[0 0 0 ; 0 0 0; 0 0 0.3 ];


%—tbs=td/K; %time between sections

constraints=[];
r_constraints=[];


i_sys=menu('Choose the following','Stochastic SF, ...
Theorem 11.3 ...
-time-varying gain [Example 13.3, td=3 sec]' ...
,'Stochastic SF, Theorem 11.3 - constant gain ');
K_value=menu('Choose the value of K ','K=1 ' ...
,'K=5 ');
if K_value==1
K=1
else
K=5
end
tbs=td/K; %time between sections
BS=K+1;
if i_sys==1
%——Theorem 11.3, Time Varying Gain

for i=1:noss
  %
    for k=1:BS
        %
        q{i,k}=sdpvar(n);
        y{i,k}=sdpvar(nu,n,'full');
        %
        constraints=[constraints,q{i,k}>en];
        r_constraints=[r_constraints,q{i,k}>0];
    end
end
gamsq=sdpvar(1);

for i=1:noss
    for k=1:K
```

```
            for v=1:pv
                  Ac=A{i,v}; %A Current
                  Bwc=Bw{i,v};
                  Buc=Bu{i,v};
                  Czc=Cz{i,v};
                  Dzwc=Dzw{i,v};
                  Dzuc=Dzu{i,v};
lm1{i,k,v}=[-(q{i,k+1}-q{i,k})/tbs+Ac*q{i,k} ...
+q{i,k}*Ac'+Buc*y{i,k} ...
+y{i,k}'*Buc',Bwc,q{i,k}*Czc'+y{i,k}'*Dzuc',q{i,k}*F{i}';
Bwc' , -gamsq*eye(nw) , Dzwc', zeros(nw,n); ...
Czc*q{i,k}+Dzuc*y{i,k},Dzwc,-eye(nz),zeros(nz,n); ...
F{i}*q{i,k},zeros(n,nw),zeros(n,nz),-q{i,k}]; ...
lm2{i,k,v}=[-(q{i,k+1}-q{i,k})/tbs+Ac*q{i,k+1} ...
+q{i,k+1}*Ac' ...
+Buc*y{i,k+1}+y{i,k+1}'*Buc' , Bwc , ...
q{i,k+1}*Czc'+y{i,k+1}'*Dzuc', q{i,k+1}*F{i}'; ...
Bwc' , -gamsq*eye(nw) , Dzwc' , zeros(nw,n); ...
Czc*q{i,k+1}+Dzuc*y{i,k+1},Dzwc,-eye(nz),zeros(nz,n); ...
F{i}*q{i,k+1},zeros(n,nw),zeros(n,nz),-q{i,k+1}]; ...
constraints=[constraints,lm1{i,k,v}<elm,lm2{i,k,v}<elm];
r_constraints=[r_constraints,lm1{i,k,v}<0,lm2{i,k,v}<0];
end
      end
      k=K+1;
      for v=1:pv
            Ac=A{i,v}; %A Current
            Bwc=Bw{i,v};
            Buc=Bu{i,v};
            Czc=Cz{i,v};
            Dzwc=Dzw{i,v};
            Dzuc=Dzu{i,v};
lm1{i,k,v}=[ Ac*q{i,k}+q{i,k}*Ac'+Buc*y{i,k} ...
+y{i,k}'*Buc' ...
,Bwc , q{i,k}*Czc'+y{i,k}'*Dzuc', q{i,k}*F{i}';
Bwc' , -gamsq*eye(nw) , Dzwc', zeros(nw,n);
Czc*q{i,k}+Dzuc*y{i,k} , Dzwc ,-eye(nz),zeros(nz,n);
F{i}*q{i,k} , zeros(n,nw), zeros(n,nz),-q{i,k}];
constraints=[constraints,lm1{i,k,v}<elm];
r_constraints=[r_constraints,lm1{i,k,v}<0];
      end
      for i2=[1:i-1,i+1:noss]
            lmsi{i,i2}=[-q{i,K+1}+q{i2,1}];
            %LM Switching Instances
            constraints=[constraints,lmsi{i,i2}>en];
            r_constraints=[r_constraints,lmsi{i,i2}>0];
      end
end
ops = sdpsettings('solver','sdpt3');
```

```
solvesdp(constraints,gamsq,ops);
checkset(constraints)
SF_LMI=all(checkset(constraints)>0);
if SF_LMI == 0
display(' No feasible solution ')
else
gamma=sqrt(double(gamsq))
end
 else
%———Theorem 11.3,  Constant Gain
for i=1:noss
    %
    for k=1:BS
        %
        q{i,k}=sdpvar(n);
        q{i,k}=q{i,1};%—————————— Note !!!
        y{i,k}=sdpvar(nu,n,'full');
       %
        constraints=[constraints,q{i,k}>en];
        r_constraints=[r_constraints,q{i,k}>0];
    end
end
gamsq=sdpvar(1);

for i=1:noss
    for k=1:K
        for v=1:pv
            Ac=A{i,v}; %A Current
            Bwc=Bw{i,v};
            Buc=Bu{i,v};
            Czc=Cz{i,v};
            Dzwc=Dzw{i,v};
            Dzuc=Dzu{i,v};
lm1{i,k,v}=[-(q{i,k+1}-q{i,k})/tbs+Ac*q{i,k} ...
+q{i,k}*Ac'+Buc*y{i,k}...
+y{i,k}'*Buc',Bwc , q{i,k}*Czc'+y{i,k}'*Dzuc', ...
q{i,k}*F{i}'; ...
Bwc',-gamsq*eye(nw) ,Dzwc',zeros(nw,n); ...
Czc*q{i,k}+Dzuc*y{i,k} ,Dzwc,-eye(nz),zeros(nz,n); ...
F{i}*q{i,k} , zeros(n,nw), zeros(n,nz),-q{i,k}]; ...
lm2{i,k,v}=[-(q{i,k+1}-q{i,k})/tbs+Ac*q{i,k+1}
+q{i,k+1}*Ac'+Buc*y{i,k+1} ...
+y{i,k+1}'*Buc',Bwc ,q{i,k+1}*Czc'+y{i,k+1}'*Dzuc', ...
q{i,k+1}*F{i}'; ...
Bwc',-gamsq*eye(nw) , Dzwc',zeros(nw,n); ...
Czc*q{i,k+1}+Dzuc*y{i,k+1},Dzwc,-eye(nz),zeros(nz,n); ...
F{i}*q{i,k+1} , zeros(n,nw), zeros(n,nz) ,-q{i,k+1}]; ...
constraints=[constraints,lm1{i,k,v}<elm,lm2{i,k,v}<elm];
r_constraints=[r_constraints,lm1{i,k,v}<0,lm2{i,k,v}<0];
```

```
            end
        end
        k=K+1;
        for v=1:pv
            Ac=A{i,v}; %A Current
            Bwc=Bw{i,v};
            Buc=Bu{i,v};
            Czc=Cz{i,v};
            Dzwc=Dzw{i,v};
            Dzuc=Dzu{i,v};
lm1{i,k,v}=[ Ac*q{i,k}+q{i,k}*Ac'+Buc*y{i,k}+y{i,k}'*Buc' ...
,Bwc , q{i,k}*Czc'+y{i,k}'*Dzuc', q{i,k}*F{i}';
Bwc' , -gamsq*eye(nw) , Dzwc', zeros(nw,n);
Czc*q{i,k}+Dzuc*y{i,k} , Dzwc , -eye(nz), zeros(nz,n);
F{i}*q{i,k} , zeros(n,nw), zeros(n,nz) ,-q{i,k}];
            constraints=[constraints,lm1{i,k,v}<elm];
            r_constraints=[r_constraints,lm1{i,k,v}<0];
        end
        for i2=[1:i-1,i+1:noss]
            lmsi{i,i2}=[-q{i,K+1}+q{i2,1}];
            %LM Switching Instances
            constraints=[constraints,lmsi{i,i2}>en];
            r_constraints=[r_constraints,lmsi{i,i2}>0];
        end
end
ops = sdpsettings('solver','sdpt3');
solvesdp(constraints,gamsq,ops);
checkset(constraints)
SF_LMI=all(checkset(constraints)>0);
%
if SF_LMI == 0
    display(' No feasible solution ')
else
gamma=sqrt(double(gamsq))
end
end
%
```

References

1. Abedor, J., Nagpal, K., Poolla, K.: A Linear matrix inequality approach to peak-to-peak gain minimization. International Journal of Robust and Nonlinear Control 6, 899–927 (1996)
2. Ackermann, J.: Longitudinal control of fighter aircraft F4E with additional canards. In: Sondergeld, K.P. (compiler) A Collection of Plant Models and Design Specifications for Robust Control. DFVLR, Oberpfaffenhoffen, Germany (1982)
3. Allerhand, L.I., Shaked, U.: Robust stability and stabilization of linear switched systems with dwell time. IEEE Transactions on Automatic Control 56, 381–386 (2011)
4. Apkarian, P., Gahinet, P.: A convex characterization of gain-scheduled H_∞ conterollers. IEEE Transactions on Automatic Control 40, 853–864 (1995)
5. Basar, T., Bernhard, P.: H_∞ Optimal Control and Related Minmax Design Problems: a Dynamic Game Approach. Birkhauser, Boston (1995)
6. Berman, N., Shaked, U.: Stochastic H_∞ control for nonlinear stochastic systems. In: Proc. Conf. Decision Control, Maui, Hawaii (2003)
7. Berman, N., Shaked, U.: H_∞-like control for nonlinear stochastic systems. Systems and Control Letters 55, 247–257 (2006)
8. Berman, N., Shaked, U.: H_∞ control for nonlinear stochastic systems, the output-feedback case. International Journal of Control 81, 1733–1746 (2008)
9. Bernt, K.: Stochastic Differential Equations: An Introduction with Applications, 6th edn. Springer, Berlin (2003)
10. Blakelock, J.H.: Automatic Control of Aircraft and Missiles, 2nd edn., pp. 176–188. Wiley, New York (1991)
11. Borkar, V., Mitter, S.: A Note on Stochastic Dissipativeness. In: Rantzer, A., Byrnes, C.I. (eds.) Directions in Mathematical Systems Theory and Optimization. LNCIS, vol. 286, pp. 41–49. Springer, Heidelberg (2003)
12. Boukas, E.K., Liu, Z.K.: Deterministic and Stochastic Time Delay Systems. Birkhäuser (2002)
13. Bouhtouri, A.E., Hinriechsen, D., Pritchard, A.J.: H_∞-type control for discrete-time stochasic systems. Int. J. Robust Nonlinear Control 9, 923–948 (1999)
14. Boyarski, S., Shaked, U.: Time-convexity and time-gain-scheduling in finite-horizon robust H_∞-control. In: Proceedings of the 48th CDC 2009, Shanghai, China (2009)

15. Boyd, S., El Ghaoui, L., Feron, E., Balakrishnan, V.: Linear Matrix Inequality in Systems and Control Theory. SIAM Frontier Series (1994)
16. Bruni, C., Dipillo, G., Koch, G.: Bilinear Systems: An appealing class of "nearly linear" systems in theory and applications. IEEE Transactions on Automatic Control 19, 334–348 (1974)
17. Charalambous, C.D.: Stochastic nonlinear minmax dynamic games with noisy measurements. IEEE Transactions on Automatic Control 48, 261–266 (2003)
18. Chen, W.H., Guan, Z., Lu, X.: Delay-dependent exponential stability of uncertain stochastic systems with multiple delays: an LMI approach. Systems and Control Letters 54, 547–555 (2005)
19. Chen, Y., Zheng, W.X., Xue, A.: A new result on stability analysis for stochastic neutral systems. Automatica 46, 2100–2104 (2010)
20. Chen, Y., Xue, A., Lu, R., Zhou, S., Wang, J.: Delay-dependent robust H_∞ control of uncertain stochastic delayed system. In: Proceedings of the 17th World Congress the International Federation of Automatic Control, Seoul, Korea, July 6-11, pp. 1:4916–1:4921 (2008)
21. Colaneri, P.: Dwell time analysis of deterministic and stochastic switched systems. European Journal of Automatic Control 15, 228–249 (2009)
22. Costa, O.L.V., Kubrusly, C.S.: State-feedback H_∞ -control for discrete-time infinite-dimensional stochastic bilinear systems. Journal of Mathematical Systems, Estimation and Control 6, 1–32 (1996)
23. Costa, O.L.V., Guilherme, R.A.M.: Linear minimum mean square filter for discrete-time linear systems with Markov jumps and multiplicative noises. Automatica 47, 466–476 (2011)
24. Dragan, V., Morozan, T., Halanay, A.: Optimal stability compensator for linear systems with state dependent noise. Stochastic Analysis and Application 10, 557–572 (1992)
25. Dragan, V., Halanay, A., Stoica, A.: A small gain theorem for linear stochastic system. System and Control Letters 30, 243–251 (1997)
26. Dragan, V., Morozan, T.: Mixed input-output optimization for time-varying Ito systems with state dependent noise. Dynamics of Continuous, Discrete and Impulsive Systems 3, 317–333 (1997)
27. Dragan, V., Stoica, A.: A γ attenuation problem for discrete-time time-varying stochastic systems with multiplicative noise. Reprint Series of the Institute of Mathematics of the Romanian Academy (10) (1998)
28. Dragan, V., Morozan, T.: The linear quadratic optimiziation problems for a class of linear stochastic systems with multiplicative white noise amd Markovian jumping. IEEE Transactions on Automatic Control 49(5), 665–675 (2004)
29. Dupuis, P., James, M.R., Petersen, I.R.: Robust properties of risk-sensitive control. Math. of Contr. Signals and Systems 13, 318–332 (2000)
30. Fen, X., Loparo, K.A., Ji, Y., Chizeck, H.J.: Stochastic Stability Properties of Jump Linear Systems. IEEE Transactions on Automatic Control 37, 38–53 (1992)
31. Florchinger, P.: A passive system approach to feedback stabilization of nonlinear control stochastic systems. SIAM Journal of Control and Optimization 37, 1848–1864 (1999)
32. Florchinger, P., Verriest, E.I.: Stabilization of nonlinear stochastic systems with delay feedback. In: Proc. Conf. Decision Control, San Antonio, TX, pp. 859–860 (December 1, 1993)

33. Fridman, E.: New Lyapunov-Krasovskii functionals for stability of linear retarded and neutral type systems. Systems and Control Letters 43, 309–319 (2001)
34. Fridman, E., Shaked, U.: Parameter dependent stability and stabilization of time-delay systems. IEEE Transactions on Automatic Control 48(5), 861–866 (2003)
35. Fridman, E., Shaked, U.: Input-output approach to stability and L_2-gain analysis of systems with time-varying delays. Systems and Control Letters 55, 1041–1053 (2006)
36. Fridman, E., Orlov, Y.: Exponential stability of linear distributed parameter systems with time-varying delays. Automatica 45(2), 194–201 (2009)
37. Srivastava, S.: A course on Borel sets, p. 112. Springer (1998)
38. Gao, H., Lam, J., Wang, C.: Robust H_∞ filtering for discrete stochastic time-delay systems with nonlinear disturbances. Nonlinear Dynamics and Systems Theory 4(3), 285–301 (2004)
39. Gao, H., Chen, T.: New results on stability of discrete-time systems with time-varying state delay. IEEE Transactions on Automatic Control 52(2), 328–334 (2007)
40. Gahinet, P.: Explicit controller formulas for LMI-based H_∞ synthesis. Automatica 32(7), 1007–1014 (1996)
41. Gahinet, P., Nimerovski, A., Laub, A.J., Chilali, M.: LMI Control Toolbox. The Mathworks Inc. (1995)
42. Geromel, J.C., Bernussou, J., Garcia, G., de Oliviera, M.C.: H_2 and H_∞robust filtering for discrete-time linear systems. SIAM Journal of Control and Optimization 38, 1353–1368 (2000)
43. Geromel, J., Colaneri, P.: Stability and stabilization of continuous-time switched linear systems. SIAM Journal of Control and Optimization 45, 1915–1930 (2006)
44. Gershon, E., Shaked, U., Yaesh, I.: H_∞ control and filtering of discrete-time stochastic systems with multiplicative noise. Automatica 37, 409–417 (2001)
45. Gershon, E., Pila, A., Shaked, U.: Difference LMIs for robust H_∞ control and filtering. In: Proceedings of the European Control Conference (ECC 2001), Porto, Portugal, pp. 3469–3474 (2001)
46. Gershon, E., Limebeer, D.J.N., Shaked, U., Yaesh, I.: Robust H_∞ filtering of stationary continuous-time linear systems with stochastic uncertainties. IEEE Transactions on Automatic Control 46, 1788–1793 (2001)
47. Gershon, E., Shaked, U., Yaesh, I.: Robust H_∞ estimation of stationary discrete-time linear processes with stochastic uncertainties. Systems and Control Letters 45, 257–269 (2002)
48. Gershon, E., Shaked, U., Yaesh, I.: H_∞ tracking of linear systems with stochastic uncertainties and preview. In: Proceedings of the IFAC World Congress, Barcelona, Spain (2002)
49. Gershon, E., Shaked, U., Yaesh, I.: Output-feedback H_∞ tracking with preview of state-multiplicative noise systems. In: Proceedings of the Mediterranean Conference on Control and Automation (MED 2002), Lisbon, Portugal (2002)
50. Gershon, E., Shaked, U., Yaesh, I.: Static H_2 and H_∞ output-feedback of linear continuous-time systems with stochastic uncertainties. In: Proceedings of the 4th IFAC Symposium on Robust Control Design (ROCOND), Milano, Italy (2003)

51. Gershon, E., Shaked, U., Yaesh, I.: Robust H_∞ tracking with preview of continuous-time linear systems with stochastic parameter uncertainties. International Journal of Robust and Nonlinear Control 14, 607–626 (2004)
52. Gershon, E., Limebeer, D.J.N., Shaked, U., Yaesh, I.: Stochastic H_∞ tracking with preview for state-multiplicative systems. IEEE Transactions on Automatic Control 49(11), 2061–2066 (2004)
53. Gershon, E., Shaked, U., Yaesh, I.: H_∞ Control and Estimation of State-multiplicative Linear Systems. LNCIS, vol. 318. Springer (2005)
54. Gershon, E., Shaked, U., Yaesh, I.: Static output-feedback of state-multiplicative systems with application to altitude control. International Journal of Control 77, 1160–1170 (2005)
55. Gershon, E., Shaked, U., Berman, N.: H_∞ control and estimation of state-multiplicative stochastic systems with delay. IEEE Transactions on Automatic Control 52, 329–334 (2007)
56. Gershon, E., Shaked, U., Berman, N.: H_∞ Control and estimation of state-multiplicative stochastic systems with delay. In: Proceedings of the American Control Conference (ACC 2007), New York, USA (2007)
57. Gershon, E., Berman, N., Shaked, U.: Robust H_∞control and estimation of state-multiplicative stochastic systems with delay. In: Proceedings of the 16th Mediterranean Conference on Control and Automation (MED 2008), Ajaccio, Corsica (2008)
58. Gershon, E., Shaked, U.: H_∞ output-feedback of discrete-time systems with state-multiplicative noise. Automatica 44, 574–579 (2008)
59. Gershon, E., Shaked, U.: Robust H_∞ output-feedback control of retarded state-multiplicative stochastic systems. International Journal of Robust and Nonlinear Control 21, 1283–1296 (2010)
60. Gershon, E., Shaked, U., Berman, N.: H_∞ Preview Tracking Control of Retarded State-Multiplicative Stochastic Systems. In: Systol 2010, Nice, France (2010)
61. Gershon, E., Shaked, U., Berman, N.: Robust Polytopic H_∞ Control of Linear Retarded State-multiplicative Stochastic Systems. In: Proceedings of the 19th Mediterranean Conference on Control and Automation (MED 2011), Corfu, Greece (2011)
62. Shaked, U., Gershon, E.: Robust H_∞ control of stochastic linear switched systems with dwell time. International Journal of Robust and Nonlinear Control (2012) (in press)
63. Gershon, E., Shaked, U.: H_∞ preview tracking control of retarded state-multiplicative stochastic systems. International Journal of Robust and Nonlinear Control (2012) (under review)
64. Gershon, E.: Robust reduced-order H_∞ output-feedback control of retarded state-multiplicative linear systems. IEEE Transactions on Automatic Control (2012) (in press)
65. Gihman, I.I., Skorohod, A.V.: Stochastic Differential equations. Springer (1972)
66. Green, M., Limebeer, D.J.N.: Linear Robust Control. Prentice Hall, N.J. (1995)
67. Has'minskii, R.H.: Stochastic Stability of Differential Equation. Sijthoff and Noordhoff (1980)
68. Hespanha, J.P.: Uniform stability of switched linear systems: extensions of LaSalle's invariance principle. IEEE Transactions on Automatic Control 49, 470–482 (2004)

69. Hinrichsen, D., Pritchard, A.J.: Stochastic H_∞. SIAM Journal of Control and Optimization 36, 1504–1538 (1998)
70. Huang, Y., Zhang, W., Feng, G.: Infinite horizon H_2-H_∞ control for stochastic systems with Markovian jumps. Automatica 44, 857–863 (2008)
71. Huang, Y.P., Zhou, K.: Robust stability of uncertain time-delay systems. IEEE Transactions on Automatic Control 45, 2169–2173 (2000)
72. Hussmann, U.G.: Optimal stationary control with state and control dependent noise. SIAM Journal of Control 9, 184–198 (1971)
73. Isidori, A.: Nonlinear Control Sytems. Springer (1995)
74. Iwasaki, T., Skelton, R.E.: The XY-centering algorithm for the dual LMI problem - A new approach to fixed order control designs. Int. J. Contr. 62, 1257–1272 (1995)
75. Jazwinsky, A.H.: Stochastic Processes and Filtering Theory. Academic Process (1970)
76. James, M.R.: Asymptotic analysis of nonlinear stochastic risk-sensitive control and differential games. Math. Control Signals Systems 5, 401–492 (1992)
77. Kao, C.Y., Lincoln, B.: Simple stability criteria for systems with time-varying delays. Automatica 40, 1429–1434 (2004)
78. Kojima, A., Ishijima, S.: Formulas on preview and delayed H_∞ control. IEEE Transactions on Automatic Control 51, 1920–1937 (2006)
79. Kolmanovskii, V., Richard, J.P.: Stability of some linear systems with delays. IEEE Transactions on Automatic Control 44, 984–989 (1999)
80. Kozin, F.: A survey of stability of stochastic systems. Automatica 5, 95–112 (1970)
81. Moon, Y.S., Park, P., Kwon, W.H., Lee, Y.S.: Delay-dependent robust stabilization of uncertain state-delayed systems. Int. J. Control 74, 1447–1455 (2001)
82. Leibfritz, F.: An LMI-based algorithm for designing suboptimal static H_2/H_∞ output-feedback controllers. SIAM Journal of Control and Optimization 39, 1711–1735 (2001)
83. Levanon, L.: Radar Principles. John Willy and Sons (1998)
84. Li, L., Ugrinovskii, V.A.: On necessary and sufficient conditions for H-infinity output feedback control of Markov jump linear systems. IEEE Transactions on Automatic Control 52, 1287–1292 (2007)
85. Li, X., de Souza, C.: Criteria for robust stability and stabilization of uncertain linear systems with state delay. Automatica 33, 1657–1662 (1997)
86. Liberzon, D.: Switching in Systems and Control. Birkhauser, Boston (2003)
87. Lin, C.F.: Modern Navigation Guidance and Control Procceing. Perentice Hall, New Jersy (1991)
88. Mao, X.: Robustness of exponential stability of stochastic differential delay equations. IEEE Transactions on Automatic Control 41, 442–447 (1996)
89. Mao, X., Shan, A.: Exponential Stability of Stochastic Delay Differential Equations. Stochastics Stochastics Rep. 60(1), 135–153 (1997)
90. Margaliot, M., Langholz, G.: Necessary and sufficient conditions for absolute stability: the case of second-order systems. IEEE Transactions on Circuits and Systems: Fundamental Theory and Applications 50, 227–234 (2003)
91. Mclane, P.J.: Optimal stochastic control of linear systems with state- and control- dependent disturbances. IEEE Transactions on Automatic Control 16, 793–798 (1971)

92. Milstein, G.N.: Design of stabilizing controller with incomplete state-data for linear stochastic system with multiplicative noise. Automat. Remote Contr. 5, 653–659 (1982)
93. Mohler, M.M., Kolodziej, W.J.: An overview of stochastic bilinear control processes. IEEE Transactions on Systems Man Cybernetics 10, 913–919 (1980)
94. Morozan, T.: Input-output operators and parametrized Riccati equations for time-varying stochastic differential equations. Reprint Series of the Institute of Mathematics of the Romanian Academy (6) (1997)
95. Morozan, T.: Parametrized Riccati equations associated to input-output operators for time-varying stochastic differential equations with state dependent noise. Preprint of Inst. Math. of Romanian Academy 37, 317–333 (1995)
96. Niculescu, S.I.: Delay effects on stability: A robust control approach. LNCIS, vol. 269. Springer, London (2001)
97. Pakshin, P.V.: State estimation and control synthesis for discrete linear systems with additive and multiplicative noise. Automat. Remote Contr. 4, 526–535 (1978)
98. Petersen, I.R.: Quadratic stabilizability of uncertain linear systems: existence of a nonlinear stabilizing control does not imply existence of a stabilizing control. IEEE Transactions on Automatic Control 30, 291–293 (1985)
99. Petersen, I.R., Ugrinovskii, V.A., Savkin, A.V.: Robust Control Design Using H_∞ Methods. Springer (2000)
100. Phillis, Y.: Controller design of systems with multiplicative noise. IEEE Transactions on Automatic Control 30, 1017–1019 (1985)
101. Prajna, S., Papachristodoulou, A., Seiler, P., Parrilo, P.A.: New developments in sum of squares optimization and SOSTOOLS. In: Proceedings of the American Control Conference (ACC), Boston, MA (2004)
102. Sagirow, P.: Stochastic methods in the dynamics of satellites. Lecture Notes, Udine, CISM (1970)
103. Scherer, C., Weiland, S.: Course on Linear Matrix Inequalities in Control, e-book, version 2.0. DISC (1999)
104. Scherer, C., Weiland, S.: Linear Matrix Inequalities in Control (2006), e-book at `http://www.st.ewi.tudelft.nl/roos/courses/WI4218/lmi052.pdf`
105. Schuss, Z.: Theory and Applications of Stochastic Differential Equations. Wiley, New-York (1980)
106. Shaked, U., de Souza, C.E.: Continuous-time tracking problems in an H_∞ setting: a game theory approach. IEEE Transactions on Automatic Control 40, 841–852 (1995)
107. Shaked, U.: A LPV approach to robust H_2 and H_∞ static output feedback design. Transactions on Automatic Control 48, 866–872 (2003)
108. Speyer, J.L., Chung, W.H.: Stochastic processes, estimation, and control. Advances in design and control. SIAM (2008)
109. Stoica, A.M., Yaesh, I.: Markovian jump delayed Hopfield networks with multiplicative noise. Automatica 44, 2157–2162 (2008)
110. Sun, Z., Ge, S.S.: Analysis and synthesis of switched linear control systems. Automatica 41, 181–195 (2005)
111. Suplin, V., Fridman, E., Shaked, U.: H_∞ control of linear uncertain time-delay systems - A projection appproach. IEEE Transactions on Automatic Control 51(4), 680–685 (2006)
112. Tadmor, G., Mirkin, L.: H_∞ control and estimation with preview - parts I and II. IEEE Transactions on Automatic Control 50(1), 19–40 (2005)

113. Thygessen, U.H.: On dissipation in stochastic systems. In: Proc. ACC Conf., San Diego, California (June 1999)
114. Urinovskii, V.A., Petersen, I.R.: Robust filtering of stochastic uncertain systems on an infinite time horizon. Int. J. Control 75, 614–626 (2002)
115. Ugrinovskii, V.A., Petersen, I.R.: Absolute stabilization and minimax optimal control of uncertain systems with stochastic uncertaity. SIAM Journal of Control and Optimization 37, 1089–1122 (1999)
116. Verriest, E.I.: Inconsistencies in systems with time-varying delays and their resolution. IMA Journal of Mathematical Control and Estimation 28, 147–162 (2011)
117. Verriest, E.I.: Asymptotic Properties of Stochastic Delay Systems. In: Niculescu, S.-I., Gu, K. (eds.) Advances in Time-Delay Systems. Lecture Notes in Computational Science and Engineering, pp. 389–435. Springer (2004)
118. Wang, Z., Ho, D.W.C., Liu, Y., LIu, X.: Robust H_∞ control for a class of nonlinear discrete time-delay stochastic systems with missing measurements. Automatica 45, 684–691 (2009)
119. Verriest, E.I., Florchinger, P.: Stability of Stochastic Systems with Uncertain Time Delays. Systems and Control Letters 24(1), 41–47 (1995)
120. Wei, G., Wang, Z., Shu, H., Fan, J.: A delay-depdendent approach to H_1 filtering for stochastic delayed jumping systems with sensor nonlinearities. International Journal of Control 80, 885–897 (2007)
121. Willlems, J.L., Willems, J.C.: Robust stabilization of uncertain systems. SIAM Journal of Control and Optimization 21, 352–373 (1983)
122. Willems, J.C.: Dissipative dynamical systems-part I: general theory. Arch. Rational Mech. Anal. 45, 321–351 (1972)
123. Wilson, D.A.: Operator norms for linear systems. IEEE Transactions on Automatic Control 34, 94–97 (1989)
124. Wonham, W.M.: Optimal stationary control of linear systems with state-dependent noise. SIAM Journal of Control 5, 486–500 (1967)
125. Wonham, W.M.: On a matrix Riccati equation of stochastic control. SIAM Journal of Control 6(4), 681–697 (1968)
126. Xia, Y., Qiu, J., Zhang, J., Gao, Z., Wang, J.: Delay-dependent robust H_∞ control for uncertain stochastic time-delay system. International Journal of Systems Science 39, 1139–1152 (2008)
127. Xie, S., Xie, L.: Stabilization of uncertain large-scale stochastic systems with time delays. Automatica 36, 161–167 (2000)
128. Xu, S., Chen, T.: Robust H_∞ control for uncertain stochastic systems with state-delay. IEEE Transactions on Automatic Control 47, 2089–2094 (2002)
129. Xu, S., Chen, T.: Robust H_∞ control for uncertain stochastic systems with state-delay. IEEE Transactions on Automatic Control 47, 2089–2094 (2002)
130. Xu, S., Lam, J., Mao, X., Zou, Y.: A new LMI condition for delay-dependent robust stability of stochastic time-delay systems. Asian J. Control 7, 419–423 (2005)
131. Xua, S., Lamb, J., Chen, T.: Robust H_∞ control for uncertain discrete stochastic time-delay systems. Systems and Control Letters 51, 203–215 (2004)
132. Yaesh, I., Shaked, U.: Robust reduced-order output-feedback H_∞ control. In: Proc. of the 6th IFAC Symposium on Robust Control, ROCOND 2009, Haifa, Israel (2009)

133. Yong, J., Zhou, X.Y.: Stochastic Controls: Hamiltonian Systems and HJB Equations. In: Applications of Mathematics. Stochastic Modelling and Applied Probability, vol. 43. Springer (1999)
134. Yue, D., Won, S.: Delay-dependent robust stability of stochastic systems with time-delay and nonlinear uncertainties. IEE Electron. Lett. 37, 992–993 (2001)
135. Yue, D., Tian, E., Zhang, Y.: Stability analysis of discrete systems with stochastic delay and its applications. International Journal of Innovative Computing, Information and Control 5(8), 2391–2403 (2009)
136. Zarchan, P.: Tactical and Strategic Missile Guidance. In: Progress in Astronautics and Aeronautics, vol. 124. AIAA, New York (1990)
137. Zhang, W., Chen, B.S.: State feedback H_∞ control for a class of nonlinear stochastic systems. SIAM Journal of Control and Optimization 44, 1973–1991 (2006)
138. Zhang, W., Chen, B.S., Tseng, C.S.: Robust H_∞ filtering for nonlinear stochastic systems. IEEE Transections on Signal Processing 53(2), 589–598 (2005)
139. Zhendong, S., Sam, S.G.: Stability Theory of Switched Dynamical Systems. Springer's Communications and Control Engineering series, Berlin (2011)
140. McRuer, D., Ashkenazi, I., Graham, D.: Aircraft Dynamics and Control, ch. 7, 8. Princeton University Press, Princeton (1973)

Index

L_2-gain switched systems, 182
$L_2 - gain$, 116, 129
$\sigma-$algebra, 219, 221
l_2-gain, 159

BRL
  continuous-time, nominal case, 30
  discrete-case, 101
  nonlinear, continuous-time, 134
  nonlinear, discrete-case, 157
  norm bounded case, 31
  polytopic case, 32
  polytopic discrete-time, 102

discrete-time NL systems, 155
dissipation, 116
dissipative stochastic systems, 9
dissipativity, 156

filtering
  continuous nominal case, 37
  continuous norm-bounded case, 41
  continuous polytopic case, 42
  continuous, peak to peak, 195
  discrete-time, 106
  polytopic discrete-time, 111
  transfer function matrix, 201
Fubini theorem, 120

GBM, 223
guidance control, example, 47

Hamilton-Jacobi inequality, 116
Hessian, 18
HJAI, 165
HJI, 117, 134, 157

infinitesimal generator, 123
input-output
  continuous-time, 3, 26, 66, 77
  discrete-time, 4, 98
inverted pendulum, 168
inverted pendulum - example, 127
Ito
  formula, 119, 120
  lemma, 17, 221
  type, 219
Ito lemma applications, 221

LMI optimization, 227
Lyapunov
  function, 222

mean square calculus, 13

nonlinear control, 5
nonlinear estimation, 130
nonlinear OF control, 140
nonlinear SF control, 117

output-feedback
  nominal case, 44
  norm-bounded case, 49
  polytopic case, 50

P2P filtering, 199
Parseval's theorem, 15
peak-to-peak, 194

reduced-order control
  nominal case, 64
  polytopic case, 72
risk-sensitive, 115

Schur's complements, 126
SOS, 167
spectral density, 15
stability
  discrete-time, 98
  discrete-time case, 11
  global, 223
  global case, 8
  in probability, 122, 133, 151
  in the mean square sense, 122
  nominal case, 25
  norm-bounded case, 28
  of tracking system, 78
  polytopic case, 29
  polytopic discrete-time, 100
  stochastic, 222, 223
  switched systems, 179
state-feedback
  continuous nominal case, 32
  continuous norm-bounded case, 34
  continuous polytopic case, 35
  continuous, peak to peak, 194
  discrete-time, 103
  polytopic discrete-time, 104
  switched systems, 187
  tracking, 80
static output-feedback
  nominal case, 53
  polytopic case, 57
stochastic differential equations, 16
stochastic differential games, 115
stochastic dissipation, 118
storage function, 116

Taylor expansion, 18
tracking
  with no preview, 84
  with preview, 84

white noise sequences, 14
Wiener process, 14, 221

MIX
Papier aus verantwortungsvollen Quellen
Paper from responsible sources
FSC® C105338

If you have any concerns about our products,
you can contact us on
ProductSafety@springernature.com

In case Publisher is established outside the EU,
the EU authorized representative is:
Springer Nature Customer Service Center GmbH
Europaplatz 3, 69115 Heidelberg, Germany

Printed by Libri Plureos GmbH
in Hamburg, Germany